Wildlife Research Monographs

The book series on "Wildlife Research Monographs" focuses on all aspects of wildlife biology to fill a vital niche at the intersection of several disciplines. The aim of the series is to summarize knowledge about interaction between wildlife species, their habitats, as well as human activities. Main areas are: wildlife ecology, diseases, toxicology, conservation, management, and the sustainable use of natural resources. The series addresses graduate students, university teachers, scientists and professionals.

More information about this series at http://www.springer.com/series/8826

Rafael Mateo • Beatriz Arroyo • Jesus T. Garcia
Editors

Current Trends in Wildlife Research

Springer

Editors

Rafael Mateo
Instituto de Investigación en Recursos
Cinegéticos (IREC, CSIC-UCLM-JCCM)
Ciudad Real
Spain

Jesus T. Garcia
Instituto de Investigación en Recursos
Cinegéticos (IREC, CSIC-UCLM-JCCM)
Ciudad Real
Spain

Beatriz Arroyo
Instituto de Investigación en Recursos
Cinegéticos (IREC, CSIC-UCLM-JCCM)
Ciudad Real
Spain

ISSN 2366-8733 ISSN 2366-8741 (electronic)
Wildlife Research Monographs
ISBN 978-3-319-27910-7 ISBN 978-3-319-27912-1 (eBook)
DOI 10.1007/978-3-319-27912-1

Library of Congress Control Number: 2016935927

Printed on acid-free paper

This Springer imprint is published by Springer Nature
The registered company is Springer International Publishing AG Switzerland

Foreword

Wildlife is a quite imprecise term for non-domestic organisms, and the taxa included into this group vary across the globe and with time. However, in most areas of the world and during most times, wildlife comprise vertebrates, namely, mammals, birds, and also fish, amphibians, or reptiles. Especially birds and mammals have a quite strong connection to humans. From the emotional point of view, we are attracted or at least fascinated by them, known as biophilia. On the other hand, many wildlife species have difficulties to survive next to humans and their activities; they are endangered and need protection. Some wildlife species have an important role in biodiversity conservation as there are keystone species. Others benefit from anthropogenic environments such as settlements or intensively used arable land. They have to be controlled to minimize damage to human infrastructure, health, or crops. Last but not least, some wildlife species are important for human nutrition, welfare, or even culture. Whatever each wildlife species means to our society, coexistence between man and wildlife deserves management in order to avoid biodiversity loss, to reduce damage caused by wildlife, or to keep wildlife as a natural resource.

However, a sustainable wildlife management needs a sound scientific basis. That is why the demand for wildlife research is growing. Consequently, research activities in scientific fields related to wildlife are increasing exponentially. The quantitative growth is characterized by a qualitative growth, too. While wildlife research was rather descriptive some decades ago, we are now able to follow a hypothesis-driven science. This fascinating development is topped by the fact that wildlife research has a broad spectrum ranging from anatomy to zoonoses and that we have nearly unlimited research avenues using inter- and transdisciplinary approaches.

As the scientific field is growing, there is the need to compile the current knowledge and to sum up the state of the art. Therefore, books like this are an important milestone on our way to find answers to current questions. This book not only provides an overview of what we have learned in the past, it also points to the future and widens our horizon to detect emerging research fields. Thus, the innovative methods and sustainable approaches described here will inspire readers and allow them to permanently improve quality in wildlife research. In addition, this book offers numerous facets for new ways to increase inter- and transdisciplinarity.

I congratulate the editors and chapter authors of this volume and look forward to see both researchers and students to refer to it as a reference and inspiration.

Klaus Hackländer
Head of the Institute of Wildlife Biology and Game Management at the University of Natural Resources and Life Sciences, Vienna (BOKU)
Deputy President of the Applied Science Division of the International Council for Game and Wildlife Conservation (CIC)

Contents

Reviewers

Nils Bunnefeld Biological and Environmental Sciences, University of Stirling, Stirling, UK
URL: http://rms.stir.ac.uk/converis-stirling/person/18125

Travis DeVault Department of Agriculture, Wildlife Services, National Wildlife Research Center, Sandusky, OH, USA
URL: http://www.aphis.usda.gov/wps/portal/aphis/ourfocus/wildlifedamage/sa_programs/sa_nwrc/sa_research/sa_scientists/ct_devault

John E. Elliott Environment Canada, Science and Technology Branch, Pacific Wildlife Research Centre, Delta, BC, Canada
URL: john.elliott@ec.gc.ca

Ezio Ferroglio Dipartimento di Scienze Veterinarie, Università degli Studi di Torino, Grugliasco (TO), Italy
URL: http://www.personalweb.unito.it/ezio.ferroglio/

Christian Gortázar SaBio, Instituto de Investigación en Recursos Cinegéticos (IREC), CSIC-UCLM-JCCM, Ciudad Real, Spain
URL: http://www.sabio-irec.com/es/equipo/christian-gortazar-schmidt/

Niels Kanstrup Dansk Jagtakademi, Rønde, Denmark
URL: nk@danskjagtakademi.dk

Jorge R. Lopez-Olvera Servei d'Ecopatologia de Fauna Salvatge (SEFaS), Facultat de Veterinària (edifici V), Universitat Autònoma de Barcelona, Bellaterra, Barcelona, Spain
URL: Jordi.Lopez.Olvera@uab.cat

Michael L. Morrison Department of Wildlife and Fisheries Sciences, Texas A&M University, College Station, TX, USA
URL: http://wfsc.tamu.edu/directory/faculty-directory/michael-l-morrison/

Raimundo Real Grupo de Biogeografía, Diversidad y Conservación, Universidad de Málaga, Málaga, Spain
URL: rrgimenez@uma.es

Judit Smits Ecosystm and Public Health, University of Calgary, Calgary, AB, Canada
URL: https://vet.ucalgary.ca/profiles/judit-smits

Giacomo Tavecchia Biodiversidad y Conservación, Institut Mediterrani d'Estudis Avançats (IMEDEA CSIC-UIB), Esporles – Illes Balears, Spain
URL: https://imedea.uib-csic.es/ficha.php?pid=387

José Luis Tella Department of Conservation Biology, Estación Biológica de Doñana (EBD-CSIC), Seville, Spain
URL: https://www.researchgate.net/profile/Jose_Tella

Carles Vilà Conservation and Evolutionary Genetics Group, Estación Biológica de Doñana (EBD-CSIC), Seville, Spain
URL: http://www.consevol.org/members/carles.html

Trends in Wildlife Research: A Bibliometric Approach

Beatriz Arroyo, Rafael Mateo, and Jesús T. García

Wildlife Research: Definitions and Concepts

"Wildlife" is a word that has different meanings for different people and in different contexts. In fact, many people use it with an unconscious attachment to a particular meaning, not necessarily aware of it being used differently by other people. According to the Oxford Advanced Learner's Dictionary, wildlife means "*the native fauna (and sometimes flora) of a region*". In many cases, however, this "native fauna" is, consciously or unconsciously, limited to vertebrate species, and it sometimes excludes fish (as implicitly implied in the names of the "Fish and Wildlife" societies and services in the US). Conversely, fish (at least fresh-water fish) is considered as "wildlife" in many countries, as they are part of the same ecosystems and their management is analogous. Likewise, butterflies and other invertebrates are usually included in "wildlife inventories" at least in the UK. Wildlife is also used as a term for "*undomesticated animals living in the wild*" (American Heritage Dictionary) or "*animals and plants that grow independently of people, usually in natural conditions*" (Cambridge Advance Learner's Dictionary & Thesaurus). Here, the emphasis is put in the "untamed" quality of species considered as wildlife. Traditionally, "wildlife" includes all game species in the US, as hunting represents, in the social discourse there, a way to approach wilderness (Good 1997). Indeed, according to the Webster's Dictionary, wildlife means "*wild animals, especially those hunted for food or sport*". On the other hand, game species are, at least in Europe, intensively managed, so they do not "grow independently of people", and some voices claim that, in these circumstances, they are livestock rather than wildlife (Díaz et al. 2009). In some European languages, there are different words for game species and non-game species, and only the latter include some reference to

B. Arroyo (✉) • R. Mateo • J.T. García
Instituto de Investigación en Recursos Cinegéticos (IREC, CSIC-UCLM-JCCM),
Ronda de Toledo 12, Ciudad Real 13005, Spain
e-mail: Beatriz.Arroyo@uclm.es; rafael.mateo@uclm.es; jesusgarcia.irec@gmail.com

R. Mateo et al. (eds.), *Current Trends in Wildlife Research*,
Wildlife Research Monographs 1, DOI 10.1007/978-3-319-27912-1_1

"wild" in the non-English term (e.g. *faune sauvage* vs *gibier* in French, or *fauna silvestre* vs *fauna cinegética* or *caza* in Spanish). The recent change of name of the "Game Conservancy Trust" in the UK to the "Game and Wildlife Conservancy Trust" somehow also confronts both terms, as if they were, if not antonyms, at least dissimilar or complementary.

Given this variety of meanings, it is equally difficult to define comprehensively and accurately the concept of "wildlife research", without making a too-wide definition like "research made on animals in a natural environment". There exist various scientific journals including the word "wildlife" in the title, many of which launched in North America (even if they have an international scope). These include, among others, the Journal of Wildlife Management, Wildlife Society Bulletin, Wildlife Monographs (all three published by The Wildlife Society in the US), Human Dimensions of Wildlife Management, Journal of Fish and Wildlife Management or the Journal of Wildlife Diseases. Other such journals exist or have appeared more recently in other geographical areas, such as the European Journal of Wildlife Research, Wildlife Research (formerly, Australian Wildlife Research), Wildlife Biology, British Wildlife, the South African Journal of Wildlife Research, or Wildlife Middle-East News. A perusal through those journals indicates that "wildlife research" is used there to refer to studies made with non-captive individuals, usually under an applied optic.

However, these journals do not necessarily encompass all studies about wildlife, or all the possible meanings of the term. Additionally, wildlife research (as many other research disciplines) has also evolved with time, modifying scope and approaches in recent decades. For example, a recent review showed that within the Wildlife Society Journals, there was a trend for an increase in papers related to non-game and multiple species, as well as those including modelling (Powell et al. 2010). Similarly, a similar exercise for papers published in the European Journal of Wildlife Research also showed an increase for papers assessing management or interventions, rather than purely descriptive ones (Gortazar 2012). As research in the field continues to grow, it may be useful to have a broader understanding of its major themes and emerging trends.

We thus aimed to provide an overview of wildlife research that contemplates its variety and changes, even if taking into account that any approach we take is likely to have biases. We opted for a wider bibliometric approach to illustrate trends about wildlife research and identify the most important or emerging research topics within this discipline.

Methods Used

We used the search engine of Scopus. We selected all documents that included the word "wildlife" in the document title, abstract or keywords, or the journal name (*source title* in Scopus) for the period 1984–2013, within Life Sciences, Health Sciences, Physical Sciences or Social Sciences and Humanities. That rendered 51,436 documents.

Within that sample, we looked for the most common specified keywords, using the "keyword" option, in two steps, first for 1984–2004, and then for 2005–2013. This was done to account for the much larger number of papers overall in the last period, and aiming to pick up the maximum possible number of keywords (as the system only shows the 160 most frequent ones). After considering those that were duplicate in both sets, this rendered 186 keywords being mentioned in at least 200 documents each. Of these, 51 referred to either the region or the taxon studied, 14 referred to methods used (e.g. "GIS", or "comparative study" or "animal tissue"), and 12 were not meaningful for the purpose of this review (e.g. "male", "female", "seasons"). The remainder 109 were grouped into categories, revising them step by step and reducing them to main categories when possible (inductive category development; Mayring 2000). This exercise rendered 14 topics, associated to a variable number of keywords (Table 1).

For evaluating the impact of each topic, we restricted the search within Scopus to those documents that contained any of the identified keywords for each topic, thus calculating the number of papers for each topic each year of the study. We subsequently used the citation overview to calculate for each year and topic the total number of citations, and the number of citations within 2 years of publication (e.g., for papers published in 1999, total number of citations up to 2001 included). We then divided this number by number of papers published, to obtain an average number of citations per paper for each topic and year. For identifying milestones in each topic, we searched, within the years where published papers in a particular topic had shown peaks in citations in the subsequent 2 years, those papers that had the highest numbers of citations within that period and assessed their contents. When clear peaks were not noticeable, we identified those papers most cited over periods where citation rate was relatively higher.

For looking at geographical trends, we grouped countries (as mentioned in the keywords) in continents, following the United Nations Statistics Division classification.

Publication Sources

As expected by our search algorithm, a majority (61 %) of the 51,436 identified documents were papers published in a wildlife research journal, with Journal of Wildlife Management and Journal of Wildlife Diseases being the two most important individual journal contributors (Fig. 1). In addition, almost 40 % of identified documents were published in more general journals, mainly journals dealing with applied ecology and management (e.g. Forest Ecology and Management, Journal of Applied Ecology or Environmental Management) or conservation journals (e.g. Biological Conservation, Conservation Biology, Oryx, Biodiversity and Conservation or Environmental Conservation among the most frequent) (Fig. 1). This highlights the fact that wildlife research has indeed a strong applied focus. Wildlife research papers also appeared, although less frequently, in interdisciplinary journals (with PloS ONE, Science and Nature being the most frequent ones).

Table 1 Identified topics in relation to keywords specified in the documents searched

Topic	Keywords
Human actions	Anthropogenic effect; disturbance; ecological impact; environmental impact assessment; environmental impact; human activity; humans; nature-society relations
Biodiversity	Biodiversity; classification; phylogeny; species difference; species specificity
Climate change	Climate change
Conservation	Conservation management; conservation of natural resources; conservation planning; conservation; endangered species; environmental protection; protected area; restoration ecology; species conservation; wildlife conservation
Demography	Abundance; demography; mortality; movement; population decline; population density; population dynamics; population estimation; population size; reproduction; survival
Disease	Animal disease; animal parasitosis; antibodies, bacteria; antibodies, viral; bacteria; bacterium antibody; bird diseases; disease carrier; disease outbreaks; disease reservoirs; disease transmission; drug effect; epidemic; epidemiology; microbiology; parasitology; pathology; prevalence; rabies; unclassified drug; vaccination; virology; virus infection; virus antibody; zoonoses; zoonosis
Ecophysiology	Immunology; metabolism, physiology
Extractive use	Harvesting; hunting
Ecology	Diet; ecology; home range; predation
Genetics	Genetics; nucleotide sequence
Habitat	Agriculture; forest management; forest; forestry; habitat conservation; habitat fragmentation; habitat management; habitat quality; habitat selection; habitat use; habitat; land use; landscape; vegetation; wetlands
Invasive species	Invasive species
Management	Decision making; management; pest control; wildlife management; forest management; habitat management
Pollution	Bioaccumulation; environmental exposure; environmental pollutants; pollution; polychlorinated biphenyl; risk assessment; toxicity; water pollutants, chemical; water pollution; water quality

Geographical Range of Studies

Almost half of the scientific literature about wildlife research published between 1984 and 2013 has been produced in North America (49 %), followed by Europe (26 %), Asia (8 %), Oceania (8 %), Africa (5 %) and Latin America and the Caribbean (3 %) (Fig. 2). Under a temporal perspective, wildlife research had in North America its initial development, and the contribution of publications from this region was predominant until mid-1990s. In this sense, it is worthwhile mentioning the significant contribution of two North American (though now international) scientific societies, The Wildlife Disease Association (publishing The Journal of Wildlife Diseases) and The Wildlife Society (publishing the Journal of Wildlife Management, Wildlife Society Bulletin and Wildlife

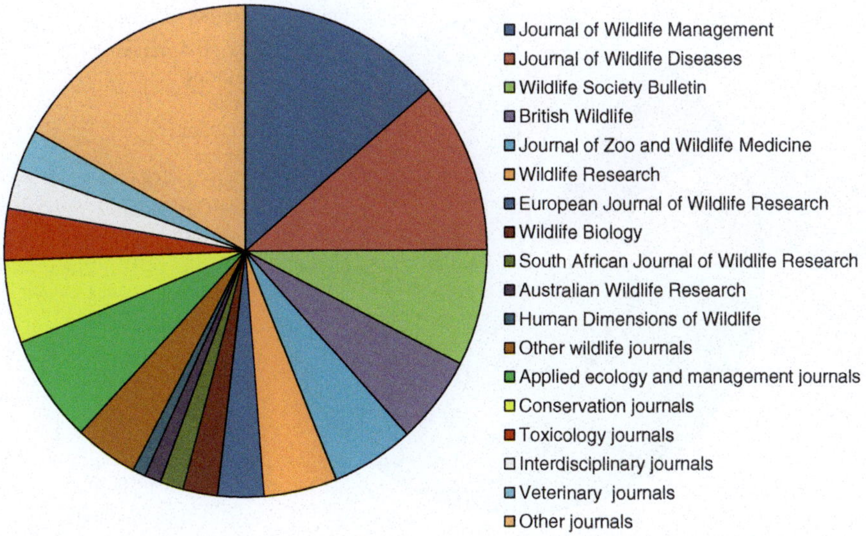

Fig. 1 Proportion of wildlife research papers (n = 51,436) published in different journals

Monographs) as pioneers and leaders in this research field. However, a significant increase in the scientific production on wildlife research in Europe occurred during the 1990s, coinciding with the launch (or restructuring) of two journals, Wildlife Biology (initiated in 1994 by the Nordic Council for Wildlife Research) and the European Journal of Wildlife Research (formerly the Zeitschrift für Jagdwissenschaft, changing name and scope in 1996). Even if the number of wildlife research papers has also strongly increased in North America during that time, the difference in contribution has decreased with time and, in 2013, scientific production in wildlife research from Europe was just 25.3 % lower than in North America (Fig. 2). Other regions have also shown a significant increase in their production, and the sum of Asia, Oceania, Africa and Latin America and the Caribbean represented 30.4 % of the wildlife research literature published in 2013.

Within each region some countries have historically contributed more significantly to wildlife research that others (Fig. 3). In the case of North America, the publications from the United States (84.8 %) exceed by large the production from Canada (15 %). Similarly, in Europe the largest contribution is from the United Kingdom (31.9 %), followed by Germany (9.7 %) and Spain (8.5 %). In Asia, India (22.4 %), China (18.7 %) and Japan (14.3 %) had the greatest contribution to wildlife research in the region. In Oceania, the largest production was from Australia (79.2 %), followed by New Zealand (19.5 %). In Africa, most of the production comes from South Africa (39.3 %), followed by the contribution of Kenia (13.0 %) and Tanzania (7.4 %). In Latin America, Brazil (32.6 %), Mexico (16.7 %) and Argentina (14.8 %) produced most of the wildlife research during our studied period of 30 years.

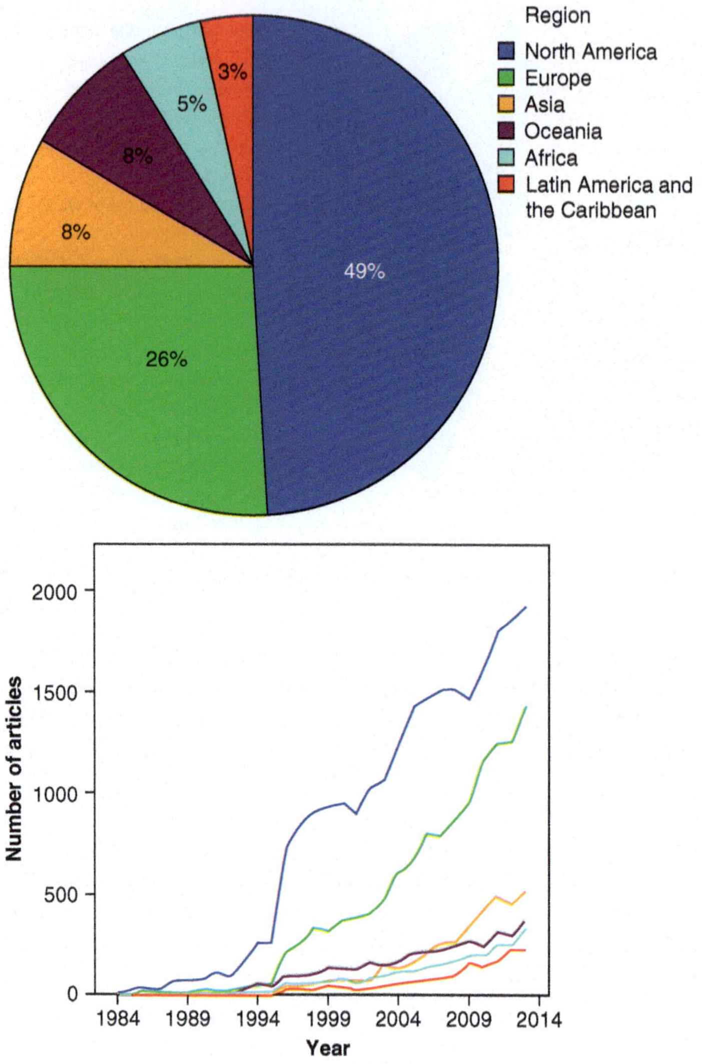

Fig. 2 Total proportion and temporal trends of wildlife research publications in relation to geographical areas

Researched Taxa

Approximately half of the publications (24,049) on wildlife research had some information in their keywords (specific information in 7311 paper keywords) about the taxonomic classification of the species under study. Mammals were the most frequent Class specified in these keywords, followed by birds, amphibians and reptiles. Within mammals, the most frequently studied groups were Cervidae, followed

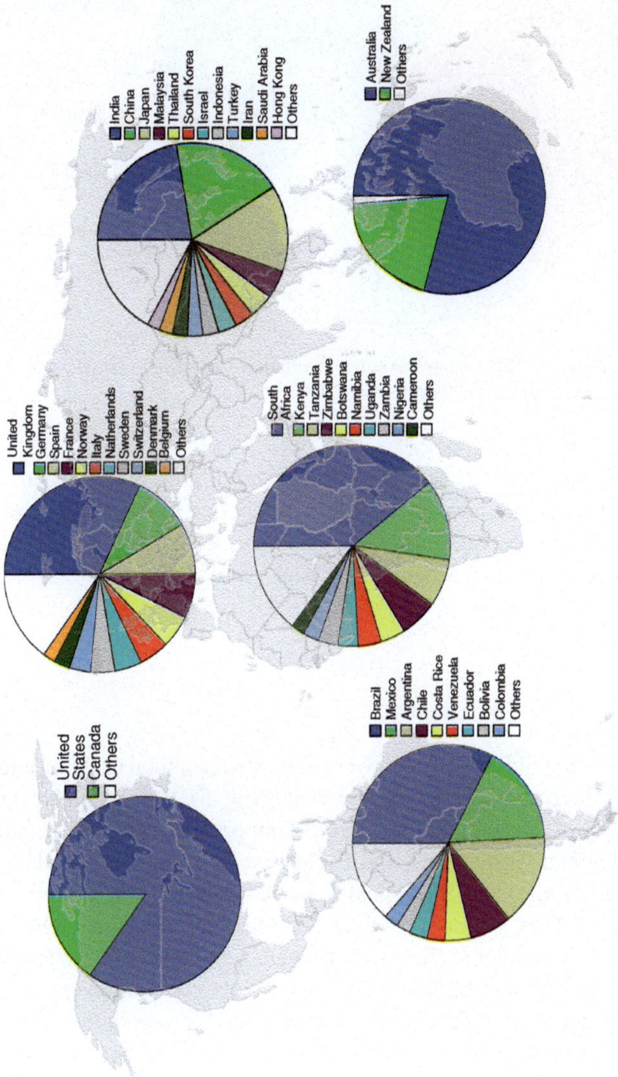

Fig. 3 Contribution of different countries (per continent) to wildlife research publications

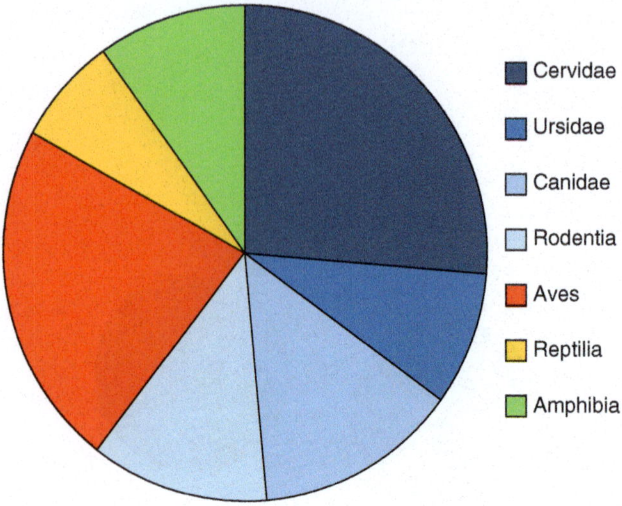

Fig. 4 Proportion of wildlife research studies in relation to studied taxa (for those documents where this was specified in the keywords, n=24,049)

by Canidae, Ursidae and Rodentia (Fig. 4). This highlights the fact, mentioned above, that there is some bias towards big vertebrates in the approach to wildlife research, and also the focus of either harvested species (such as deer) or conflictive species (both carnivores and rodents) as model studies.

Knowledge Areas

The publications on wildlife research were classified in a total of 27 different knowledge areas, of which nine are the most relevant (Fig. 5), the remainder representing each less than 1 % of all publications. The most important areas are the Environmental Sciences (35 %) and Agricultural and Biological Sciences (31 %). The important presence in Agricultural and Biological Sciences was something expected because of the role of wildlife as a natural resource. On the other hand, the inclusion of wildlife research in Environmental Sciences, as well as in Earth and Planetary Sciences (5 %), denotes that the wildlife concept has a wide application in many areas of research, including those focused in the relationship between humans and the environment. This arises because of the use of wildlife as bioindicator of global issues like environmental pollution or climate change.

Wildlife is also relevant in the areas of Veterinary Sciences, Medicine, Immunology and Microbiology and Biochemistry, Genetics and Molecular Biology. The sum of these areas than can be included in the broader group of Health Sciences represent 16 % of publications under the wildlife research concept. This arises not only from the role of wildlife species as bioindicators, but also

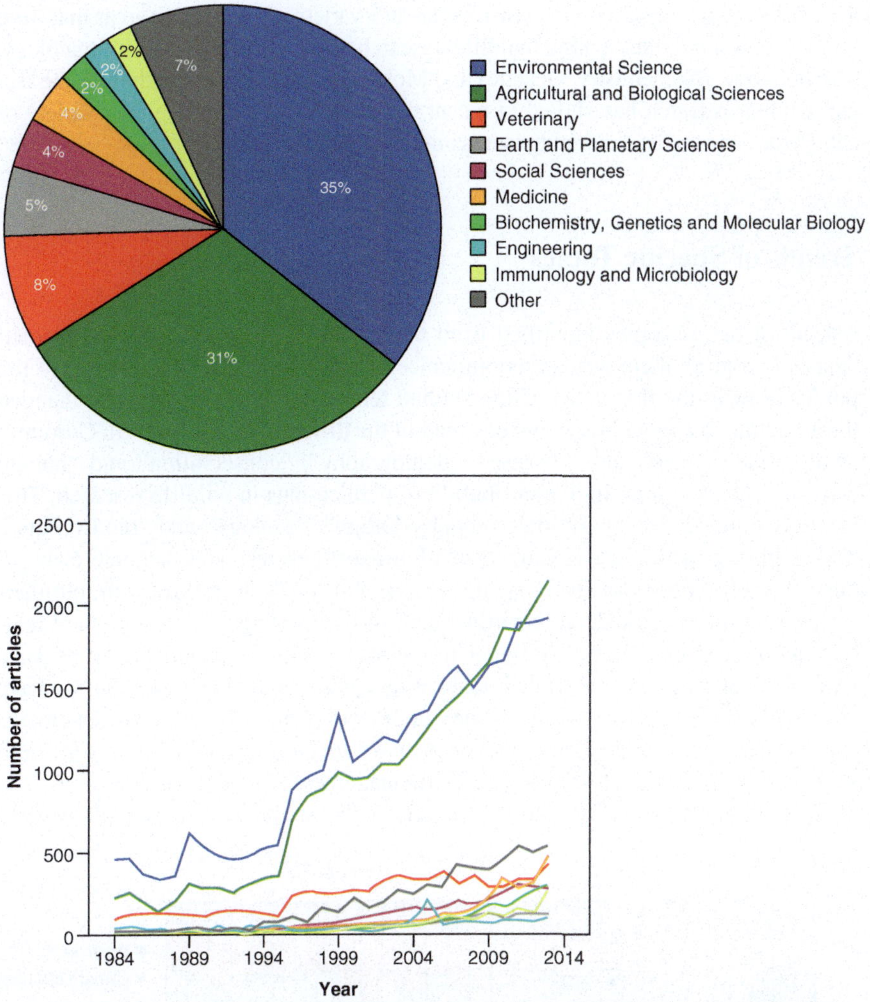

Fig. 5 Total proportion and temporal trends of wildlife research publications in relation to knowledge areas

because of the importance of wildlife as reservoir of infections that can also affect humans and livestock. Finally, the presence of wildlife research in areas like Engineering (2 %) and Social Science (2 %) highlights the relevance of the interactions between humans and wildlife in many aspects of life, like economy, policy and leisure, and the need of applied and technological approaches to face these interactions.

From a temporal perspective, scientific production of wildlife research in Veterinary Science journals has been more or less constant (Fig. 5). In contrast, a sharp increase in wildlife research publications being categorized within Enviromental Sciences or Agricultural and Biological Sciences is evidenced since mid-1990s (Fig. 5). Additionally, there has been an increase in the last 15 years in a variety of knowledge areas that were initially less important within wildlife research, such as Medicine, Immunology, Microbiology, Biochemistry, Genetics and Molecular Biology. This highlights the fact that wildlife research has shifted in recent decades to incorporate a wider variety of approaches and topics, becoming increasingly multidisciplinary.

Trends of Specific Topics

Indeed, of the 14 topics identified from the keywords extracted from the studied dataset (Table 1), there was a predominance of "diseases" among the topics of the publications in the first years of the studied temporal series, but this has changed through time to a more homogeneous load of the different topics (Fig. 6). Currently "management", "habitat", "disease", "demography", "conservation" and "human activity" have a similar high contribution to publications in wildlife research. This group is followed by "pollution", "ecophysiology", "ecology" and "biodiversity". Topics like "genetics", "extractive use", "invasive species" and "climate change" have currently a comparatively smaller contribution to wildlife research, but their increase has been marked, and their impact is also high (Figs. 7 and 8), so their relative importance may be much higher in the near future. Logically, some of these topics may overlap, and individual publications may contain keywords that we have classified in different topics (e.g. a study on how to control the spread of an emerging disease introduced in the wild by an invasive species may well be included both in "invasive species", "diseases" and in "management"). In any case, these results again highlight the fact that wildlife research studies are interested in a wider variety

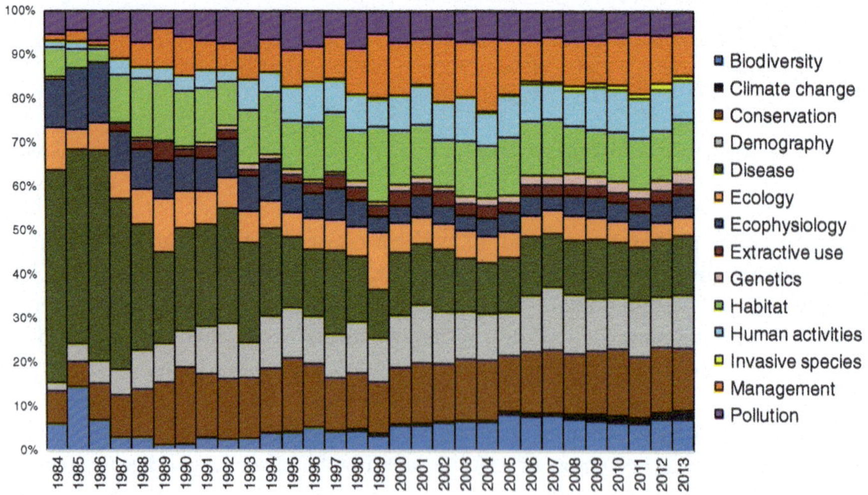

Fig. 6 Proportion of wildlife research publications in relation to topic and publication year

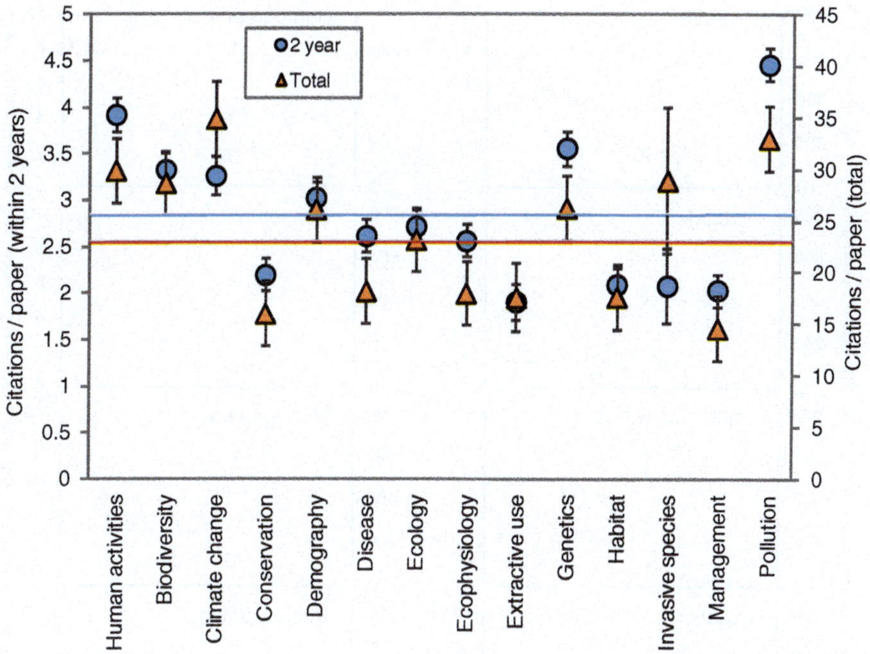

Fig. 7 Average (± SE) citations per paper within 2 years of publication and in total (up to July 2015) for wildlife research publications according to topic. Means calculated as LSMeans from a General Linear Model including "year of publication" as a continuous variable, and "topic" as a categorical variable. Horizontal lines represent the mean for all topics (*in red*, for total citations; in blue, for citations within 2 years of publication)

of issues, and incorporate a wider range of approaches in current times. Multidisciplinarity is thus a marked current trend for wildlife research, as also suggested by the study of scientific impact of different topics.

Scientific Impact by Topic: Trends and Milestones

As in all other disciplines, number of citations per paper has overall increased with time for all topics, in line with the increasing number of papers published (Fig. 8). Taking this overall increase into account, the mean number of citations per paper for studies published during the last 30 years in the area of wildlife research is 2.8 within the first 2 years after publication (i.e. year of publication + 2), and 23.8 in total (i.e., up to July 2015). Five topics had more impact both short-term (within 2 years) and long-term (in total) than the average; these were "pollution", "human activities", "biodiversity", "climate change" and "genetics" (Fig. 7). In addition, the topic of "invasive species" had a strong impact when considering total citations, which is particularly notable considering that most of those papers are relatively

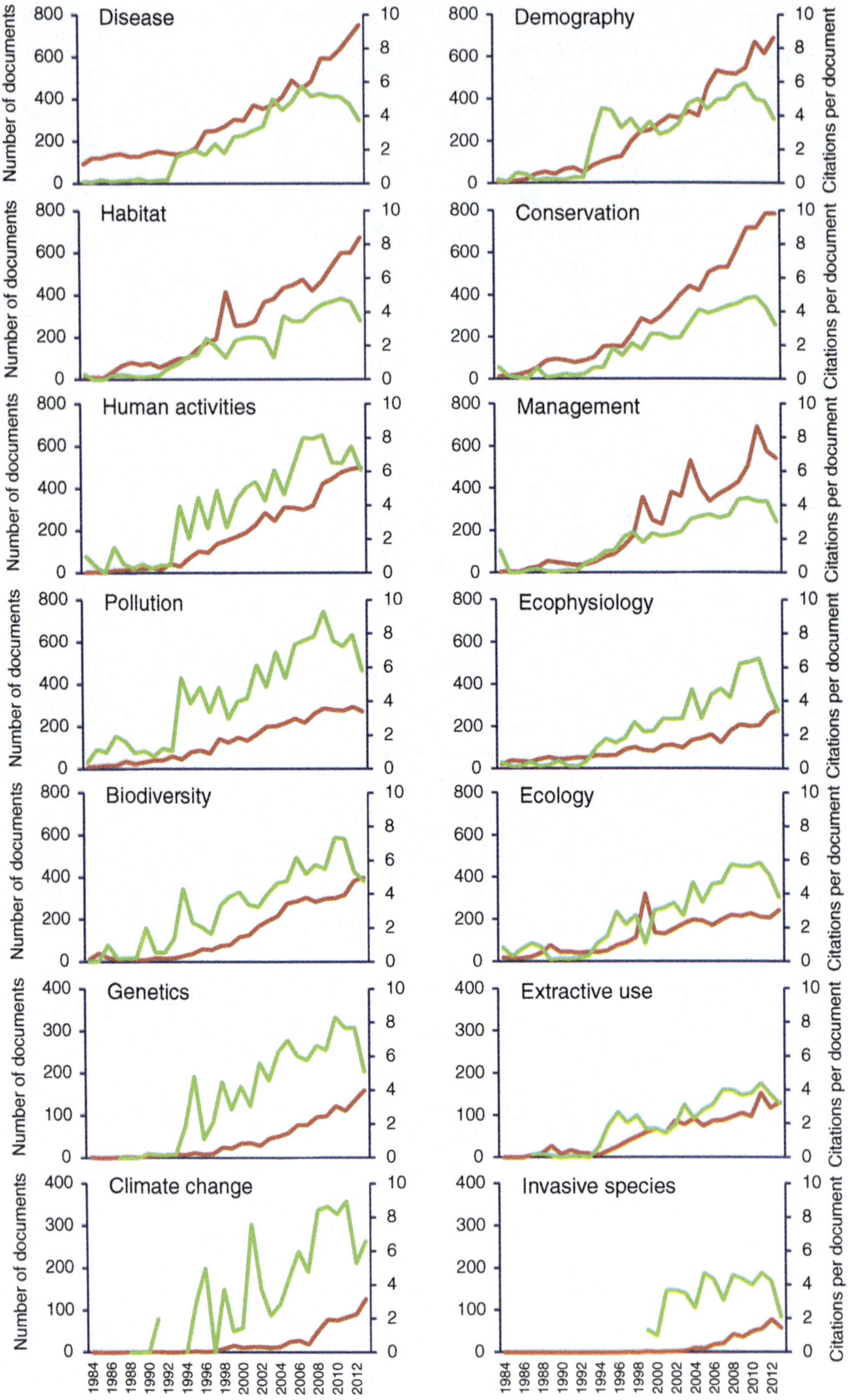

Disease
Demography
Habitat
Conservation
Human activities
Management
Pollution
Ecophysiology
Biodiversity
Ecology
Genetics
Extractive use
Climate change
Invasive species
Number of documents
Citations per document
800
600
400
200
0
300
100
10
8
6
4
2
1984
1986
1988
1990
1992
1994
1996
1998
2000
2002
2004
2006
2008
2010
2012

recent (there were no papers published on this topic before 1997 and regular numbers only appear in the last 10 years, Fig. 8).

Beyond the overall temporal increase in citations, several topics showed marked peaks in the number of citations within 2 years after their publication (Fig. 8), which relate to themes that were timely and relevant and may have helped to increase the visibility of the area at that time. The analysis of the temporal trend of citations together with the number of published papers in each topic permits to outline some of those themes and milestones, which we detail below.

Disease

This topic is one of the most relevant within wildlife research from the beginning, and the number of papers published has increased constantly since the mid-1990s. It showed a remarkable peak of citations in 1995–1996 and then smaller ones in 2004 and 2007 (Fig. 8).

The first citation peak reflected growing interest in wildlife diseases with a zoonotic potential, such as rabies or tuberculosis (O'Reilly and Daborn 1995; Rupprecht et al. 1995), including development of vaccines to prevent some diseases (Ertl and Xiang 1996). Additionally, a study modelling how even moderately severe diseases could increase probability of metapopulation extinctions (Hess 1996) also attracted large attention. In 2004, high impact was partly related to the identification of diclofenac, a veterinary drug used in cattle, as the cause of the decline of Asian vultures (Oaks et al. 2004). Other emerging topics that year leading to high impact of research were the highly pathogenic strain of avian influenza H5N1, not only because of the potential threat for humans, but also by the impact on wildlife conservation (Keawcharoen et al. 2004), and the epidemiology of Ebola virus in wildlife (Leroy et al. 2004). Finally, in 2007 attention was again focused on the role of wildlife in maintaining infections affecting domestic animals or humans (Wolfe et al. 2007; Dubey et al. 2007; Chomel et al. 2007; Gortazar et al. 2007).

Demography

Evaluation of population size, trends and demographic parameters is at the basis of population ecology, and thus wildlife management. The number of papers in this topic within wildlife research is thus large and has increased constantly since mid-1990s (Fig. 8).

Fig. 8 Number of publications (*in red*) and citations per paper within 2 years of publication (*in green*) according to year of publication and topic. Note that the scale for number of publications in the four graphs in the bottom (genetics, extractive use, climate change, invasive species) is different

In terms of citations, it showed a remarkable peak in 1994–1996, and then smaller ones in 2004–2005 and 2009–2010 (Fig. 8). The first one was associated to the publication of papers relating genetic parameters and effective population sizes in wildlife (Frankham 1995c), as well as the effects of contaminants or diseases on reproduction and population declines of wildlife (Facemire et al. 1995; Jobling et al. 1996; Villafuerte et al. 1994). At this time, the concept of virus-vectored contraception as a management tool for wildlife was also in the spotlight (Tyndale-Biscoe 1994). The second one highlights studies on the effects of farming on wildlife declines (Green et al. 2005), and it also picks up those relating the effects on survival or reproduction of pharmaceuticals (Oaks et al. 2004; Nash et al. 2004) or the impact of diseases on population declines (Leroy et al. 2004) already highlighted in other topics. In more recent years, highest impact has been for studies assessing new techniques for estimating population size or habitat-performance relationships (Gaillard et al. 2010; Luikart et al. 2010; Thomas et al. 2010), as well as others highlighting long-term negative fitness effects of captive breeding as a conservation management tool (Araki et al. 2009); impacts of plasticizers (Oehlmann et al. 2009); or the high mortality caused by fungi in amphibians (Harris et al. 2009), which has led to dramatic declines in many places. These examples highlight the transversality of this topic within wildlife research, with regular links to most of the other identified topics.

Habitat

The study of habitat-wildlife relationships is also central to wildlife research, being among the most important numerically at present (Fig. 8). The number of documents dealing with wildlife habitat has increased constantly during the 30-year study period. The marked peak in 1999 was due to the inclusion within this topic of 196 publications of a NCASI Technical Bulletin (a bulletin published by the National Council for Air and Stream Improvement, a non-profit research institute that focuses on environmental topics of interest to the forest products industry) which, that particular year, focused on the relationships between forestry and wildlife.

In terms of scientific impact, some early publications about the importance of habitat fragmentation (Fahrig 1997) and agriculture intensification (Krebs et al. 1999) became influential in the following years. Later, highly cited publications about wildlife habitat dealt with impacts of global changes of land use (Foley et al. 2005; Pickett et al. 2011), and especially with the dilemma of reconciling food production and wildlife conservation and defining appropriate farming strategies (Green et al. 2005; Power 2010).

Conservation

Conservation is also a core topic in wildlife research, with an important and steadily increasing number of papers (Fig. 8). Citation rate has regularly increased, reaching maxima in recent years. Papers highlighted at that time include a variety of themes, reflecting some important current wildlife conservation problems.

For example, papers highlighting the importance of fungus spread for the decline and extinction of frogs worldwide (Skerratt et al. 2007; Kilpatrick et al. 2010) have had strong impact. Conservation problems associated to climate change also feature in this topic (Post et al. 2009; Mawdsley et al. 2009), as well as those associated to other human activities like accumulation of plastic debris (Barnes et al. 2009) or genetic problems associated to harvest or release of plants or animals (Allendorf et al. 2008; Laikre et al. 2010). Also featured are the themes of human-wildlife conflicts (Inskip and Zimmermann 2009), or conservation in urban environments (Goddard et al. 2010). Other studies focus on how to find solutions to conservation problems, including the importance and caveats of assisted colonization or reintroductions to mitigate species extinctions (Seddon et al. 2007; Hoegh-Guldberg et al. 2008); the promotion of citizen science as a tool for conservation (Cooper et al. 2007); or the development of strategies that allow nature conservation and economic development or food production (Tallis et al. 2008; Power 2010). As for demography, these examples underline the transversality of this topic within wildlife research, and the links to most other identified topics.

Human Activities

The number of papers in this topic has been increasing constantly since 1994. Citation rate has been overall high for this topic, but there were marked peaks in the mean number of citations for papers published in the years 1994, 1996, 1998, 2004 and in the period 2007–2010 (Fig. 8).

Many of these high rates of citation are due to publications about environmental pollutants produced as a consequence of human activities. In particular, the peak in 1994 is related to some original and review papers about xenobiotics (i.e. chemical substances found within an organism not normally present there) with endocrine disruption activity, such as polychlorinated biphenyls (PCBs) (Bergman et al. 1994; Bimbaum 1994; Safe 1994). The peak in 1996 also relates to papers considering environmental pollutants because of human activity, but here the endocrine disruptors, including some emerging pollutants like alkylphenol ethoxylates widely used as surfactants, seem to have a leading role (Toppari et al. 1996; Kavlock et al. 1996; Nimrod and Benson 1996; Shelby et al. 1996). This year, research on the effect of human disturbance on animal populations was also on the spotlight (Gill et al. 1996), and had increasing impact subsequently. In the year 1998, research on persistent halogenated pollutants (i.e. dioxin-like compounds) had again a relevant impact due to the development of methods for their risk assessment in humans and wildlife (Van Den Berg et al 1998). More recently, in the period 2007–2010, research on other emerging pollutants originated from a wide range of manufactured products, like the perfluorinated compounds (Lau et al. 2007) or bisphenol A (Wetherill et al. 2007), was still in the focus of researchers, but in this period there was a higher diversity of subjects having high impact: the emergence of new wildlife diseases as a consequence of human actions (Dubey et al. 2007; Chomel et al. 2007); the impacts of biomass production (Semere and Slater 2007); effects of human-induced climate change in the arctic (Post et al. 2009); environmental impact of plastics (Barnes et al. 2009; Thompson et al. 2009; Oehlmann et al. 2009); impacts on protected species of predator control used for sport hunting (Packer et al. 2009); the adverse effects of widely used herbicides such as glyphosate-based products (Lushchak et al. 2009); or effects on wildlife of infrastructures for energy production or distribution (Kuvlesky et al. 2007; Benítez-López et al. 2010).

Management

As occurs with other broad topics, this one overlaps with many of our identified ones, and the total number of publications including it is elevated. Peaks in the citations of the publications of this topic are, however, not very evident (Fig. 8).

If we focus our analysis on the last part of the study period (2009 and afterwards), we can detect important issues like the transport and release to wildlife of chemical pollution by the plastics present in the environment (Teuten et al. 2009); concepts and methods for the joint analyses of spatial and genetic data (Guillot et al. 2009); the genetic consequences of plant and animal releases (Laikre et al. 2010); the spread of infectious and non-infectious diseases as a consequence of the alteration of global nutrient cycles of phosphorus and nitrogen caused by global changes (Johnson et al. 2010); management plans to cope with climate change effects of biodiversity (Mawdsley et al. 2009) or other types of adaptive management strategies addressed to current challenges (i.e. fire management, food demand) (Driscoll et al. 2010; Phalan et al. 2011). Other ecological aspects less related to human activity have been also on the focus on wildlife management research. In this sense, the perception by prey species of predation risk was found to be important for wildlife population dynamics (Zanette et al. 2011), so the accurate measurement of stress in wildlife was a relevant subject in recent years (Sheriff et al. 2011).

Pollution

The topic of pollution has many similarities with that of human activities in its trend over time (although the overall number of published documents is lower) and in some of the citation peaks. These occurred in 1994, 2002, 2004 and 2009 (Fig. 8).

The peak of citations in 1994 is mostly explained by work on PCBs (Safe 1994) and in particular a paper about their impact on birds from the Great Lakes region (Giesy et al. 1994). By the same time, endocrine disruptors like alkylphenolic compounds used as surfactants were an emerging topic (White et al. 1994). Later in 2002, research on other pollutants like brominated flame retardants (de Wit 2002), alkylphenol ethoxylates (Ying et al. 2002) and other xenoestrogens (Hong et al. 2002; Rajapakase et al. 2002), were of interest for the scientific community. Water pollution with pathogens (i.e *Toxoplasma gondii*) was another issue highlighted in this topic (Miller et al. 2002). By 2004, high relevance was obtained by some papers about the effect on fish of nanomaterials (i.e. fullerenes) (Oberdörster 2004) or pharmaceuticals such as contraceptive pills (Nash et al. 2004), as well as the mentioned review on brominated flame retardants (Bimbaum and Staskal 2004). Finally in 2009, the relevance is shared among bisphenol A and other endocrine disrupting-chemicals (Diamanti-Kandarakis et al. 2009; Vandenberg et al. 2009), pharmaceuticals (Kümmerer 2009) and plastics (Barnes et al. 2009).

Ecophysiology

The rate of increase in number of papers published in this topic has been more marked in the last 10 years, coinciding with a peak of citations in 2004 and 2009–2011 (Fig. 8).

High impact studies in 2004 were quite diverse, including a review about toxins of plants (i.e. pyrrolizidine alkaloids) (Fu et al. 2004), the application of novel molecular technologies in ecotoxicological studies (Snape et al. 2004), or the use of faecal glucocorticoids (an indicator of physiological stress) in ecological and conservation biology studies (Millspaugh and Washburn 2004). The latter issue (glucocorticoid analyses as a measure of stress in wildlife) was also a highly cited issue later in the period 2010–2011 (Sheriff et al. 2011), as well as the physiological effects of different types of persistent organic pollutants and metals (Chen and Hale 2010; Koivula and Eeva 2010; Letcher et al. 2010) and the global impact of wildlife diseases in the ecosystems (Tompkins et al. 2011).

Biodiversity

The study of biodiversity as part of wildlife research has also increased steadily throughout our study period. The mean number of citations per paper showed a marked peak in mid-1990s, as well as in 2006 and more recently (Fig. 8), related mainly to methods to study biodiversity and to the identified threats for biodiversity conservation.

Thus, in the 1990s, publications about persistent organic pollutants deserved great attention in the scientific community (Tilson et al. 1990; Murk et al. 1994; Safe 1994).

Later, the development of methods to study spatial distribution of wildlife at different scales, including models used to predict species presence, had a strong impact on wildlife research (Calenge 2006; Hirzel et al. 2006). Biodiversity is an ever-present topic in the challenge to harmonize food production and farming with the conservation of common and endangered species, and studies discussing strategies published in this period had also strong impact (Kleijn et al. 2006). At that time, the concept of ecosystem services as a means to value biodiversity gained relevance

(Losey and Vaughan 2006; Christie et al 2006; Power 2010). The increase and aggregation of human population is probably behind the rising interest for urban biodiversity in the last years (Chace and Walsh 2006; Goddard et al. 2010; Pickett et al. 2011). Finally, the concern about chemical pollutants highlighted in the 1990s has turned into an interest to identify the impact of emerging diseases on the conservation of biodiversity (i.e. bats or amphibians) (Frick et al. 2010; Kilpatrick et al. 2010; Altizer et al. 2011).

Ecology

The number of wildlife research publications within the topic of "ecology" is lower than the ones above, but has regularly increased throughout the study period. The marked peak in 1999 was due, as explained above, to the inclusion within this topic of the NCASI Technical Bulletin focusing on the relationships between forestry and wildlife.

Citation rate of papers within this topic did not show marked peaks throughout the study period, but maxima in 1996, 2002 and 2009–2011. The first one included highly cited methodological publications, including one for estimating animal home ranges (Kie et al. 1996) and the already mentioned one assessing how to quantify effects of human disturbance on animal populations (Gill et al. 1996). In 2002 appeared high impact publications on the ecology of emerging pathogens (Woolhouse 2002), but also a paper reviewing the economic reasons for conserving wild nature (Balmford et al. 2002), which further developed into the concept of ecosystem services. More recently, papers highlighted in this topic include ones about urban ecosystems (Chamberlain et al. 2009; Pickett et al. 2011); ecological effects of organohalogen contaminants (Letcher et al. 2010); and the already-mentioned ecological impacts of climate change in the arctic (Post et al. 2009).

Genetics

The field of genetics within wildlife research has only taken importance since 1998, according to the number of papers published in this topic (Fig. 8).

The increase in the occurrence of this topic followed the high impact of two publications in 1995 introducing the concept of population genetics and conservation genetics (Frankham 1995a, b). Subsequently, there has been a constant increase of this topic within wildlife research, and several peaks in citations have occurred. For instance, the concept of genetics became more present in highly cited publications about wildlife diseases (Hanlon et al. 1998; Tryland et al. 1998; Chua et al. 2002; Robinson et al. 2010). Several studies integrating the concept of population genetics in the conservation of species (Maudet et al. 2002; Bowen et al. 2005) and in strategies of extractives uses (Harris et al. 2002; Laikre et al. 2010) also had strong impact. Moreover, using genetics to assess the potential of adaptation of wildlife to our changing world has also gained relevance (Nussey et al. 2005; Charmantier and Garant 2005). In recent years, the impact of landscape fragmentation on the genetics of the species has been addressed in some influential papers (Goddard et al. 2010; Shirk et al. 2010).

Extractive Use

The concept of wildlife has been historically linked to game animals, but the number of publications specifically mentioning hunting or extractive use is much lower than those on conservation or management (Fig. 8). Peaks in citation rates were not marked, but several issues and publications can be highlighted during the last 30 years.

Some early influential publications proposed the extractive use of some game species as a method to reduce overpopulation, with examples as geese species in North America (Ankney 1996). This contrasts with the publications based on other scenarios where the regulation of harvest is necessary to avoid the overharvest of some populations and the consequent spatial extinctions (McCullough 1996; Milner-Gulland and Bennet 2003; Corlett 2007). Another topic has been the study of the consequence of trophy hunting of genetic shifts and demography of the populations under pressure (Coltman et al. 2003; Milner et al. 2007; Allendorf et al. 2008; Packer et al. 2011). The behavioural response of game animals to human disturbance in relation to hunting has been another issue studied (Stankowich 2008), as well as the importance of game meat (bushmeat) in human nutrition of developing countries and its conflicts with conservation and sustainable use of such natural resource (Golden et al. 2011). Finally, the consequences of other extractive uses (i.e. fisheries or seal and whale hunting) on wildlife species have been the subject of some impact studies (Hall and Harding 1997; Trivelpiece et al. 2011). Moreover, the growing concern about climate change introduces new aspects in the decision making process for wildlife management, including extractive uses (Nichols et al. 2011).

Climate Change

This issue is of great interest for the general public currently, but the number of documents relating climate changes and wildlife just peaked very recently, after 2007 (Fig. 8).

Publications on this topic had already a strong scientific impact from 1995, with a study of the impact on wildlife of reforestation with the purpose of sequestering carbon (Englin and Callaway 1995). This highlights the interest of not only the impact of climate change on wildlife, but also of the measures adopted to cope with the causes (i.e. wind power plants to reduce fossil fuel use). A second peak in 1998 related to another issue of great public attention: the conservation of the Amazonia forest (Laurance 1998). The quality of collected data for evaluating impacts was another concern in those early years with strong repercussion thereafter (Anderson 2001), as well as the effect of climate change on the ecology of diseases, especially those transmitted by vectors (Randolph 2001). After the take-off of this topic, in the period 2008–2011, impact of research has been related to very different research subjects, like the conservation of the arctic (Post et al. 2009); how climate changes modulate, both directly and indirectly, host–parasite interaction and disease persistence (Harvell et al. 2009; Altizer et al. 2011); the influence on the global cycle of contaminants (Noyes et al. 2009); and management and conservation tools to address the effects of climate change on wildlife (Mawdsley et al. 2009; Robinson et al. 2009). This highlights that this is a very transversal topic, interacting with many others.

Invasive Species

The topic of invasive species only appeared in the keywords of wildlife research publications in 1997, and only a few papers per year were published subsequently until 2004, where it took off exponentially (Fig. 8). Even if, in number of papers, it is only a minor topic right now, its impact (particularly observed in total number of citations per paper, Fig. 7) highlights that it will strongly increase in the near future.

Assessment of citations rates indicates that topics having higher impact included initially the assessment of methods to control invasive vertebrates endangering protected wildlife (Engeman and Vice 2001; Savarie et al. 2001) and the effects of invasive plants on wildlife (Herrera and Dudley 2003; Scheiman et al. 2003; Shafroth et al. 2005). More recently, focus has been on the impacts of invasive vertebrates on other wildlife (Doody et al. 2006; Dolman and Wäber 2008); modelling range expansion of invasive species to foresee management actions (Muñoz and Real 2006; Tattoni et al. 2006); or assessing the problems and costs associated to manage invasive species (Hayward and Kerley 2009; Cruz et al. 2009; Britton et al. 2011).

Conclusions and Perspectives

This short (and thus necessarily superficial) review shows that wildlife research is an increasingly productive and eclectic discipline, with progressively varied research subjects, leading to a more holistic approach. The depiction of some of the papers that had impact at different times in each identified topic also highlights that wildlife research studies are mostly focused not just on the animals, but on the relationship between animals and humans, and has thus a strong applied approach. This is shown in the increasing importance and impact of studies relating to the interactions with humans in each identified topic (consequences of pollution produced by humans on wildlife, the importance of zoonotic diseases, farming and wildlife, human-wildlife conflicts, etc.), as well as those evaluating ways of mitigating problems (conservation and management studies, which are somewhat transversal to all other topics). A significant result in this bibliometric review is that the highest scientific impact of wildlife research publications is usually achieved by studies that touch several topics. This highlights the importance of pluridisciplinarity for current wildlife research, which will undoubtedly increase in the future.

This bibliometric exercise has identified a number of topics that seem fundamental to describe current wildlife research trends. Studies on *demography* (including the study of factors affecting population trends), *wildlife diseases* (including the importance of wildlife to maintain infections that are relevant for domestic animals or humans) and *habitat* are core to the discipline of wildlife research, based on numbers of papers. *Pollution* (i.e. effects of different types of pollutants on wildlife) is, in addition, the one with highest impact at present, and one that appears also mentioned in many other topics (human activities, conservation, biodiversity), and particularly in ecophysiology because of the need of understanding the modes of action of pollutants. *Invasive species* and *genetics* are topics with strong recent increase and impact. Hunting and the impacts of *extractive use* of wildlife are not only traditional topics, but also increasing in the light of sustainable approaches. Although research themes identified by our search in the topics of "biodiversity" or "human activities" are mostly mentioned in other topics, *species distribution* and the impact of *human infrastructures* are not mentioned anywhere else and could be representative issues of both topics. Finally, research on wildlife conservation is frequently associated to minimizing problems about diseases, pollution or habitat loss, which usually conflicts with human development or other activities. *Conservation conflicts* are thus very relevant in terms of conservation and management of wildlife. These ten themes represent the topics of the following chapters that represent an updated overview of the knowledge attained on wildlife research, as well as a guide to identify current trends in this discipline and the remaining gaps to fill according to the new requirements of our changing world.

Acknowledgements We are thankful to C. Gortazar for insightful comments on the MS.

References

Allendorf FW, England PR, Luikart G, Ritchie PA, Ryman N (2008) Genetic effects of harvest on wild animal populations. Trends Ecol Evol 23:327–337

Altizer S, Bartel R, Han BA (2011) Animal migration and infectious disease risk. Science 331:296–302

Anderson DR (2001) The need to get the basics right in wildlife field studies. Wildl Soc Bull 29:1294–1297

Ankney CD (1996) An embarrassment of riches: too many geese. J Wildl Manage 60: 217–223
Araki H, Cooper B, Blouin MS (2009) Carry-over effect of captive breeding reduces reproductive fitness of wild-born descendants in the wild. Biol Lett 5:621–624
Balmford A, Bruner A, Cooper P, Costanza R, Farber S, Green RE, Jenkins M, Jefferiss P, Jessamy V, Madden J, Munro K, Myers N, Naeem S, Paavola J, Rayment M, Rosendo S, Roughgarden J, Trumper K, Turner RK (2002) Ecology: economic reasons for conserving wild nature. Science 297:950–953
Barnes DKA, Galgani F, Thompson RC, Barlaz M (2009) Accumulation and fragmentation of plastic debris in global environments. Philos Trans R Soc Lond B Biol Sci 364:1985–1998
Benítez-López A, Alkemade R, Verweij PA (2010) The impacts of roads and other infrastructure on mammal and bird populations: a meta-analysis. Biol Conserv 143:1307–1316
Bergman A, Lasson-Wehler E, Kuroki H (1994) Selective retention of hydroxylated PCB metabolites in blood. Environ Health Perspect 102:464–469
Bimbaum LS (1994) Endocrine effects of prenatal exposure to PCBs, dioxins, and other xenobiotics: implications for policy and future. Environ Health Perspect 102:676–679
Bimbaum LS, Staskal DF (2004) Brominated flame retardants: cause for concern? Environ Health Perspect 112:9–17
Bowen BW, Bass AL, Soares L, Toonen RJ (2005) Conservation implications of complex population structure: lessons from the loggerhead turtle (Caretta caretta). Mol Ecol 14:2389–2402
Britton JR, Gozlan RE, Copp GH (2011) Managing non-native fish in the environment. Fish Fish 12:256–274
Calenge C (2006) The package "adehabitat" for the R software: a tool for the analysis of space and habitat use by animals. Ecol Model 197:516–519
Chace JF, Walsh JJ (2006) Urban effects on native avifauna: a review. Landsc Urban Plan 74:46–69
Chamberlain DE, Cannon AR, Toms MP, Leech DI, Hatchwell BJ, Gaston KJ (2009) Avian productivity in urban landscapes: a review and meta-analysis. Ibis 151:1–18
Charmantier A, Garant D (2005) Environmental quality and evolutionary potential: lessons from wild populations. Proc Roy Soc B Biol Sci 272:1415–1425
Chen D, Hale RC (2010) A global review of polybrominated diphenyl ether flame retardant contamination in birds. Environ Int 36:800–811
Chomel BB, Belotto A, Meslin F-X (2007) Wildlife, exotic pets, and emerging zoonoses. Emerg Infect Dis 13:6–11
Christie M, Hanley N, Warren J, Murphy K, Wright R, Hyde T (2006) Valuing the diversity of biodiversity. Ecol Econ 58:304–317
Chua KB, Lek Koh C, Hooi PS, Wee KF, Khong JH, Chua BH, Chan YP, Lim ME, Lam SK (2002) Isolation of Nipah virus from Malaysian Island flying-foxes. Microbes Infect 4:145–151
Coltman DW, O'Donoghue P, Jorgenson JT, Hogg JT, Strobeck C, Festa-Blanchet M (2003) Undesirable evolutionary consequences of trophy hunting. Nature 426:655–658
Cooper CB, Dickinson J, Phillips T, Bonney R (2007) Citizen science as a tool for conservation in residential ecosystems. Ecol Soc 12: Article number 11
Corlett RT (2007) The impact of hunting on the mammalian fauna of tropical Asian forests. Biotropica 39:292–303
Cruz F, Carrion V, Campbell KJ, Lavoie C, Donlan CJ (2009) Bio-economics of large-scale eradication of feral goats from Santiago Island, Galapagos. J Wild Manage 73:191–200
de Wit CA (2002) An overview of brominated flame retardants in the environment. Chemosphere 46:583–624
Diamanti-Kandarakis E, Bourguignon J-P, Giudice LC, Hauser R, Prins GS, Soto AM, Zoeller RT, Gore AC (2009) Endocrine-disrupting chemicals: an Endocrine Society scientific statement. Endocr Rev 30:293–342
Díaz M, Campos P, Pulido F (2009) Importancia de la caza en el desarrollo sustentable y en la conservación de la biodiversidad. In: Sáez de Buruaga M, Carranza J (eds) Gestión cinegética en los ecosistemas mediterráneos. Junta de Andalucía, Sevilla, pp 21–33
Dolman PM, Wäber K (2008) Ecosystem and competition impacts of introduced deer. Wildl Res 35:202–214

Doody JS, Green B, Sims R, Rhind D, West P, Steer D (2006) Indirect impacts of invasive cane toads (*Bufo marinus*) on nest predation in pig-nosed turtles (*Carettochelys insculpta*). Wildl Res 33:349–354

Driscoll DA, Lindenmayer DB, Bennett AF, Bode M, Bradstock RA, Cary GJ, Clarke MF, Dexter N, Fensham R, Friend G, Gill M, James S, Kay G, Keith DA, MacGregor C, Russell-Smith J, Salt D, Watson JEM, Williams RJ, York A (2010) Fire management for biodiversity conservation: key research questions and our capacity to answer them. Biol Conserv 143:1928–1939

Dubey JP, Schares G, Ortega-Mora LM (2007) Epidemiology and control of neosporosis and Neospora caninum. Clin Microbiol Rev 20:323–367

Engeman RM, Vice DS (2001) Objectives and integrated approaches for the control of brown tree snakes. Int Pest Manage Rev 6:59–76

Englin J, Callaway JM (1995) Environmental impacts of sequestering carbon through forestation. Clim Change 31:67–78

Ertl HCJ, Xiang Z (1996) Novel vaccine approaches. J Immunol 156:3579–3582

Facemire CF, Gross TS, Guillette LJ Jr (1995) Reproductive impairment in the Florida panther: nature or nurture? Environ Health Perspect 103:79–86

Fahrig L (1997) Relative effects of habitat loss and fragmentation on population extinction. J Wildl Manage 61:603–610

Foley JA, DeFries R, Asner GP, Barford C, Bonan G, Carpenter SR, Chapin FS, Coe MT, Daily GC, Gibbs HK, Helkowski JH, Holloway T, Howard EA, Kucharik CJ, Monfreda C, Patz JA, Prentice IC, Ramankutty N, Snyder PK (2005) Global consequences of land use. Science 309:570–574

Frankham R (1995a) Effective population size/adult population size ratios in wildlife: a review. Genet Res 66:95–107

Frankham R (1995b) Conservation genetics. Annu Rev Genet 29:305–327

Frankham R (1995c) Relationship of genetic variation to population size in wildlife. Conserv Biol 10:1500–1508

Frick WF, Pollock JF, Hicks AC, Langwig KE, Reynolds DS, Turner GG, Butchkoski CM, Kunz TH (2010) An emerging disease causes regional population collapse of a common North American bat species. Science 329:679–682

Fu PP, Xia Q, Lin G, Chou MW (2004) Pyrrolizidine alkaloids – genotoxicity, metabolism enzymes, metabolic activation, and mechanisms. Drug Metab Rev 36:1–55

Gaillard J-M, Hebblewhite M, Loison A, Fuller M, Powell R, Basille M, Van Moorter B (2010) Habitat-performance relationships: finding the right metric at a given spatial scale. Philos Trans R Soc Lond B Biol Sci 365:2255–2265

Giesy JP, Ludwig JP, Tillitt DE (1994) Deformities in birds of the Great Lakes region: assigning causality. Environ Sci Tech 28:128A–135A

Gill JA, Sutherland WJ, Watkinson AR (1996) A method to quantify the effects of human disturbance on animal populations. J Appl Ecol 33:786–792

Goddard MA, Dougill AJ, Benton TG (2010) Scaling up from gardens: biodiversity conservation in urban environments. Trends Ecol Evolut 25:90–98

Golden CD, Fernald LCH, Brashares JS, Rasolofoniaina BJR, Kremen C (2011) Benefits of wildlife consumption to child nutrition in a biodiversity hotspot. Proc Natl Acad Sci U S A 108:19653–19656

Good SP (1997) Wilderness and the hunting experience: what it means to be a hunter. Wildl Soc Bull 25:563–567

Gortazar C (2012) Wildlife research – science for a changing environment. Eur J Wildl Res 58:1–4

Gortazar C, Ferroglio E, Höfle U, Frölich K, Vicente J (2007) Diseases shared between wildlife and livestock: a European perspective. Eur J Wildl Res 53:241–256

Green RE, Cornell SJ, Scharlemann JPW, Balmford A (2005) Farming and the fate of wild nature. Science 307:550–555

Guillot G, Leblois R, Coulon A, Frantz AC (2009) Statistical methods in spatial genetics. Mol Ecol 18:4734–4756

Hall SJ, Harding MJC (1997) Physical disturbance and marine benthic communities: the effects of mechanical harvesting of cockles on non-target benthic infauna. J Appl Ecol 34:497–517

Hanlon CA, Niezgoda M, Hamir AN, Schumacher C, Koprowski H, Rupprecht CE (1998) First North American field release of a vaccinia-rabies glycoprotein recombinant virus. J Wildl Dis 34:228–239

Harris RB, Wall WA, Allendorf FW (2002) Genetic consequences of hunting: what do we know and what should we do? Wildl Soc Bull 30:634–643

Harris RN, Brucker RM, Walke JB, Becker MH, Schwantes CR, Flaherty DC, Lam BA, Woodhams DC, Briggs CJ, Vredenburg VT, Minbiole KPC (2009) Skin microbes on frogs prevent morbidity and mortality caused by a lethal skin fungus. ISME J 3:818–824

Harvell D, Altizer S, Cattadori IM, Harrington L, Weil E (2009) Climate change and wildlife diseases: when does the host matter the most? Ecology 90:912–920

Hayward MW, Kerley GIH (2009) Fencing for conservation: restriction of evolutionary potential or a riposte to threatening processes? Biol Conserv 142:1–13

Herrera AM, Dudley TI (2003) Reduction of riparian arthropod abundance and diversity as a consequence of giant reed (*Arundo donax*) invasion. Biol Invasions 5:167–177

Hess G (1996) Disease in metapopulation models: implications for conservation. Ecology 77:1617–1632

Hirzel AH, Le Lay G, Helfer V, Randin C, Guisan A (2006) Evaluating the ability of habitat suitability models to predict species presences. Ecol Model 199:142–152

Hoegh-Guldberg O, Hughes L, McIntyre S, Lindenmayer DB, Parmesan C, Possingham HP, Thomas CD (2008) Ecology: assisted colonization and rapid climate change. Science 321:345–346

Hong H, Tong W, Fang H, Shi L, Xie Q, Wu J, Perkins R, Walker JD, Branham W, Sheehan DM (2002) Prediction of estrogen receptor binding for 58,000 chemicals using an integrated system of a tree-based model with structural alerts. Environ Health Perspect 110:29–36

Inskip C, Zimmermann A (2009) Human-felid conflict: a review of patterns and priorities worldwide. Oryx 43:18–34

Jobling S, Sheahan D, Osborne JA, Matthiessen P, Sumpter JP (1996) Inhibition of testicular growth in rainbow trout (Oncorhynchus mykiss) exposed to estrogenic alkylphenolic chemicals. Environ Toxicol Chem 15:194–202

Johnson PTJ, Townsend AR, Cleveland CC, Glibert PM, Howarth RW, Mckenzie VJ, Rejmankova E, Ward MH (2010) Linking environmental nutrient enrichment and disease emergence in humans and wildlife. Ecol Appl 20:16–29

Kavlock RJ, Daston GP, DeRosa C, Fenner-Crisp P, Gray LE, Kaattari S, Lucier G, Luster M, Mac MJ, Maczka C, Miller R, Moore J, Rolland R, Scott G, Sheehan DM, Sinks T, Tilson HA (1996) Research needs for the risk assessment of health and environmental effects of endocrine disrupters: a report of the U.S. EPA-sponsored workshop. Environ Health Perspect 104:715–740

Keawcharoen J, Oraveerakul K, Kuiken T, Fouchier RAM, Amonsin A, Payungporn S, Noppornpanth S, Wattanodorn S, Theamboonlers A, Tantilertcharoen R, Pattanarangsan R, Arya N, Ratanakorn P, Osterhaus ADME, Poovorawan Y (2004) Avian influenza H5N1 in tigers and leopards. Emerg Infect Dis 10(12):2189–2191

Kie JG, Baldwin JA, Evans CJ (1996) CALHOME: a program for estimating animal home ranges. Wildl Soc Bull 24:342–344

Kilpatrick AM, Briggs CJ, Daszak P (2010) The ecology and impact of chytridiomycosis: an emerging disease of amphibians. Trends Ecol Evolut 25:109–118

Kleijn D, Baquero RA, Clough Y, Díaz M, De Esteban J, Fernández F, Gabriel D, Herzog F, Holzschuh A, Jöhl R, Knop E, Kruess A, Marshall EJP, Steffan-Dewenter I, Tscharntke T, Verhulst J, West TM, Yela JL (2006) Mixed biodiversity benefits of agri-environment schemes in five European countries. Ecol Lett 9:243–254

Koivula MJ, Eeva T (2010) Metal-related oxidative stress in birds. Environ Pollut 158:2359–2370

Krebs JR, Wilson JD, Bradbury RB, Siriwardena GM (1999) The second silent spring? Nature 400:611–612

Kümmerer K (2009) The presence of pharmaceuticals in the environment due to human use – present knowledge and future challenges. J Environ Manage 90:2354–2366

Kuvlesky WP Jr, Brennan LA, Morrison ML, Boydston KK, Ballard BM, Bryant FC (2007) Wind energy development and wildlife conservation: challenges and opportunities. J Wildl Manage 71:2487–2498

Laikre L, Schwartz MK, Waples RS, Ryman N (2010) Compromising genetic diversity in the wild: unmonitored large-scale release of plants and animals. Trends Ecol Evolut 25:520–529

Lau C, Anitole K, Hodes C, Lai D, Pfahles-Hutchens A, Seed J (2007) Perfluoroalkyl acids: a review of monitoring and toxicological findings. Toxicol Sci 99:366–394

Laurance WF (1998) A crisis in the making: responses of Amazonian forests to land use and climate change. Trends Ecol Evolut 13:411–415

Leroy EM, Rouquet P, Formenty P, Souquière S, Kilbourne A, Froment J-M, Bermejo M, Smit S, Karesh W, Swanepoel R, Zaki SR, Rollin PE (2004) Multiple Ebola virus transmission events and rapid decline of central African wildlife. Science 303:387–390

Letcher RJ, Bustnes JO, Dietz R, Jenssen BM, Jørgensen EH, Sonne C, Verreault J, Vijayan MM, Gabrielsen GW (2010) Exposure and effects assessment of persistent organohalogen contaminants in arctic wildlife and fish. Sci Total Environ 408:2995–3043

Losey JE, Vaughan M (2006) The economic value of ecological services provided by insects. BioScience 56:311–323

Luikart G, Ryman N, Tallmon DA, Schwartz MK, Allendorf FW (2010) Estimation of census and effective population sizes: the increasing usefulness of DNA-based approaches. Conserv Genet 11:355–373

Lushchak OV, Kubrak OI, Storey JM, Storey KB, Lushchak VI (2009) Low toxic herbicide Roundup induces mild oxidative stress in goldfish tissues. Chemosphere 76:932–937

Maudet C, Miller C, Bassano B, Breitenmoser-Würsten C, Gauthier D, Obexer-Ruff G, Michallet J, Taberlet P, Luikart G (2002) Microsatellite DNA and recent statistical methods in wildlife conservation management: Applications in Alpine ibex [Capra ibex (ibex)]. Mol Ecol 11:421–436

Mawdsley JR, O'Malley R, Ojima DS (2009) A review of climate-change adaptation strategies for wildlife management and biodiversity conservation. Conserv Biol 23:1080–1089

Mayring P (2000) Qualitative content analysis. Forum Qualitative Sozial- forschung/Forum: Qualitative Social Research, 1, art. 20. Available at http://www.qualitative-research.net/index.php/fqs/article/view/1089/2385

McCullough DR (1996) Spatially structured populations and harvest theory. J Wildl Manage 60: 1–9

Miller MA, Gardner IA, Kreuder C, Paradies DM, Worcester KR, Jessup DA, Dodd E, Harris MD, Ames JA, Packham AE, Conrad PA (2002) Coastal freshwater runoff is a risk factor for *Toxoplasma gondii* infection of southern sea otters (Enhydra lutris nereis). Int J Parasitol 32:997–1006

Millspaugh JJ, Washburn BF (2004) Use of fecal glucocorticoid metabolite measures in conservation biology research: considerations for application and interpretation. Gen Comp Endocrinol 138:189–199

Milner JM, Nilsen EB, Andreassen HP (2007) Demographic side effects of selective hunting in ungulates and carnivores: review. Conserv Biol 21:36–47

Milner-Gulland EJ, Bennet EL (2003) Wild meat: the bigger picture. Trends Ecol Evol 18: 351–357

Muñoz A-R, Real R (2006) Assessing the potential range expansion of the exotic monk parakeet in Spain. Divers Distrib 12:656–665

Murk A, Morse D, Boon J, Brouwer A (1994) In vitro metabolism of 3,3′,4,4′-tetrachlorobiphenyl in relation to ethoxyresorufin-O-deethylase activity in liver microsomes of some wildlife species and rat. Eur J Pharmacol 270:253–261

Nash JP, Kime DE, Van der Ven LTM, Wester PW, Brion F, Maack G, Stahlschmidt-Allner P, Tyler CR (2004) Long-term exposure to environmental concentrations of the pharmaceutical ethynylestradiol causes reproductive failure in fish. Environ Health Perspect 112:1725–1733

Nichols JD, Koneff MD, Heglund PJ, Knutson MG, Seamans ME, Lyons JE, Morton JM, Jones MT, Boomer GS, Williams BK (2011) Climate change, uncertainty, and natural resource management. J Wildl Manage 75:6–18

Nimrod AC, Benson WH (1996) Environmental estrogenic effects of alkylphenol ethoxylates. Crit Rev Toxicol 26:335–364

Noyes PD, McElwee MK, Miller HD, Clark BW, Van Tiem LA, Walcott KC, Erwin KN, Levin ED (2009) The toxicology of climate change: environmental contaminants in a warming world. Environ Int 35:971–986

Nussey DH, Postma E, Gienapp P, Visser ME (2005) Evolution: selection on heritable phenotypic plasticity in a wild bird population. Science 310:304–306

O'Reilly LM, Daborn CJ (1995) The epidemiology of Mycobacterium bovis infections in animals and man: a review. Tuber Lung Dis 76(Suppl 1):1–46

Oaks JL, Gilbert M, Virani MZ, Watson RT, Meteyer CU, Rideout BA, Shivaprasad HL, Ahmed S, Chaudhry MJI, Arshad M, Mahmood S, Ali A, Khan AA (2004) Diclofenac residues as the cause of vulture population decline in Pakistan. Nature 427:630–633

Oberdörster E (2004) Manufactured nanomaterials (fullerenes, C60) induce oxidative stress in the brain of juvenile largemouth bass. Environ Health Perspect 112:1058–1062

Oehlmann J, Schulte-Oehlmann U, Kloas W, Jagnytsch O, Lutz I, Kusk KO, Wollenberger L, Santos EM, Paull GC, VanLook KJW, Tyler CR (2009) A critical analysis of the biological impacts of plasticizers on wildlife. Phil Trans R Soc B Biol Sci 364(1567):2047–2062

Packer C, Brink H, Kissui BM, Maliti H, Kushnir H, Caro T (2011) Effects of trophy hunting on lion and leopard populations in Tanzania. Conserv Biol 25:142–153

Packer C, Kosmala M, Cooley HS, Brink H, Pintea L, Garshelis D, Purchase G, Strauss M, Swanson A, Balme G, Hunter L, Nowell K (2009) Sport hunting, predator control and conservation of large carnivores. PLoS One 4:e5941

Phalan B, Balmford A, Green RE, Scharlemann JPW (2011) Minimising the harm to biodiversity of producing more food globally. Food Policy 36(Suppl 1):S62–S71

Pickett STA, Cadenasso ML, Grove JM, Boone CG, Groffman PM, Irwin E, Kaushal SS, Marshall V, McGrath BP, Nilon CH, Pouyat RV, Szlavecz K, Troy A, Warren P (2011) Urban ecological systems: scientific foundations and a decade of progress. J Environ Manage 92:331–362

Post E, Forchhammer MC, Bret-Harte MS, Callaghan TV, Christensen TR, Elberling B, Fox AD, Gilg O, Hik DS, Høye TT, Ims RA, Jeppesen E, Klein DR, Madsen J, McGuire AD, Rysgaard S, Schindler DE, Stirling I, Tamstorf MP, Tyler NJC, Van Der Wal R, Welker J, Wookey PA, Schmidt NM, Aastrup P (2009) Ecological dynamics across the arctic associated with recent climate change. Science 326:1355–1358

Powell RA, Ransom D Jr, Slack RD, Silvy NJ (2010) Dynamics of content and authorship patterns in the Wildlife Society Journals (1937–2007). J Wildl Manage 74:816–827

Power AG (2010) Ecosystem services and agriculture: tradeoffs and synergies. Phil Trans R Soc B Biol Sci 365:2959–2971

Rajapakase N, Silva E, Kortenkamp A (2002) Combining xenoestrogens at levels below individual no-observed-effect concentrations dramatically enhances steroid hormone action. Environ Health Perspect 110:917–921

Randolph SE (2001) The shifting landscape of tick-borne zoonoses: tick-borne encephalitis and Lyme borreliosis in Europe. Phil Trans R Soc B Biol Sci 356:1045–1056

Robinson RA, Crick HQP, Learmonth JA, Maclean IMD, Thomas CD, Bairlein F, Forchhammer MC, Francis CM, Gill JA, Godley BJ, Harwood J, Hays GC, Huntley B, Hutson AM, Pierce GJ, Rehfisch MM, Sims DW, Santos MB, Sparks TH, Stroud DA, Visser ME (2009) Travelling through a warming world: climate change and migratory species. Endanger Species Res 7:87–99

Robinson RA, Lawson B, Toms MP, Peck KM, Kirkwood JK, Chantrey J, Clatworthy IR, Evans AD, Hughes LA, Hutchinson OC, John SK, Pennycott TW, Perkins MW, Rowley PS, Simpson VR, Tyler KM, Cunningham AA (2010) Emerging infectious disease leads to rapid population declines of common British birds. PLoS One 5:e12215

Rupprecht CE, Smith JS, Makonnen Fekadu MS, Childs JE (1995) The ascension of wildlife rabies: a cause for public health concern or intervention? Emerg Infect Dis 1:107–114
Safe SH (1994) Polychlorinated biphenyls (PCBs): environmental impact, biochemical and toxic responses, and implications for risk assessment. Crit Rev Toxicol 24:87–149
Savarie PJ, Shivik JA, White GC, Hurley JC, Clark L (2001) Use of acetaminophen for large-scale control of brown treesnakes. J Wildl Manage 65:356–365
Scheiman DM, Bollinger EK, Johnson DH (2003) Effects of leafy spurge infestation on grassland birds. J Wild Manage 67:115–121
Seddon PJ, Armstrong DP, Maloney RF (2007) Developing the science of reintroduction biology. Conserv Biol 21:303–312
Semere T, Slater FM (2007) Ground flora, small mammal and bird species diversity in miscanthus (*Miscanthus × giganteus*) and reed canary-grass (*Phalaris arundinacea*) fields. Biomass Bioenergy 31:20–29
Shafroth PB, Cleverly JR, Dudley TL, Taylor JP, Van Riper C III, Weeks EP, Stuart JN (2005) Control of *Tamarix* in the western United States: implications for water salvage, wildlife use, and riparian restoration. Environ Manage 35:231–246
Shelby MD, NEwbold RR, Tully DB, Chae K, Davis VL (1996) Assessing environmental chemicals for estrogenicity using a combination of in vitro and in vivo assays. Environ Health Perspect 104:1296–1300
Sheriff MJ, Dantzer B, Delehanty B, Palme R, Boonstra R (2011) Measuring stress in wildlife: techniques for quantifying glucocorticoids. Oecologia 166:869–887
Shirk AJ, Wallin DO, Cushman SA, Rice CG, Warheit KI (2010) Inferring landscape effects on gene flow: a new model selection framework. Mol Ecol 19:3603–3619
Skerratt LF, Berger L, Speare R, Cashins S, McDonald KR, Phillott AD, Hines HB, Kenyon N (2007) Spread of chytridiomycosis has caused the rapid global decline and extinction of frogs. Ecohealth 4:125–134
Snape JR, Maund SJ, Pickford DB, Hutchinson TH (2004) Ecotoxicogenomics: the challenge of integrating genomics into aquatic and terrestrial ecotoxicology. Aquat Toxicol 67:143–154
Stankowich T (2008) Ungulate flight responses to human disturbance: a review and meta-analysis. Biol Conserv 141:2159–2173
Tallis H, Kareiva P, Marvier M, Chang A (2008) An ecosystem services framework to support both practical conservation and economic development. Proc Natl Acad Sci U S A 105:9457–9464
Tattoni C, Preatoni DG, Lurz PWW, Rushton SP, Tosi G, Bertolino S, Martinoli A, Wauters LA (2006) Modelling the expansion of a grey squirrel population: implications for squirrel control. Biol Invasions 8:1605–1619
Teuten EL, Saquing JM, Knappe DRU, Barlaz MA, Jonsson S, Björn A, Rowland SJ, Thompson RC, Galloway TS, Yamashita R, Ochi D, Watanuki Y, Moore C, Viet PH, Tana TS, Prudente M, Boonyatumanond R, Zakaria MP, Akkhavong K, Ogata Y, Hirai H, Iwasa S, Mizukawa K, Hagino Y, Imamura A, Saha M, Takada H (2009) Transport and release of chemicals from plastics to the environment and to wildlife. Phil Trans R Soc B Biol Sci 364:2027–2045
Thomas L, Buckland ST, Rexstad EA, Laake JL, Strindberg S, Hedley SL, Bishop JR, Marques TA, Burnham KP (2010) Distance software: design and analysis of distance sampling surveys for estimating population size. J Appl Ecol 47:5–14
Thompson RC, Moore CJ, Saal FSV, Swan SH (2009) Plastics, the environment and human health: current consensus and future trends. Phil Trans R Soc B Biol Sci 364:2153–5166
Tilson HA, Jacobson JL, Rogan WJ (1990) Polychlorinated biphenyls and the developing nervous system: cross-species comparisons. Neurotoxicol Teratol 12:239–248
Tompkins DM, Dunn AM, Smith MJ, Telfer S (2011) Wildlife diseases: from individuals to ecosystems. J Wildl Ecol 80:19–38
Toppari J, Larsen JC, Christiansen P, Giwercman A, Grandjean P, Guillette LJ Jr, Jégou B, Jensen TK, Jouannet P, Keiding N, Leffers H, McLachlan JA, Meyer O, Müller J, Rajpert-De Meyts E, Scheike T, Sharpe R, Sumpter J, Skakkebæk NE (1996) Male reproductive health and environmental xenoestrogens. Environ Health Perspect 104:741–803

Trivelpiece WZ, Hinke JT, Miller AK, Reiss CS, Trivelpiece SG, Watters GM (2011) Variability in krill biomass links harvesting and climate warming to penguin population changes in Antarctica. Proc Natl Acad Sci U S A 108:7625–7628

Tryland M, Sandvik T, Mehl R, Bennett M, Traavik T, Olsvik O (1998) Serosurvey for orthopoxviruses in rodents and shrews from Norway. J Wildl Dis 34:240–250

Tyndale-Biscoe CH (1994) Virus-vectored immunocontraception of feral mammals. Reprod Fertil Dev 6:281–287

Van Den Berg M, Birnbaum L, Bosveld ATC, Brunström B, Cook P, Feeley M, Giesy JP, Hanberg A, Hasegawa R, Kennedy SW, Kubiak T, Larsen JC, Van Leeuwen FXR, Liem AKD, Nolt C, Peterson RE, Poellinger L, Safe S, Schrenk D, Tillitt D, Tysklind M, Younes M, Wærn F, Zacharewski T (1998) Toxic equivalency factors (TEFs) for PCBs, PCDDs, PCDFs for humans and wildlife. Environ Health Perspect 106:775–792

Vandenberg LN, Maffini MV, Sonnenschein C, Rubin BS, Soto AM (2009) Bisphenol-a and the great divide: a review of controversies in the field of endocrine disruption. Endocr Rev 30:75–95

Villafuerte R, Calvete C, Gortázar C, Moreno S (1994) First epizootic of rabbit hemorrhagic disease in free living populations of Oryctolagus cuniculus at Doñana National Park, Spain. J Wildl Dis 30:176–179

Wetherill YB, Akingbemi BT, Kanno J, McLachlan JA, Nadal A, Sonnenschein C, Watson CS, Zoeller RT, Belcher SM (2007) In vitro molecular mechanisms of bisphenol A action. Reprod Toxicol 24:178–198

White R, Jobling S, Hoare SA, Sumpter JP, Parker MG (1994) Environmentally persistent alkylphenolic compounds are estrogenic. Endocrinology 135:175–182

Wolfe ND, Dunavan CP, Diamond J (2007) Origins of major human infectious diseases. Nature 447:279–283

Woolhouse MEJ (2002) Population biology of emerging and re-emerging pathogens. Trends Microbiol 10:S3–S7

Ying G-G, Williams B, Kookana R (2002) Environmental fate of alkylphenols and alkylphenol ethoxylates – a review. Environ Int 28:215–226

Zanette LY, White AF, Allen MC, Clinchy M (2011) Perceived predation risk reduces the number of offspring songbirds produce per year. Science 334:1398–1401

Wildlife Demography: Population Processes, Analytical Tools and Management Applications

Jean-Dominique Lebreton and Jean-Michel Gaillard

Introduction

Although the term "wildlife" should literally encompass all wild living forms (including e.g. arthropods, plants...), it is generally taken in the more restricted meaning of wild terrestrial vertebrates, primarily birds and mammals. In modern societies, the interactions between humans and wildlife are many and diverse: for instance, wildlife is nowadays a recreational resource (Duffus and Dearden 1990) and induces risks of disease transmission to domestic animals (Frölich et al. 2002). In terms of space and time, human-wildlife interactions range from small-scale issues, e.g. impacts of foxes on poultry (Moberly et al. 2004), to large scale and even global issues, such as the impact of the various components of global change on wildlife and possible mitigation measures (Inkley et al. 2004; Mawdsley et al. 2009).

The consequences for the two partners – humans and wildlife – of these various interactions depend largely on the number of animals concerned and change in these numbers, whether under natural mechanisms or through human harvest or management actions. Hence, wildlife demography necessarily plays a central role in answering any question on human – wildlife interactions. For instance the sustain-

J.-D. Lebreton (✉)
CEFE, UMR 5175, Campus CNRS,
1919 Route de Mende, 34293 Montpellier cedex 5, France
e-mail: jean-dominique.lebreton@cefe.cnrs.fr

J.-M. Gaillard
LBBE, UMR 5558, Université Claude Bernard Lyon I,
43 Boulevard du 11 Novembre 1918, Villeurbanne 69622, France

R. Mateo et al. (eds.), *Current Trends in Wildlife Research*,
Wildlife Research Monographs 1, DOI 10.1007/978-3-319-27912-1_2

ability of harvest of quarry species as well as the effectiveness of conservation plans deal with the impact of man-induced changes in wildlife demography, which can be addressed through the theory of exploited populations (Lebreton 2005a). Wildlife demography is strongly rooted in human demography with similar processes (such as density-dependence, stochastic variation, or individual heterogeneity) and models (such as logistic population growth or Gompertz law of mortality). Recent research on senescence provides the best evidence supporting the strong connexion existing between human and wildlife demography (Jones et al. 2014; Caswell 2014). Thanks to the accumulation of demographic data at the individual level collected during long-term research programs (Clutton-Brock and Sheldon 2010), wildlife demography is increasingly used to solve management problems (see e.g., Gamelon et al. 2012 for a case study on wild boar, *Sus scrofa*). One main limitation of wildlife demography is the strong data requirement. It is not feasible to obtain detailed demographic data for all species involved in management plans. However, both the development of demographic approaches based on incomplete data (e.g. Niel and Lebreton 2005) and the identification of broad types of populations or species with similar demographic patterns based on analyses of life history variation (e.g. Gamelon et al. 2014) should allow to generalize wildlife demography to address management issues.

Another connection between wildlife and human studies lies in the similarity of statistical tools used to analyse flows, such as death rate, or rates of transition among various states. The tools used for wildlife, mostly within the capture-mark- recapture framework, have common roots with those used in human epidemiology (in particular cancer research) (Lebreton 1995), with a specific attention on non-exhaustive registration that was only progressively transferred to human studies.

The purpose of this chapter is to provide a broad overview of wildlife demography and explain how demographic approaches shed light on wildlife conservation and management issues. We will first summarize the main interactions between humans and wildlife and the questions they raise (2), and then briefly review the history of research on wildlife demography (3) and modern tools for wildlife demography (4). In a comparative perspective, we will next show how the diversity of wildlife life cycles and demography is organized on a slow-fast continuum (5) and how the theory of exploited populations can be used to understand different sensitivities to extra sources of mortality along this continuum. We then present some key aspects of variation in demographic parameters (6) such as environmental variation and density-dependence. In a general discussion (7) we will briefly present what we think are the dominant trends for future research on wildlife demography, in particular in face of global change. We will not cover issues dealing with fish (see e.g. Reynolds et al. 2001), wildlife being again taken in the restricted meaning of terrestrial vertebrates, but we will see in discussion that many aspects of wildlife demography converges with what is known of fisheries.

Table 1 Some typical cases of interactions between human activities and wildlife, arranged by species groups and type of activity, with references that discuss wildlife demographic aspects in the interaction concerned

	Waterbirds, Seabirds	Ungulates	Predators	Others
Conservation	Impact of climate change on high latitude and elevation species (Le Bohec et al. 2008) Longline fishery seabird bycatch (Weimerskirch et al. 1997)	Endangered species (Hoffmann et al. 2011) Impact of climate change on high latitude and elevation species (Post and Forchhammer 2008) and on lowland and temperate species (Plard et al. 2014)	Managements of populations (Treves and Karanth 2003) Impact of fragmentation of distribution (Dolrenry et al. 2014) Climate change (Regehr et al. 2010)	Endangered species recovery plans (Morris et al. 2002) Impact of climate change (Araujo et al. 2006)
Hunting	Sustainability of waterbird hunting, regulations (Nichols et al. 2007)	Sustainability of hunting (Robinson and Bennett 2004) Impact of trophy hunting (Palazy et al. 2012)	Safaris (Loveridge et al. 2007)	Hunting regulations (Sandercock et al. 2011)
Recreational use	Disturbance (McLeod et al. 2013)	Impact of recreational cabins (Nellemann et al. 2010) Implementation of recreational areas (Bischof et al. 2012)	Ecotourism (Hayward and Hayward 2009)	Ecotourism (Müllner et al. 2004)
Indirect interactions	Overabundance of geese (Koons et al. 2013)	Impact on forests (Rooney and Waller 2003)	Impact on livestock (Kissui 2008; Reza et al. 2002)	Overabundant species, bird pests (Dolbeer 1998)
Epidemiology	Emerging or re-merging diseases (Kuiken et al. 2006)	Disease reservoir (Gortázar et al. 2008; Godfroid 2002)	Disease reservoir (Anderson et al. 1981)	Emerging diseases (Daszak et al. 2001)

Interactions Between Wildlife and Humans

Table 1 summarizes some interactions between humans and wildlife in which demographic aspects are prominent.

This table is far from exhaustive, but is sufficient to realize that most aspects can be grouped in a few major themes:

- Determining if some level of harvest or extra-mortality is sustainable or not;
- More generally, determining the impact of some change in demographic parameters, which can be dispersal or reproductive parameters as well as mortality ones, on the fate of a population;

- Determining the impact of habitat fragmentation, in relation with role of dispersal and population subdivision (metapopulation aspects, Hanski and Gilpin 1997);
- Translate such changes in demography into changes in distribution, in particular as a result of climate change.

When dealing with such questions, our knowledge is always incomplete and "*even the most general mathematical model is a plaything relative to the complexities of an animal population*" (Cormack 1968). Uncertainty at all levels should be taken as much as possible into account when developing and implementing management rules. One way of accounting for uncertainty in each particular case is to place any result in a comparative perspective, leading to quite a general question: are some wildlife species more sensitive than others to interactions with human activities, because of specific demographic characteristics?

A Brief History of Wildlife Demography Research

For the sake of brevity, we restrict this historical account to a few major landmarks. Malthus (1798), by first proposing in the context of human populations that "*population, when unchecked, increases in a geometrical ratio*" and assuming by contrast that resources only increase arithmetically, set the foundation for the environmental context of population growth. Much later, Leopold (1933) defined the bases of wildlife management while Lotka (1939), based on both Euler's theoretical works and Malthus' works, launched the fundamental Equation of Demography. These seminal works lead to the development of empirical approaches to the demographic analysis of wildlife populations still in use. Based on the life table approach, first developed in the seventeenth century in the context of insurance policies and used by Pearl and Miner (1935) to compare mortality curves among some invertebrates, mice and cars, Deevey (1947) proposed the first comprehensive empirical analysis of life tables of wildlife species. While such a demographic approach of wildlife populations has deeply improved our understanding of wildlife population biology, population management has been mostly relying on population counts and abundance indices rather than on demography. As population dynamics corresponds to variation in abundance over time and across space, managers have strongly focused on direct assessments of population size and its changes over time to develop management rules. As a consequence, an impressive amount of work has concentrated on reliable ways of estimating the size of wildlife populations (see reviews in Seber 1973, 1982, 1986, 1992; Caughley 1977; Schwarz and Seber 1999; Skalski et al. 2005). However, management only based on abundance monitoring is unlikely to be reliable, even for large conspicuous species such as large mammals (Gaillard et al. 2003). Population size estimates generally have low accuracy and low precision; moreover, even when reliably assessed, absolute changes in numbers between years are difficult to interpret in terms of population mechanisms (Caughley 1977; Lebreton and Gimenez 2013), especially in the absence of information on the environmental context. These problems have led wildlife biologists to investigate relationships between a population and its environment based on indicators of ecological changes (Morellet et al. 2007). Such approaches strongly rely on the concept of density-dependence and, as such, involve wildlife demography. However, this approach is entirely retrospective and does not allow population forecast. Eberhardt (1985) was among the first to show that accounting for age-dependence beyond the simple juvenile vs. adult dichotomy (first noted by Lack 1943) was required for a reliable assessment of population dynamics of long-lived species and proposed to develop simple demographic models to manage those populations (Eberhardt 1991). As discussed below, matrix models (Caswell 2001) using field estimates of demographic parameters have then progressively become the gold standard for wildlife demography studies.

Modern Tools for Wildlife Demography

Population Counts and Individual Demographic Performance: From Population Patterns to Population Processes

Population dynamics is the general study of population change, and pays attention for instance to patterns of changes in numbers using time series approaches (e.g. Stenseth et al. 2003; Brook and Bradshaw 2006). However, as just recalled, time series of counts generally fail to identify reliable demographic patterns and prevent any information on the exact demographic mechanisms to be gained, in particular because counts do not include age-structure that shape population dynamics in most wildlife populations (Coulson et al. 2001). Moreover uncertainty in counts is often strong and needs to be accounted for in analyses (Knape and de Valpine 2012; Lebreton and Gimenez 2013). Within population dynamics, demography, a field primarily developed for human populations, provides a useful way to account for age structure and puts an emphasis on processes by assuming some empirical and statistical knowledge of population flows, usually expressed as demographic parameter estimates (mortality, fecundity, etc…). Process-oriented approaches like demography are definitely more powerful to understand population dynamics than pattern-oriented approaches like abundance time series (Coulson et al. 2000; Gaillard et al. 2010; Lebreton and Gimenez 2013).

Translating demographic knowledge into predictions of population change is essential for any question dealing with man-wildlife interactions, such as the sustainability of a man-induced demographic change (Reynolds et al. 2001). However, the multiplicative nature of population processes prevents a direct translation, and, as usual "*when you are faced with a difficulty or a controversy in science, an ounce of algebra is worth a ton of verbal argument*" (J.B.S Haldane circa 1920, quoted by Maynard Smith 1965). Mathematical models based on the species life cycle are thus essential, and are used to understand past population change, in so-called retrospective analyses, or predict future change under more or less restrictive assumptions, in prospective analyses. In prospective analyses, one always work under some restrictive assumptions on future parameters, and one generally speaks of "population projections" rather than of "population predictions" (Caswell 2000, 2001). The general framework is that of birth-immigration-death-emigration ("bide") models, under a variety of demographic parameter representations and a variety of mathematical model forms (Williams et al. 2002 ch.8; Lebreton 2006), a most convenient one being matrix models (Caswell 2001). In the two components of wildlife demography, using statistical models to acquire empirical knowledge of individual demographic performance, and using dynamical models to translate this knowledge into population level consequences, individuals are usually classified in discrete biologically relevant states, closely following the "i-state philosophy" (Diekmann 2005). These components of wildlife demographic studies have developed since the 60s in parallel with a large number of long-term field programs on Vertebrates (see e.g. Perrins et al. 1991). Wildlife demography has thus been part of a general move in ecology from pattern to process (Coulson et al. 2000; Swihart et al. 2002).

A Key-Step in Wildlife Demography: Estimating Demographic Parameters

As it is extremely difficult and risky to move back from pattern to process, transversal data such as estimates of age structure are less and less used (unless combined with some other data; Udevitz and Gogan 2010) to estimate mortality patterns. Longitudinal individual data should ideally consist of dates of birth and death of individuals together with data on their reproduction output and dates. While such data are commonly available in a nearly exhaustive fashion for human populations, at least in developed countries, they are with few exceptions very difficult to obtain in animal populations, even for a subsample of the population (Clutton-Brock and Sheldon 2010). Obtaining longitudinal data in animal populations requires first that individuals receive a unique mark, or are uniquely identified by unambiguous features such as coat marks, often in combination with "camera trapping" when elusive species are targeted (e.g., for tiger, Karanth and Nichols 1998). Then the detection of marked individuals, whether by physical recapture or by indirect means such as resightings, is rarely if ever exhaustive. The "individual histories" obtained are thus far from complete. The statistical models for the analysis of such data should thus consider probabilities of detection besides the demographic parameters of main interest (e.g. annual survival probabilities, a common parameterization of death/survival process). This is the rationale for the so-called methods of capture-mark-recapture (CMR) (Nichols 1992; Lebreton et al. 1992; Gimenez et al. 2008). Survival and detection parameters can be considered as constant, varying with time, age, cohort, environmental covariates, or combinations of such effects, as in generalized linear models (Lebreton et al. 1992; Frederiksen et al. 2014), the most appropriate model being usually selected using information criteria which brings some protection again the risks of multiple statistical tests (Burnham and Anderson 2002).

After an early phase of development from the 1930s to the 1960s largely devoted to the estimation of population size (for the early history of CMR, see Lebreton et al. 2009), CMR methods progressively put increasingly more emphasis on the estimation of demographic parameters (Burnham et al. 1987; Clobert and Lebreton

1987; Lebreton et al. 1992) and developed explosively over the last 30 years, wildlife being the main biological material of application.

In the modern view of CMR methods (Lebreton et al. 2009; Gimenez et al. 2012), individuals are assumed to move among the states of a Markov chain. These states can simply be dead/alive, but can also cover a number of other features (sites, Hestbeck et al. 1991; body mass classes, Gamelon et al. 2012; reproductive status, Lebreton et al. 2009). Such general states being chosen to be relevant with respect to the species biology and life cycle, the Markov chain representation just implements the i-state philosophy (Diekmann 2005), which indeed represents demography at the individual level as transitions between life-cycle stages (Tuljapurkar et al. 2009). The transitions probabilities between states are the demographic parameter of interest: dispersal probabilities, survival probabilities, recruitment and breeding probabilities, and various combinations of these basic parameters.

Non-exhaustive detection is represented as an observation process on top of the Markov chain. Multistate CMR models (Lebreton et al. 2009) account for non-exhaustive detection assuming that when an individual is detected its state is exactly known. Multi-event models relax the latter assumption by considering there can be further uncertainty on the state occupied when detected (Pradel 2005). The probabilistic models resulting from the combination of a Markov chain for individual fates and an incomplete observation process are called hidden Markov models (HMM). In HMM, all parameters are not necessarily separately estimable (or "identifiable"). The use of HMM as a general framework for CMR models has been reviewed in greater detail by Gimenez et al. (2012). The most general models can currently be fitted using program E-SURGE (Choquet et al. 2009), which in particular provides reliable diagnostics of parameter identifiability. Modern CMR methodology is particularly adequate for analysing CMR data arising from long-term programs on wildlife: when investigating relationships between a demographic parameter and an environmental covariate (Frederiksen et al. 2014), each year is a point, and decent power for detecting such relationships usually requires 30 years or more (Grosbois et al. 2008). Moreover, whatever the intensity of data collection, estimating survival requires a sufficient number of death events: covering one or several generations is a prerequisite. The difficulty is extreme for long-lived species such as cetaceans: "*de mémoire de rose on n'a jamais vu mourir un jardinier*" ("*in the memory of roses, one never saw a gardener die*", Fontenelle, seventeenth century). Even under such stringent constraints, not only estimates of demographic parameters for wildlife focus on processes rather than on patterns, but, as in human populations (Keyfitz and Caswell 2005), they are usually less biased and more precise than population size estimates.

Linking Individual and Population Performances: Dynamical Models

Matrix models represent the changes in a population vector over time through a matrix product. The model matrix summarizes the demographic flows and can easily be obtained from graphical representations of the life cycle summarizing the transitions between a set of relevant states. They can be studied numerically, to iterate a population size vector over time, but a variety of theoretical results are available. The most classical result is the existence under mild conditions of an asymptotic exponential growth rate, for constant parameter models as well as in a random environment context (Caswell 2001; Tuljapurkar 1990). Matrix models are excellently covered in the literature in particular in Caswell's book (2001). They can be generalized in a variety of directions (Lebreton 2006).

While matrix models are often based on a categorization of individuals in age classes, they can often be best based on a set of biologically relevant states, in so-called "stage-based models" (Caswell 2001; Ch. 3 and 4). For instance, Gamelon et al. (2012) developed a matrix model with 3 body mass classes for each sex in a wild boar population. Not only did demography depend more on body mass than age in this species, but, moreover, animals killed by hunters were systematically sexed and weighed. Gamelon et al. (2012) took advantage of this situation by making explicitly appear in the model specific states for individuals harvested by hunting for each sex and body mass category (Fig. 1, from Gamelon et al. 2012).

They could then easily predict from the model numbers of individuals killed by hunting, just extracting from model runs the components corresponding to the 6 "dead" states (2 sexes X 3 body weight classes). They ran then a cross-validation by comparing these predicted numbers to the observed ones (Fig. 2, after Gamelon et al. 2012).

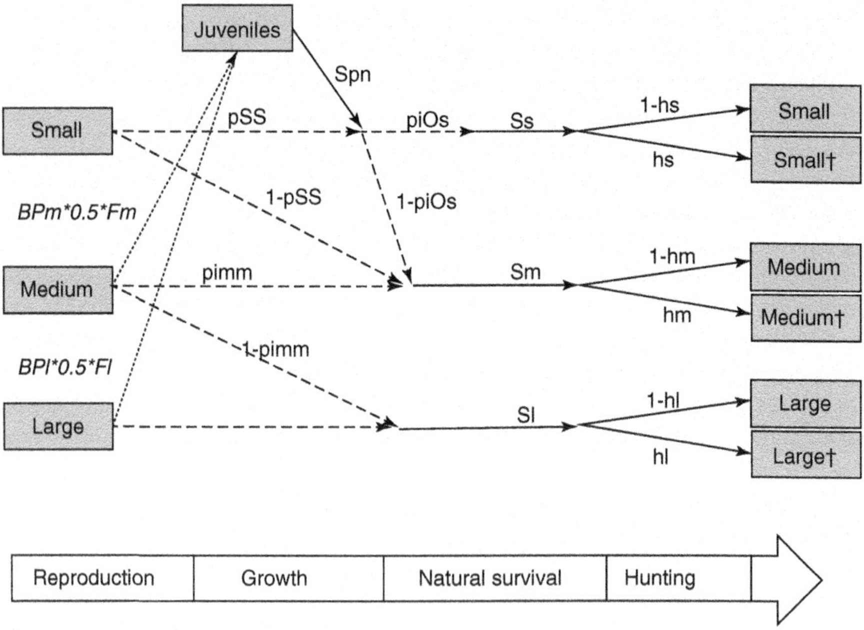

Fig. 1 Sex and size based model of female wild boar life cycle. The sex ratio is balanced. Individuals killed by hunting are represented by † (From Gamelon et al. 2012)

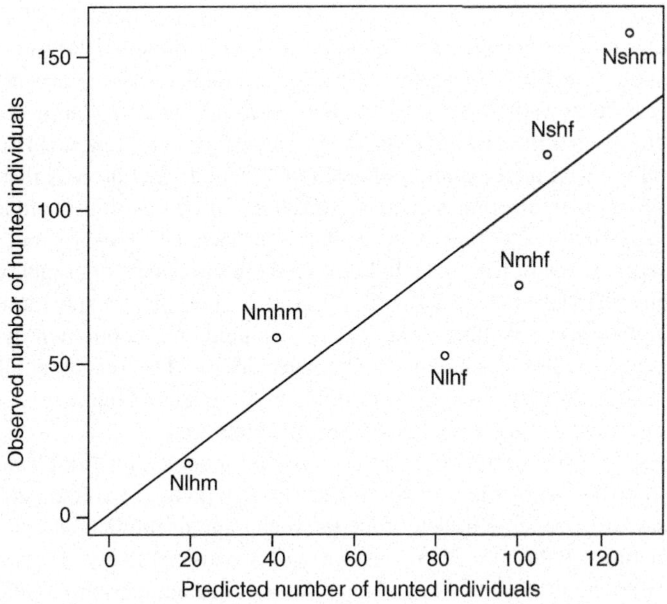

Fig. 2 Relationship between the mean numbers of wild boar for each sex and weight class hunted in an Eastern France estate overs the years 1993–1999 and those predicted by a sex and size-based matrix model (R2=0.73). Nshf, Nmhf and Nlhf=number of small, medium-sized and large females, respectively, shot by hunters; Nshm, Nmhm and Nlhm=number of small, medium-sized and large males, respectively, killed by hunting (From Gamelon et al. 2012)

The traditional use of matrix models has thus been made of two steps: (1) injecting estimates of demographic parameters in the matrix models; (2) studying the model properties under the estimated parameter scenario, comparing trajectories with population counts, and considering alternative scenarios such as changes in harvest intensity. Rather than this two-step approach, Integrated Population Modelling (Besbeas et al. 2005) now offers the possibility of combining all the pieces of information available within a same framework through a combined likelihood. When for instance population size surveys and CMR demographic parameters are combined, model trajectories are immersed in a statistical framework and come with confidence intervals, while the precision of demographic parameter estimates is increased by the combination of information. Gauthier et al. (2007) provide an example on the Greater snow Goose, *Chen caerulescens atlanticus* of this extremely promising approach. Even when extra parameters are needed in the information combination, the precision can never decrease (Barker and Kavalieris 2001). A limitation of present Integrated Population Models is that the sources of data should be independent to be easily combined (as a product of likelihoods). This limitation could be circumvented in the future through the use of more sophisticated state-space models (Pradel, pers. comm.)

Recently, age-stage structured models have been extended to measuring demographic consequences of selective pressures on life-history traits. First introduced by Easterling et al. (2000), and developed by Ellner and Rees (2006), Vindenes et al. (2008), and Coulson (2012), Integral Projection Models (IPM, not to be confounded with Integrated Population Models mentioned above) are increasingly commonly used nowadays to study the dynamics of continuous trait distributions and their fitness consequences. IPMs are based on four main functions that link the variation of the trait under study with each of the biological process that influence trait change between one time step to the next one. These functions include survival, recruitment, growth and inheritance. This approach has recently been used to address a large range of topics, including assessing the effects of climate change on demography (Vindenes et al. 2014; Plard et al. 2014), of size-specific harvest on population dynamics (Wallace et al. 2013), or elucidating population dynamics of invasive plant species (Dauer and Jongejans 2013). The recent publication of practical guides (Merow et al. 2014; Rees et al. 2014) should make the use of IPMs growing in the next years to analyse wildlife demography.

IPMs nicely illustrate that even in stage-based models, individuals do age, and the aging process can be made explicit (Caswell 2001, ch. 7; Lebreton 2005b). As a consequence it is possible to determine the mean age of mothers at childbirth, i.e. generation time (Lebreton 1996). As we will see immediately, generation time, although often neglected, is a key statistic in wildlife demography (Gaillard et al. 2005) as it varies from a few weeks in small rodents to more than 30 years in large cetaceans and large turtles, in a 300-fold order of magnitude.

An Overview of Wildlife Demography: The Slow-Fast Continuum and Its Consequences

The Slow-Fast Continuum

Even in inter-tropical environments with weaker seasonality than in temperate areas, most terrestrial vertebrates have a well-defined reproductive season during which births occur with variable level of synchrony. With the exception of rodents (but sciurids) and lagomorphs among mammals, which are often multivoltine and have several reproductive attempts across seasons within the year, and nearly no exception among birds, terrestrial vertebrates tend to have a well-defined annual reproductive season (birth-pulse populations *sensu* Caughley 1977), sometimes with several consecutive litters or broods. In Hares (*Lepus spp.*) and some rodents like microtines (e.g. *Microtus montanus*; Negus and Berger 1988), some individuals born in spring seem to reproduce in the autumn of their first year of life, i.e. around a few months of age.

Virtually all females in terrestrial vertebrates are iteroparous, contrary e.g. to some fish as Salmons. In medium-sized and large species reproduction is annual. It becomes biennial in very large seabirds like albatross *Diomedea spp.* (Tickell 1968) as the full cycle from egg laying to fledging takes more than 365 days. The frequency of reproduction also decreases to one every second, third, or even more year in megaherbivores (*sensu* Owen-Smith 1988), whales, and some primates like apes due to the constraints of very long gestation (e.g. 22 months in African elephants, *Loxodonta africana*) and lactation periods. In parallel brood/litter size decreases from small to large species when calculated to account for the frequency of reproductive events within a year (annual fecundity *sensu* Allainé et al. 1987). Altogether, the inverse of fecundity, age at first reproduction and life expectancy all corresponds to biological times (*sensu* Lindstedt and Calder 1981) and are intimately positively correlated (Calder 1984; Gaillard et al. 1989), leading to rank species along a slow-fast continuum (Gaillard et al. 1989; Promislow and Harvey 1990; Stearns 1992; Bielby et al. 2007; Jeschke and Kokko 2009).

Gaillard et al. (2005) (see also Oli and Dobson 2003) analysed the demography of 126 mammal species, based on age at first reproduction, age at last reproduction, first year survival probability, after first-year annual survival probability, and annual fecundity. The first Principal Component of these 4 demographic variables (log-transformed) explained 69 % of their variability, and its correlation with generation time was equal to 0.903. We ran a similar analysis based on 3 log-transformed demographic variables (age at first reproduction, after first-year annual survival probability and annual fecundity) for 199 bird species in the bird demographic database (BIDDABA, Lebreton et al. 2012) already studied by Desprez (2009). The first principal component accounted for 78 % of the variance, and its correlation with generation time (also log-transformed) was equal to 0.797. The lower correlation between generation time and the first principal component obtained in birds is likely to be explained by the absence of consideration of first year survival probability, which contributes to changes in generation time for a given set of other demographic parameters. Despite this limitation, variation in generation time accounted for almost two-third of the observed variation in the first principal component. While a Principal Component of demographic parameters is a very neutral statistical way of representing the position of a species or a population on the slow-fast continuum, generation time has the advantage of being indeed expressed in time units, of being biologically interpretable and of bearing strong relationships with essential demographic concepts, driving in particular patterns of sensitivity of population growth rates to changes in demographic parameters (Hamilton 1966; Gaillard et al. 2005).

Like most life history traits, generation time is subject to strong evolutionary allometric constraints (Peters 1983; Calder 1984). However, while the position on the slow-fast continuum is clearly related to body size, it varies with the phylogenetic group concerned, as it depends on the general design (the "bauplan"; Stearns 1992) of the species. Thus, generation time increases with size both among bats as well as other mammal orders, but the linear relationships between body mass and generation time has a very different intercept because bats have to be light to fly (Fig. 3 after Gaillard et al. 2005). Such "grade shifts effects" (*sensu* Harvey and Pagel 1991) in evolutionary allometry have been repeatedly reported (e.g. Ferguson et al. 1996, in gestation length and neonatal mass of North American carnivores; Isler et al. 2008, in encephalization of Primates).

The slow-fast continuum has been mostly used at the inter-specific level to rank species over a gradient of pace of life. However, a slow-fast continuum also occurs

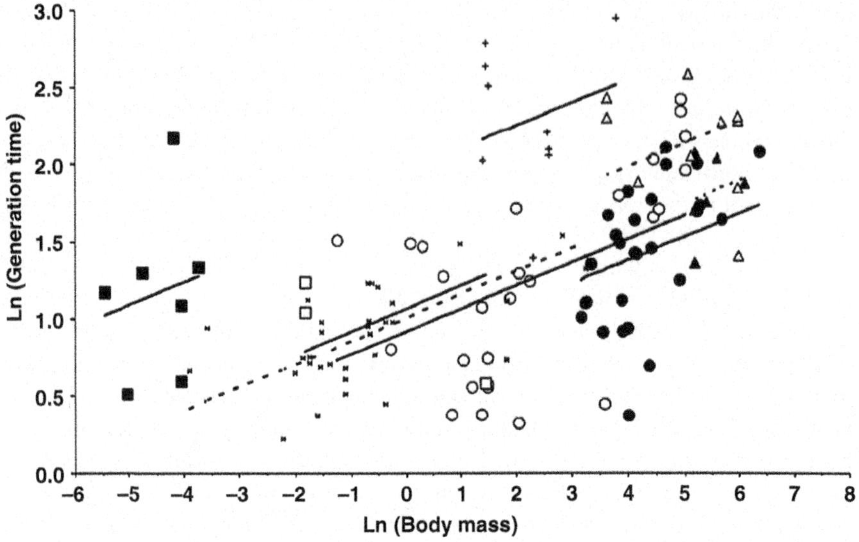

Fig. 3 Relationship between generation time and female adult body mass according to the order level for 123 populations of mammals: artiodactyls (filled circles, solid line), perissodactyls (filled triangles, dashed line), pinnipeds (open triangles, dashed line), primates (plus signs, solidline), fissipeds (open circles, solid line), lagomorphs (open squares, solid line), bats (filled squares, solid line), rodents (crosses, dashed line) (from Gaillard et al. 2005). Both scales are logarithmic

at the population level. When subjected to markedly different environmental conditions, populations within a same species can display about twofold variation in generation time (Crampe et al. 2006; Nilsen et al. 2009). In long-lived iteroparous species with low fecundity like large mammalian herbivores, high hunting and/or predation pressure cause populations to decline in absence of compensatory increase of recruitment, as observed in bighorn sheep (Festa-Bianchet et al. 2006) or woodland caribou (Wittmer et al. 2005). These declining populations consistently show much longer generation times, involving a slower life history (Nilsen et al. 2009).

Population Growth Rate Along the Slow-Fast Continuum and Its Consequences

Compared with "fast" populations, populations on the slow side of the continuum have less young produced per year and a longer reproductive life, starting later. Unavoidably, they have a lower maximum population growth rate r_{max}, as well shown for all living forms by Fenchel (1974). The relationship is particularly narrow when considering only homeotherms. Among birds, r_{max} varies approximately as 1/T where T is the generation time (Fig. 4, after Niel and Lebreton 2005), leading the product r_{max}. T to be a "dimensionless number" (*sensu* Charnov 1993) The same

Fig. 4 Relationship between generation time (abscissa) and maximum population growth rate (ordinate) in 13 bird species (From Niel and Lebreton 2005). Both scales are logarithmic

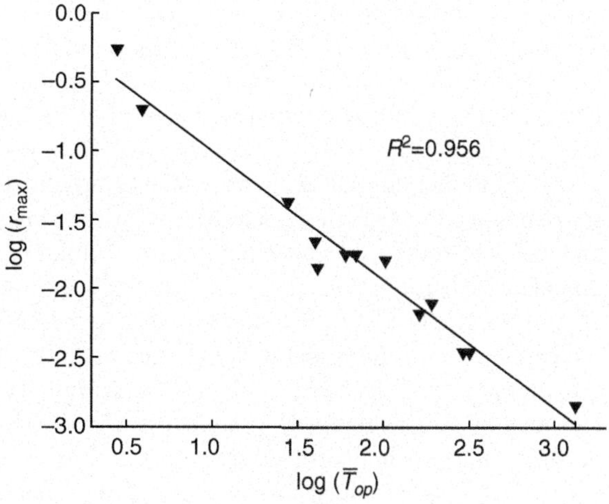

relationship with similar values holds for mammals (Gaillard et al., unpublished results).

A first consequence concerns the evolutionary mechanisms for the diversification of life history strategies along the slow-fast continuum. Certainly, the evolution to a large time and space scale is linked to access to distant and/or coarse-grained resource, as well shown by Jouventin and Mougin (1981) for seabirds. However, evolving towards a large time and space scale requires evolving towards a lower maximum population growth rate r_{max}, in apparent contradiction with the fitness maximization principle. The only plausible mechanism to date is r-K selection (MacArthur and Wilson 1967) according to which genotypes with a lower r_{max} can be selected because they do better in situations of competition, mostly intraspecific competition. r-K selection is selection for genotypes that do better under density-dependence (Boyce 1984). In this view, the link with space and time scale remains obscure and has indeed been only stated in vague and descriptive terms (Pianka 1970). Evolving towards a large time/space scale makes it possible to avoid diffuse interspecific competition, and thus to outnumber small time/space scale genotypes with a higher r_{max} but strongly submitted to interspecific competition. Avoidance of diffuse interspecific competition seems a more plausible explanation for the demographic diversification towards slow demographic strategy than r-K selection in its narrower sense (Lebreton unpublished).

The second consequence of being "slow" is a greater demographic sensitivity to increases in mortality, as a result of the lower r_{max}. When a proportion *h* of a population is harvested uniformly irrespective of age or stage, the maximum sustainable harvest proportion h_{max} increases with r_{max} as 1-exp(−r_{max}), well approximated in birds and mammals by the inverse of generation time (Niel and Lebreton 2005). It comes to no surprise that the slow, long-lived species are more impacted by hunting, exploitation, and even diffuse mortality induced by human activities than short-lived ones. There are many examples of this "malediction of long-lived species" (Lebreton 2006, 2011), not restricted to terrestrial animals as strong declines in fish stocks have been linked with low r_{max} and associated demographic characteristics (Jennings et al. 1998). The impact of changes in fecundity is markedly different, as the relative sensitivity (or "elasticity") of population growth rate to an overall change in fecundity, again irrespective of age or stage, is the inverse of generation time (Gaillard et al. 2005). For an albatross species with T being approximately 25 years, the maximum rate of increase will be around 4 %. The bycatch of Albatross at sea by long line fisheries can be of that order of magnitude (Anderson et al. 2011) and has indeed induced severe population declines (e.g. Weimerskirch et al. 1997). Albatross species lay a single egg annually or every other year in the largest *Diomedea* species. Even if they were able to increase their fecundity by 10 % by increasing the hatching and fledging success, that would in turn only change the growth rate from 4 % a year to 4.04 %, a negligible change: while short-lived species may be able to compensate some extra-mortality by changes in fecundity, this cannot be the case in long-lived species. A comparative analysis of transient dynamics has recently provided a new facet of this malediction of long-lived species (Gamelon et al. 2014). Slow populations tend to decrease in size after a disturbance, whereas fast populations can respond by an increase in size. Lindberg et al. (2013), based on reproductive values, give simple formulas that generalize the approach above to changes in mortality not uniformly distributed among the different stages or age-classes in the population.

We discuss below the possible sources of compensation, i.e. of attenuation of the effect of an increase in mortality, but the link between r_{max} and generation time clearly gives to the slow-fast gradient a key role in deciphering the effect of human impacts on wildlife: in any sharp of sustained decrease in a long-lived wildlife population, one should first suspect a change in mortality.

Variation in Demographic Parameters

While each species or population can be positioned on the slow-fast continuum based on average demographic flows, in the dynamics of a particular population these demographic flows change under a variety of mechanisms: environmental variation, demographic stochasticity, density-dependence, heterogeneity among individuals… Each type of variation has been the subject of entire books, and we briefly review here only general ideas with an emphasis on recent research trends.

Environmental variation is usually taken in the sense of environmental variation over time, although spatial variation has been receiving increasing attention during the two last decades (Bjornstad et al. 1999). Matrix models with parameters varying over time in a predetermined random fashion can easily be built (Caswell 2001; ch. 14). Studies relating variation over time in a demographic parameter to environmental covariates are of special interest in the context of climate change, as climate models can provide future scenarios for covariates driving key demographic flows. As the number of years remains moderate (<40 in general), and as they are many potential candidate covariates, special care must be exercised not to detect spurious relationships (Grosbois et al. 2008), using in general a linear regression, often embedded in a CMR model (Lebreton et al. 1992). In a linear regression between Y and X_i, if there are 20 candidate covariates X_i, in absence of any relationship between Y and the covariates, there will be on the average one covariate declared as significant at the 0.05 level. Contrary to a common belief, information indices such as the AIC do not bring any kind of protection against the risk of detecting spurious relationships. A straightforward technique consists of using principal components analysis (PCA) to reduce the covariates to a few uncorrelated ones (Grosbois et al. 2008). Of special interest is then PCA with missing values (Josse and Husson 2011) that makes it possible to consider relevant covariates even if some values are missing.

While multistate CMR and matrix models can be used with geographical strata as states (Hestbeck et al. 1991; Hénaux et al. 2007; Péron et al. 2010), there are few studies considering asynchronous environmental variation over several sites, i.e. space and time environmental heterogeneity, while it is a dominant feature in plant and animal populations. However, there is accumulating evidence that spatial heterogeneity in habitat quality is strong for most populations and markedly influences body mass and thereby demographic parameters (e.g. Pettorelli et al. 2003 in roe deer, *Capreolus capreolus*). Some demographic analyses have even reported that in Soay sheep (*Ovis aries*) the sensitivity of population growth to demographic parameters varied across habitats (Ezard et al. 2008). We thus expect spatial variation in habitat quality to be increasingly considered when analysing wildlife demography.

Demographic stochasticity is the ultimate variability that remains for fixed parameters, notably the binomial variation inherent in a death/survival process. While it is often treated in an ad hoc and blind fashion on canned software for Population Viability Analysis (PVA), mathematical tools exist (Gosselin and Lebreton 2000; Caswell 2001, ch. 15). A common thought is that demographic stochasticity has only to be considered in small populations, especially to model extinction risk, as it rapidly becomes negligible when population size increases. This is not the case in subdivided populations as noted long ago by Chesson (1978) because demographic stochasticity can remain prevalent within each subpopulation. In extinction or PVA models, environmental stochasticity plays a major role in interaction with demographic stochasticity as it may induce excursions to low population size with huge effects on the extinction probability. Evidence of the key role of demographic stochasticity in population demography has been recently reported in several vertebrate populations including e.g. lemon sharks, *Nagaprion brevirostris* (White et al. 2014), European squirrel, *Sciurus vulgaris* (Rezouki et al. 2014), or ibex, *Capra ibex* (Grotan et al. 2008). In particular, for a moose *Alces alces* population studied in a Norwegian Island, demographic stochasticity accounted for most of the temporal variation observed in demography (Saether et al. 2007).

Density-dependence has been a major topic in wildlife demography (e.g. McCullough 1979; Fowler 1987; Bonenfant et al. 2009) for various reasons. First, in relation with Malthus (1798)'s statement recalled earlier and the classical logistic growth law (Verhulst 1838), density-dependence, arising through a shortage of per capita resource, is the long-term mechanism of population stabilization. Second, it has been considered as a major mechanism of compensation of the effects of any harvest: the harvest induces a drop in population size, which in turn induces a release in the per capita resource shortage, i.e. induces an increase in demographic performance bringing the population size to a higher level that it would have had under the effect of harvest in the absence of density-dependence. Seeking empirical evidence for compensation has been a major issue in quarry species management (Burnham and Anderson 1984). While incorporating some level of density-dependence in population projection models is relatively straightforward, obtaining evidence for density-dependence and estimating its intensity is difficult: simple *ad hoc* methods based on series of population size estimates are biased and tend to detect too often density-dependence (Solow 1998; Knape and de Valpine 2012; Lebreton and Gimenez 2013). On the contrary, using an estimate of population size as a covariate in a CMR analysis is conservative because of the unavoidable uncertainty in population size estimates, leading to a situation known as "regression with errors". Modern tools make it possible to circumvent these difficulties (Lebreton and Gimenez 2013; De Valpine and Hastings 2002; Barker et al. 2002), and new results and views on density-dependence are likely to appear. Density-dependence as a compensatory mechanism appears as less universal than thought a few years ago: assessments of hunting mortality based on estimation of band reporting rates in ducks seem less conclusive than in the past as new studies with reward bands showed that reporting rates were underestimated (Nichols et al. 1991), and that, as a consequence, hunting mortality estimates were biased. Moreover, the variation in survival with density needed for compensation (Lebreton 2005a) seems too high to make density-dependence a strong compensatory mechanism.

Individual heterogeneity in demographic parameters has often been implicit in demographic analyses or models: when considering and age effect in a CMR model one simply recognizes some heterogeneity in demographic parameter that can be measured by age. Beside age, which is the most commonly used in wildlife demography, individual heterogeneity can be measured by either fixed (like birth date, birth mass, cohort, or longevity) or dynamic (like mass, size, or reproductive success) traits (e.g. Hill et al. 1999; Plard et al. 2012). However, a variable proportion of individual heterogeneity remains once the effects of those fixed and dynamic traits have been accounted for. Such hidden individual differences can now routinely be accounted for when estimating demographic parameters in wildlife populations (van de Pol and Verhulst 2006). Individual differences are relatively easily evaluated for reproduction because repeated measures for a same individual over time are possible. It is more involved for survival, and CMR models explicitly considering heterogeneity (Pledger et al. 2003), which, in turn, are specific multi-event models (Gimenez et al. 2012, p. 310) had to be developed. For instance, the number of survivors among N individuals rather than being considered as a binomial distribution Bin(N,p) with a fixed probability p applying to all individuals (the demographic stochasticity view) is viewed as a mixture of binomials $Bin(N_1,p_1)$ and Bin (N_2, p_2) with N_1 (and $N_2=N-N_1$) unknown. The mixture model above can be viewed as adding for each individual the value of a discrete two-level random variable to a mean survival probability. Models with a Gaussian individual random effect have been considered by Gimenez and Choquet (2010). However, the demographic consequences of unmeasured individual heterogeneity have not been fully investigated. Case studies performed to date seem to suggest that the effect of unmeasured individual differences on population growth might be weak (Rees et al. 1999 on plants; Coulson et al. 2010 on Soay sheep; Plard et al. 2015 on roe deer).

While neglecting heterogeneity of capture has devastating effects on the estimation of population size by CMR, considering heterogeneity in survival raises a number of questions of interest. Variability in survival among individuals has been a topic of interest among human demographers for a long time (Vaupel et al. 1979; Vaupel and Yashin 1985): its effect on the detection of senescence, its links with individual quality and microhabitat variation, potential mechanisms for maintaining heterogeneity over time (while it should be counter-selected if it is even weakly heritable) are all subjects of great interest that directly impact our understanding of wildlife population mechanisms and management decisions. In particular heterogeneity in demographic performance can be a strong compensatory mechanism if the "poor" individuals are more sensitive to hunting than the "good" ones (Lindberg et al. 2013), a likely assumption in many cases (Greenwood et al. 1986). It currently seems to many experts the potentially dominant mechanism for partial compensation of exploitation (by hunting, incidental man-induced deaths) in wildlife, density-dependence seeming less susceptible to play a major role as noted above.

"Variation in demographic parameters", briefly covered here, thus encompasses a wide array of processes that are essential to decipher for efficient wildlife management. The development of "Individual-based models" (IBM) (DeAngelis and Gross 1992) can be viewed as a way of accounting for variation among individuals, by representing variation in demographic parameters through individual level rules (dealing e.g. with distance to food resources, availability of partners, etc.). Unfortunately, if all the rules in such models had to be empirically documented, even more data would be needed than in classical demographic models, and, although heuristically interesting, IBM are difficult to use for wildlife management (see broader comments by Grimm 1999). A comparison of IBM with more classical demographic models is proposed about the marmot *Marmota marmota* by Stephens et al. (2002).

Discussion

The emergence of modern statistical and modelling tools reviewed above, largely the result of computer development, coincided with the development of many long-term empirical studies of vertebrate populations, clearly visible in journals such as Journal of Wildlife Management, Journal of Applied Ecology, Ecological Applications, Wildlife Biology etc.... As a result, the number of reliable demographic assessments of wildlife population dynamics has widely increased over the last thirty years and participated to the general move from pattern to process in ecology (Swihart et al. 2002). This development provided the material for the comparative look we privileged, which gives a central role in each demographic study of a species or population to its position on the slow-fast continuum. We strongly recommend that in any management or conservation issue of a wildlife population, a clear view is taken of its position on the slow-fast continuum. Advantage can be taken in particular of the current development of open demographic data bases (Lebreton et al. 2012; Jones and Salguero Gomez 2014): when the information is scarce, one may in particular produce, by appropriate linear regressions within an homogeneous group of species, predictions of demographic parameters than can be used as prior information in further models. A general principle arising from the comparative approach when applied to wildlife demography is that long-lived species are especially sensitive to man-induced impacts, and require specific attention. Another key point in the methodological developments is that many different population structures besides age-classes can now be easily considered if biologically relevant.

Our review of mechanism of change in demographic parameters (environmental and demographic stochasticity, density-dependence, heterogeneity in space and time, heterogeneity among individuals…) has been necessarily superficial. Our main message is that some difficult issues such as density-dependence are opened to new approaches and can now be efficiently revisited. State-space models, by making it possible to account for uncertainty in population surveys, to combine different pieces of information such as individual-based data and counts, have started playing a central role.

Currently, heterogeneity among individuals is receiving a lot of attention, and bears a deep relationship to studies of heterogeneity in other traits, related e.g. to energy acquisition and behaviour (Réale et al. 2010). However, in spite of its focus on flows of individuals, the current dominant demographic approach remains largely correlative and phenomenological. The diversification of modelling tools should in our opinion progressively induce a move to more and more causal and biologically inspired models. The challenge is to keep the population level attention to flows of individuals inherent to demography while incorporating an attention to heterogeneity, traits, and interaction with resource at the individual level, in the spirit of individual-based models. A good discussion of the various types of models and of their complementarity is provided by Stephens et al. (2002) based on examples on the alpine marmot. This trend will benefit from the development of data-loggers of all kinds which bridge classical CMR with studies of individual performance.

In conclusion, it is fortunate that wildlife demography has been and is still evolving to such a pace, if one hopes to mitigate to some extent the effect of global change on wildlife.

Acknowledgements We warmly thank an anonymous reviewer for insightful comments on a previous draft of that work.

References

Allainé D, Pontier D, Gaillard JM, Lebreton JD, Trouvilliez J, Clobert J (1987) The relationship between fecundity and adult body weight in homeotherms. Oecologia 73:478–480
Anderson RM, Jackson HC, May RM, Smith AM (1981) Population dynamics of fox rabies in Europe. Nature 289:765–771
Anderson ORJ, Small CJ, Croxall JP, Dunn EK, Sullivan BJ, Yates O, Black A (2011) Global seabird bycatch in longline fisheries. Endanger Species Res 14:91–106
Araujo MB, Thuiller W, Pearson RG (2006) Climate warming and the decline of amphibians and reptiles in Europe. J Biogeogr 33:1712–1728
Barker RJ, Kavalieris L (2001) Efficiency gain from auxiliary data requiring additional nuisance parameters. Biometrics 57:563–566
Barker R, Fletcher D, Scofield P (2002) Measuring density dependence in survival from mark-recapture data. J Appl Stat 29:305–313
Besbeas P, Freeman SN, Morgan BJT (2005) The potential of integrated population modelling. Aust N Z J Stat 47:35–48
Bielby J, Mace GM, Bininda-Emonds ORP, Cardillo M, Gittleman JL, Jones KE, Orme CDL, Purvis A (2007) The fast-slow continuum in mammalian life history: an empirical reevaluation. Am Nat 169:748–757
Bischof R, Nilsen EB, Broseth H, Mannil P, Ozolins J, Linnell JDC (2012) Implementation uncertainty when using recreational hunting to manage carnivores. J Appl Ecol 49:824–832
Bjornstad ON, Ims RA, Lambin X (1999) Spatial population dynamics: analysing patterns and processes of population synchrony. Trends Ecol Evol 14:427–432
Bonenfant C, Gaillard JM, Coulson T, Festa-Bianchet M, Loison A, Garel M, Loe LE, Blanchard P, Pettorelli N, Owen-Smith N, Du Toit J, Duncan P (2009) Empirical evidence of density-dependence in populations of large herbivores. Adv Ecol Res 41:313–357
Boyce MS (1984) Restitution of r-selection and K-selection as a model of density-dependent natural selection. Annu Rev Ecol Syst 15:427–447
Brook BW, Bradshaw CJA (2006) Strength of evidence for density dependence in abundance time series of 1198 species. Ecology 87:1445–1451
Burnham KP, Anderson DR (1984) Tests of compensatory vs additive hypotheses of mortality in mallards. Ecology 65:105–112
Burnham KP, Anderson DR (2002) Model selection and multimodel inference; a practical information-theoretic approach. Springer, New York
Burnham KP, Anderson DR, White GC, Brownie C, Pollock KH (1987) Design and analysis methods for fish survival experiments based on release-recapture. American Fisheries Society Monograph 5
Calder WA (1984) Size, function, and life history. Harvard University Press, Harvard
Caswell H (2000) Prospective and retrospective perturbation analyses: their roles in conservation biology. Ecology 81:619–627
Caswell H (2001) Matrix population models. Construction, analysis, and interpretation. 2nd edn. Sinauer Associates, Sunderland
Caswell H (2014) A matrix approach to the statistics of longevity in heterogeneous frailty models. Demogr Res 31:553–592
Caughley G (1977) Analysis of wildlife populations. Wiley, Chichester
Charnov EL (1993) Life history invariants. Oxford University Press, Oxford
Chesson P (1978) Predator-prey theory and variability. Annu Rev Ecol Syst 9:323–347
Choquet R, Rouan L, Pradel R (2009) E-SURGE: a software application for fitting multievent models. In: Thomson DKL, Cooch EG, Conroy MJ (eds) Modeling Demographic Processes in Marked Populations. Springer, New-York, p 845–865
Clobert J, Lebreton JD (1987) Recent models for mark recapture and mark resighting data – response. Biometrics 43:1019–1022
Clutton-Brock TH, Sheldon BC (2010) Individuals and populations: the role of long-term, individual-based studies of animals in ecology and evolutionary biology. Trends Ecol Evol 25:562–573

Cormack RM (1968) The statistics of capture-recapture methods. Oceanogr Mar Biol Annu Rev 6:455–506
Coulson T (2012) Integral projections models, their construction and use in posing hypotheses in ecology. Oikos 121:1337–1363
Coulson T, Milner-Gulland EJ, Clutton-Brock TH (2000) The relative roles of density and climatic variation on population dynamics and fecundity rates in three contrasting ungulate species. Proc R Soc Lond Ser B 267:1771–1779
Coulson T, Catchpole EA, Albon SD, Morgan BJT, Pemberton JM, Clutton-Brock TH, Crawley MJ, Grenfell BT (2001) Age, sex, density, winter weather, and population crashes in Soay sheep. Science 292:1528–2531
Coulson T, Tuljapurkar S, Childs DZ (2010) Using evolutionary demography to link life history theory, quantitative genetics and population ecology. J Anim Ecol 79:1226–1240
Crampe JP, Loison A, Gaillard JM, Florence E, Caens P, Appolinaire J (2006) Monitoring of the reproduction in isard females (*Rupicapra pyrenaica pyrenaica*) in a non-hunted population and demographic consequences. Can J Zool 84:1263–1268
Daszak P, Cunningham AA, Hyatt AD (2001) Anthropogenic environmental change and the emergence of infectious diseases in wildlife. Acta Trop 78:103–116
Dauer JT, Jongejans E (2013) Elucidating the population dynamics of Japanese Knotweed using integral projection models. PLoS One 8:e75181
De Valpine P, Hastings A (2002) Fitting population models incorporating process noise and observation error. Ecol Monogr 72:57–76
DeAngelis DL, Gross LJ (eds) (1992) Individual-based models and approaches in ecology: populations, communities and ecosystems. Chapman Hall, New York
Deevey ES (1947) Life tables for natural populations of animals. Quart Rev Biol 22:283–314
Desprez M (2009) Démographie comparée et statut de conservation chez les oiseaux. Unpublished Master report. Université Montpellier 2
Diekmann O (2005) On the Mathematical synthesis of physiological and behavioural mechanisms and population dynamics. Centre for Mathematics and Computer Science, Amsterdam
Dolbeer RA (1998) Population dynamics: the foundation of wildlife damage management for the 21st century. Proceedings of the 18th vertebrate pest conference. Paper 9. http://digitalcommons.unl.edu/vpc18/9
Dolrenry S, Stenglein J, Hazzah L, Lutz RS, Frank L (2014) A metapopulation approach to African lion (*Panthera leo*) conservation. PLoS One 9:e88081
Duffus DA, Dearden P (1990) Nonconsumptive wildlife-oriented recreation – a conceptual framework. Biol Conserv 53:213–231
Easterling MR, Ellner SP, Dixon PM (2000) Size-specific sensitivity: applying a new structured population model. Ecology 81:694–708
Eberhardt LL (1985) Assessing the dynamics of wild populations. J Wildl Manage 49:997–1012
Eberhardt LL (1991) Models of ungulate population dynamics. Rangifer 7:24–29
Ellner SP, Rees M (2006) Integral projection models for species with complex demography. Am Nat 167:410–428
Ezard THG, Gaillard JM, Crawley MJ, Coulson T (2008) Habitat dependence and correlations between elasticities of long-term growth rates. Am Nat 172:424–430
Fenchel T (1974) Intrinsic rate of natural increase – relationship with body size. Oecologia 14:317–326
Ferguson SH, Virgl JA, Larivière S (1996) Evolution of delayed implantation and associated grade shifts in life history traits of North American carnivores. Ecoscience 3:7–17
Festa-Bianchet M, Coulson T, Gaillard JM, Hogg JT, Pelletier F (2006) Stochastic predation events and population persistence in bighorn sheep. Proc R Soc B 273:1537–1543
Fowler CW (1987) A review of density dependence in populations of large mammals. Curr Mammal 1:401–441
Frederiksen M, Lebreton JD, Pradel R, Choquet R, Gimenez O (2014) Identifying links between vital rates and environment: a toolbox for the applied ecologist. J Appl Ecol 51:71–81

Frölich K, Thiede S, Kizikowski T, Jakob W (2002) A review of mutual transmission of important infectious diseases between livestock and wildlife in Europe. Ann N Y Acad Sci 969:4–13
Gaillard JM, Pontier D, Allainé D, Lebreton JD, Trouvilliez J, Clobert J (1989) An analysis of demographic tactics in birds and mammals. Oikos 56:59–76
Gaillard JM, Loison A, Toigo C (2003) Variation in life-history traits and realistic population models for wildlife management: the case of ungulates. In: Festa-Bianchet M, Apollonio M (eds) Animal behavior and wildlife conservation. Island Press, Washington, DC, pp 115–132
Gaillard JM, Yoccoz NG, Lebreton JD, Bonenfant C, Devillard S, Loison A, Pontier D, Allainé D (2005) Generation time: a reliable metric to measure life-history variation among mammalian populations. Am Nat 166:119–123
Gaillard JM, Coulson T, Festa-Bianchet M (2010) Chapter 4. Demographic processes: lessons from long-term, individual-based studies. In: Owen-Smith N (ed) Dynamics of large herbivore populations in changing environments: towards appropriate models. Princeton University Press, Princeton, pp 98–116
Gamelon M, Gaillard JM, Servanty S, Gimenez O, Toigo C, Baubet E, Klein F, Lebreton JD (2012) Making use of harvest information to examine alternative management scenarios: a body weight-structured model for wild boar. J Appl Ecol 49:833–841
Gamelon M, Gimenez O, Baubet E, Coulson T, Tuljapurkar S, Gaillard JM (2014) Influence of life-history tactics on transient dynamics: a comparative analysis across populations. Am Nat 184:673–683
Gauthier G, Besbeas P, Lebreton JD, Morgan BJT (2007) Population growth in snow geese: a modelling approach integrating demographic and survey information. Ecology 88:1420–1429
Gimenez O, Choquet R (2010) Individual heterogeneity in studies on marked animals using numerical integration: capture-recapture mixed models. Ecology 91:951–957
Gimenez O, Viallefont A, Charmantier A, Pradel R, Cam E, Brown CR, Anderson MD, Bomberger-Brown M, Vovas R, Gaillard JM (2008) The risk of flawed inference in evolutionary studies when detectability is less than one. Am Nat 172:441–448
Gimenez O, Lebreton JD, Gaillard JM, Choquet R, Pradel R (2012) Estimating demographic parameters using hidden process dynamic models. Theor Popul Biol 82:307–316
Godfroid J (2002) Brucellosis in wildlife. Revue Scientifique et Technique de l'Office International des Epizooties 21:277–286
Gortázar C, Torres MJ, Vicente J, Acevedo P, Reglero M et al (2008) Bovine tuberculosis in Doñana Biosphere Reserve: the role of wild ungulates as disease reservoirs in the last Iberian Lynx strongholds. PLoS One 3(7):e2776. doi:10.1371/journal.pone.0002776
Gosselin F, Lebreton JD (2000) The potential of branching processes as a modeling tool for conservation biology. In: Ferson S (ed) Quantitative methods in conservation biology. Springer, New York
Greenwood H, Clark RG, Weatherhead PJ (1986) Condition bias of hunter-shot mallards (*Anas platyrhynchos*). Can J Zool 64:599–601
Grimm V (1999) Ten years of individual-based modelling in ecology: what have we learned and what could we learn in the future? Ecol Model 115:129–148
Grosbois V, Gimenez O, Gaillard JM, Pradel R, Barbraud C, Clobert J, Moller AP, Weimerskirch H (2008) Assessing the impact of climate variation on survival in vertebrate populations. Biol Rev 83:357–399
Grotan V, Saether BE, Filli F, Engen S (2008) Effects of climate on population fluctuations of ibex. Glob Chang Biol 14:218–228
Hamilton WD (1966) The moulding of senescence by natural selection. J Theor Biol 12:12–45
Hanski I, Gilpin ME (1997) Metapopulation biology: ecology, genetics, and evolution. Academic Press, San Diego
Harvey PH, Pagel MD (1991) The comparative method in evolutionary biology. Oxford University Press, Oxford
Hayward MW, Hayward GJ (2009) The impact of tourists on lion *Panthera leo* behavior, stress and energestics. Acta Theriol 54:219–224

Hénaux V, Bregnballe T, Lebreton JD (2007) Dispersal and recruitment during population growth in a colonial bird, the great cormorant *Phalacrocorax carbo sinensis*. J Avian Biol 38:44–57
Hestbeck JB, Nichols JD, Malecki RA (1991) Estimates of movement and site fidelity using mark-resight data of wintering Canada geese. Ecology 72:523–533
Hill JA, Enstrom DA, Ketterson ED, Nolan V, Ziegenfus C (1999) Mate choice based on static versus dynamic secondary sexual traits in the dark-eye junco. Behav Ecol 10:91–96
Hoffmann M, Belant JL, Chanson JS, Cox NA, Lamoreux J, Rodrigues ASL, Schipper J, Stuart SN (2011) The changing fates of the world's mammals. Phil Trans R Soc B 366:2598–2610
Inkley DB, Anderson MG, Blaustein AR, Burkett VR, Felzer B, Griffith B, Price J, Root TL (2004) Climate-change adaptation strategies and wildlife in North America. Technical review 04–2. The Wildlife Society, Bethesda
Isler K, Kirk EC, Miller JMA, Albrecht GA, Gelvin BR, Martin RD (2008) Endocranial volumes of primate species: scaling analyses using a comprehensive and reliable data set. J Hum Evol 55:967–978
Jennings S, Reynolds JD, Mills SC (1998) Life history correlates of response to fisheries exploitation. Proc R Soc Lond Ser B 265:333–339
Jeschke JM, Kokko H (2009) The roles of body size and phylogeny in fast and slow life histories. Evol Ecol 23:867–878
Jones OR, Salguero Gomez R (2014) The COMPADRE and COMADRE population matrix databases. http://dx.doi.org/10.6084/m9.figshare.1102955
Jones OR, Scheuerlain A, Salguero Gomez R, Camarda CG, Schaible R, Casper BB, Dahlgren JP, Ehrlen J, Garcia MB, Menges ES, Quintana-Ascencio PF, Caswell H, Baudisch A, Vaupel JW (2014) Diversity of ageing across the tree of life. Nature 505:169–173
Josse J, Husson F (2011) Multiple imputation in principal component analysis. Adv Data Anal Classific 5:231–246
Jouventin P, Mougin JL (1981) Les stratégies adaptatives des oiseaux de mer. Revue d'Ecologie (Terre et Vie) 35:217
Karanth KU, Nichols JD (1998) Estimation of tiger densities in India using photographic captures and recaptures. Ecology 79:2852–2862
Keyfitz N, Caswell H (2005) Applied mathematical demography. Springer, New York
Kissui BM (2008) Livestock predation by lions, leopards, spotted hyenas, and their vulnerability to retaliatory killing in the Maasai steppe, Tanzania. Anim Conserv 11:422–432
Knape J, de Valpine P (2012) Are patterns of density dependence in the Global Population Dynamics Database driven by uncertainty about population abundance? Ecol Lett 15:17–23
Koons DN, Rockwell RF, Aubry LM (2013) Effects of exploitation on an overabundant species: the lesser snow goose predicament. J Anim Ecol. doi:10.1111/1365-2656.12133
Kuiken T, Fouchier RAM, Rimmelzwaan GF, Osterhaus ADME (2006) Emerging viral diseases in waterbirds. In: Boere GC, Galbraith CA, Stroud DA (eds) Waterbirds around the world. The Stationery Office, Edinburgh, UK, pp 418–421
Lack D (1943) The life of the Robin. Witherby, London
Le Bohec C, Durant JM, Gauthier-Clerc M, Stenseth NC, Park YH, Pradel R, Gremillet D, Gendner JP, Le Maho Y (2008) King penguin population threatened by Southern Ocean warming. Proc Nat Acad Sci U S A 105:2493–2497
Lebreton JD (1995) The future of population dynamic studies using marked individuals: a statistician's perspective. J Appl Stat 22:1009–1030
Lebreton JD (1996) Demographic models for subdivided populations: the renewal equation approach. Theor Popul Biol 49:291–313
Lebreton JD (2005a) Dynamical and Statistical models for exploited populations. Aust N Z J Stat 47:49–63
Lebreton JD (2005b) Age, stages and the role of generation time in matrix models. Ecol Model 188:22–29
Lebreton JD (2006) Dynamical and statistical models of Vertebrate population dynamics. C R Biol 329:804–812

Lebreton JD (2011) The impact of global change on terrestrial vertebrates. C R Biol 334:360–369
Lebreton JD, Gimenez O (2013) Detecting and estimating Density-Dependence in wildlife populations. J Wildl Manage 77:12–23
Lebreton JD, Burnham KP, Clobert J, Anderson DR (1992) Modeling survival and testing biological hypotheses using marked animals: a unified approach with case studies. Ecol Monogr 62:67–118
Lebreton JD, Nichols JD, Barker RJ, Pradel R, Spendelow JA (2009) Modeling individual animal histories with multistage capture-recapture models. Adv Ecol Res 41:87–173
Lebreton JD, Devillard S, Popy S, Desprez M, Besnard A, Gaillard JM (2012) Towards a vertebrate demographic data bank. J Ornithol 152:617–624
Leopold A (1933) Game management. Charles Scribner's Sons, New York
Lindberg MS, Sedinger JS, Lebreton JD (2013) Individual heterogeneity in black brant survival and recruitment with implications for harvest dynamics. Ecol Evol 3:4045–4056
Lindstedt SL, Calder WA (1981) Body size, physiological time, and longevity of homeothermic animals. Quart Rev Biol 56:1–16
Lotka AJ (1939) On an integral equation in population analysis. Ann Math Stat 10:144–161
Loveridge AJ, Searle AW, Murindagomo F, Macdonald DW (2007) The impact of sport-hunting on the population dynamics of an African lion population in a protected area. Biol Conserv 134:548–558
MacArthur RH, Wilson EO (1967) The theory of island biogeography. Princeton University Press, Princeton/Oxford
Malthus TR (1798) An essay on the principle of population, as it affects the future improvement of society. J. Johnson, St Paul's Church-Yard, London
Mawdsley JR, O'Malley R, Ojima DS (2009) A Review of climate-change adaptation strategies for wildlife management and biodiversity conservation. Conserv Biol 23:1080–1089
Maynard Smith J (1965) Professor J. B. S. Haldane, FRS. Nature 206:239–240
McCullough DR (1979) The George Reserve deer herd. Population ecology of a k-selected species. University of Michigan Press, Ann Arbor
McLeod EM, Guay PJ, Taysom AJ, Robinson RW, Weston MA (2013) Buses, cars, bicycles and walkers: the influence of the type of human transport on the flight responses of waterbirds. PLOS One. doi:10.1371/journal.pone.0082008
Merow C, Dahlgren JP, Metcalf CJE, Childs DZ, Evans MEK, Jongejans E, Record S, Rees M, Salguero-Gomez R, McMahon SM (2014) Advancing population ecology with integral projection models: a practical guide. Meth Ecol Evol 5:99–110
Moberly RL, White PCL, Harris S (2004) Mortality due to fox predation in free-range poultry flocks in Britain. Vet Rec 155:48–52
Morellet N, Gaillard JM, Hewison AJM, Ballon P, Boscardin Y, Duncan P, Klein F, Maillard D (2007) Indicators of ecological change: new tools for managing populations of large herbivores. J Appl Ecol 44:634–643
Morris WF, Bloch PL, Hudgens BR, Moyle LC, Stinchcombe JR (2002) Population viability analysis in endangered species recovery plans: past use and future improvements. Ecol Appl 12:708–712
Müllner A, Linsenmair E, Wikelski M (2004) Exposure to ecotourism reduces survival and affects stress response in hoatzin chicks (*Opisthocomus hoazin*). Biol Conserv 118:549–558
Negus NC, Berger PJ (1988) Cohort analysis: environmental cues and diapause in microtine rodents. In: Boyce MS (ed) Evolution of life histories in mammals. Yale University Press, New Haven, pp 65–74
Nellemann C, Vistnes I, Jordhoy P, Stoen OG, Kaltenborn BP, Hanssen F, Helgesen R (2010) Effects of recreational cabins, trails and their removal for restoration of reindeer winter ranges. Restor Ecol 18:873–881
Nichols JD (1992) Capture-recapture models: using marked animals to study population dynamics. BioScience 42:94–102

Nichols JD, Blohm RJ, Reynolds RE, Trost RE, Hines JE, Bladen JP (1991) Band reporting rates for mallards with reward bands of different dollar values. J Wildl Manage 55:119–126
Nichols JD, Runge MC, Johnson FA, Williams BK (2007) Adaptive harvest management of North American waterfowl populations: a brief history and future prospects. J Ornithol 148(Suppl 2):S343–S349
Niel C, Lebreton JD (2005) Using demographic invariants to detect overharvested bird populations from incomplete data. Conserv Biol 19:826–835
Nilsen EB, Gaillard JM, Andersen R, Odden J, Delorme D, van Laere G, Linnell JDC (2009) A slow life in hell and a fast life in heaven: demographic analyses of contrasting roe deer populations. J Anim Ecol 78:585–594
Oli MK, Dobson FS (2003) The relative importance of life-history variables to population growth rate in mammals: Cole's prediction revisited. Am Nat 161:422–440
Owen-Smith N (1988) Megaherbivores. The influence of very large body size on ecology. Cambridge University Press, Cambridge, UK
Palazy L, Bonenfant C, Gaillard JM, Courchamp F (2012) Rarity, trophy hunting and ungulates. Anim Conserv 15:4–11
Pearl R, Miner JR (1935) Experimental studies on the duration of life. XIV. The comparative mortality of certain lower organisms. Quart Rev Biol 10:60–79
Péron G, Crochet PA, Doherty PF, Lebreton JD (2010) Dispersal at the landscape scale: efficient combination of population surveys and capture-recapture data. Ecology 91:3365–3375
Perrins CM, Lebreton JD, Hirons GJM (eds) (1991) Bird population studies: their relevance to conservation and management. Oxford University Press, Oxford
Peters RH (1983) The ecological implications of body size. Cambridge University Press, Cambridge
Pettorelli N, Gaillard JM, Duncan P, Maillard D, Van Laere G, Delorme D (2003) Age and density modify the effects of habitat quality on survival and movements of roe deer. Ecology 84:3307–3316
Pianka ER (1970) On r and K selection. Am Nat 104:592–597
Plard F, Bonenfant C, Delorme D, Gaillard JM (2012) Modeling reproductive trajectories of roe deer females: fixed or dynamic heterogeneity? Theor Popul Biol 82:317–328
Plard F, Gaillard JM, Coulson T, Hewison AJM, Delorme D, Warnant C, Bonenfant C (2014) Mismatch between birth date and vegetation phenology slows the demography of roe deer. Plos Biol 12:e1001828
Plard F, Gaillard JM, Coulson T, Delorme D, Warnant C, Michallet J, Tuljapurkar S, Krishnakurmar S, Bonenfant C (2015) Quantifying the influence of measured and unmeasured individual differences on demography. J Anim Ecol 84(5):1434–1445
Pledger S, Pollock KH, Norris JL (2003) Open capture-recapture models with heterogeneity: I. Cormack-Jolly-Seber model. Biometrics 59:786–794
Post E, Forchhammer MC (2008) Climate change reduces reproductive success of an Arctic herbivore through trophic mismatch. Philos Trans R Soc Lond B Biol Sci 363:2369–2375
Pradel R (2005) Multievent: an extension of multistate capture-recapture models to uncertain states. Biometrics 61:442–447
Promislow DEL, Harvey PH (1990) Living fast and dying young: a comparative analysis of life-history variation among mammals. J Zool 220:417–437
Réale D, Garant D, Humphries MM, Bergeron P, Careau V, Montiglio PO (2010) Personality and the emergence of the pace-of-life syndrome concept at the population level. Phil Trans R Soc B 365:4051–4063
Rees M, Sheppard A, Briese D, Mangel M (1999) Evolution of size-dependent flowering in *Onopordum Illyricum*: a quantitative assessment of the role of stochastic selection pressures. Am Nat 154:628–651
Rees M, Childs DZ, Ellner SP (2014) Building integral projection models: a user's guide. J Anim Ecol 83:528–545
Regehr EV, Hunter CM, Caswell H, Amstrup SC, Stirling I (2010) Survival and breeding of polar bears in the southern Beaufort Sea in relation to sea ice. J Anim Ecol 79:117–127

Reynolds JD, Jennings S, Dulvy NK (2001) Life histories of fishes and population responses to exploitation. Conserv Biol 6:147–168
Reza AHMA, Feeroz MM, Islam MA (2002) Man-tiger interactions in the Bangladesh Sundarbans. Bangladesh J Life Sci 14:75–82
Rezouki C, Dozières A, Le Coeur C, Thibaut S, Pisanu B, Chapuis JL, Baudry E (2014) A viable population of the European red squirrel in an urban park. PLoS One 9:e105111
Robinson JG, Bennett EL (2004) Having your wildlife and eating it too: an analysis of hunting. Anim Conserv 7:397–408
Rooney TP, Waller DM (2003) Direct and indirect effects of white-tailed deer in forest ecosystems. For Ecol Manage 181:165–176
Saether BE, Engen S, Solberg E, Heim M (2007) Estimating the growth of a new established moose population using reproductive value. Ecography 30:417–421
Sandercock BK, Martin K, Segelbacher G (2011) Ecology, conservation, and management of grouse. Stud Avian Biol 39:1–376
Schwarz CJ, Seber GAF (1999) A review of estimating animal abundance III. Stat Sci 14:427–456
Seber GAF (1973) The estimation of animal abundance and related parameters. Griffith, London
Seber GAF (1982) The estimation of animal abundance and related parameters, 2nd edn. Griffith, London
Seber GAF (1986) A review of estimating animal abundance. Biometrics 42:267–292
Seber GAF (1992) A review of estimating animal abundance II. Int Stat Rev 60:129–166
Skalski JR, Ryding KE, Millspaugh JJ (2005) Wildlife demography. Analysis of sex, age, and count data. Elsevier Academic Press, Boston
Solow AR (1998) On fitting a population model in the presence of observation error. Ecology 79:1463–1466
Stearns SC (1992) The evolution of life histories. Oxford University Press, Oxford, UK
Stenseth NC, Viljugrein H, Saitoh T, Hansen TF, Kittilsen MO, Bølviken E, Glöckner F (2003) Seasonality, density dependence, and population cycles in Hokkaido voles. Proc Natl Acad Sci U S A 100:11478–11483
Stephens PA, Frey-Roos F, Arnold W, Sutherland WJ (2002) Model complexity and population predictions. The alpine marmot as a case study. J Anim Ecol 71:343–361
Swihart RK, Dunning JB Jr, Waser PM (2002) Gray matters in ecology: dynamics of pattern, process, and scientific progress. Bull Ecol Soc Am 83:149–155
Tickell WLN (1968) The biology of the great albatrosses *Diomedea exulans* and *Diomedea epomophora*. Antarc Res Serv 12:1–55
Treves A, Karanth KU (2003) Human-Carnivore conflict and perspectives on Carnivore management worldwide. Conserv Biol 17:1491–1499
Tuljapurkar S (1990) Population dynamics in variable environments. Springer, New York
Tuljapurkar S, Steiner UK, Orzack SH (2009) Dynamic heterogeneities in life histories. Ecol Lett 12:93–106
Udevitz MS, Gogan PJ (2010) Estimating survival rates with time series of standing age-structure data. Ecology 93:726–732
Van de Pol M, Verhulst S (2006) Age-dependent traits: a new statistical model to separate within- and between-individual effects. Am Nat 167:766–773
Vaupel JW, Yashin AI (1985) Heterogeneity ruses – some surprising effects of selection on population dynamics. Am Stat 39:176–185
Vaupel JW, Manton KG, Stallard E (1979) The impact of heterogeneity in individual frailty on the dynamics of mortality. Demography 16:439–454
Verhulst PF (1838) Notice sur la loi que la population suit dans son accroissement. Corresp Math Phys 10:113–121
Vindenes Y, Engen S, Saether BE (2008) Individual heterogeneity in vital parameters and demographic stochasticity. Am Nat 171:455–467
Vindenes Y, Edeline E, Ohlberger J, Langangen O, Winfield IJ, Stenseth NC, Vollestad LA (2014) Effects of climate change on trait-based dynamics of a top predator in freshwater ecosystems. Am Nat 183:243–256

Wallace K, Leslie A, Coulson T (2013) Re-evaluating the effect of harvesting regimes on Nile crocodiles using an integral projection model. J Anim Ecol 82:155–165

Weimerskirch H, Brothers N, Jouventin P (1997) Population dynamics of wandering albatross *Diomedea exulans* and Amsterdam albatross *D. amsterdamensis* in the Indian Ocean and their relationships with long-line fisheries: conservation implications. Biol Conserv 79:257–270

White ER, Nagy JD, Gruber SH (2014) Modeling the population dynamics of lemon sharks. Biol Direct 9:23

Williams BK, Nichols JD, Conroy MJ (2002) Analysis and management of animal populations. Academic Press, San Diego

Wittmer HU, McLellan BN, Seip DR, Young JA, Kinley TA, Watts GS, Hamilton D (2005) Population dynamics of the endangered mountain ecotype of woodland caribou (*Rangifer tarandus caribou*) in British Columbia, Canada. Can J Zool 83:407–418

Research Priorities and Trends in Infections Shared with Wildlife

Christian Gortázar, Joaquín Vicente, Margarita Villar, Francisco Ruiz-Fons, Ursula Höfle, and José de la Fuente

Introduction

This chapter intends to identify research priorities and current trends in wildlife diseases. This is an ever-growing field where veterinary science has important overlaps and interactions with many related scientific fields such as molecular biology, wildlife ecology, human medicine and animal science, among others. This emerging research field is in the core of the One Health idea.

The concept of diseases shared with wildlife, or "shared diseases", indicates situations where a wildlife disease does also occur in other species, such as human beings or livestock. Shared diseases have been the focus of several reviews (e.g. Gortazar et al. 2007). However, the correct term is rather "shared infections", since not all infections do actually cause disease in each of the infected host species. Shared infections are important because they can constitute zoonoses, i.e. infections that are transmissible from animals to humans; because they can affect livestock production and economy; and because they can affect wildlife management and conservation. . This creates a specific need for our understanding of wildlife diseases with view to prevention, management and disease control, and thus a completely new scenario for wildlife disease research such as the application of new methods and techniques and experimental research in addition to surveillance. The following sections address each of the fields and identify specific research needs.

C. Gortázar (✉) • J. Vicente • M. Villar • F. Ruiz-Fons • U. Höfle • J. de la Fuente
SaBio, Instituto de Investigación en Recursos Cinegéticos (IREC, CSIC-UCLM-JCCM), Ronda de Toledo 12, Ciudad Real 13005, Spain
e-mail: christian.gortazar@uclm.es

R. Mateo et al. (eds.), *Current Trends in Wildlife Research*,
Wildlife Research Monographs 1, DOI 10.1007/978-3-319-27912-1_3

Wildlife Diseases

Pathogens are natural components of all ecosystems. As such, their effect on their host populations is as natural as other interactions between living beings, such as predation. However, human-driven changes have created situations where diseases can pose a major direct or indirect threat to the conservation of endangered species; can severely affect game production; or can compromise (domestic) animal and human health and wellbeing.

Examples of pathogens with effects on conservation are well known, for instance the sometimes-disastrous consequences of distemper, rabies and other viral diseases in endangered carnivores. In addition, diseases of keystone prey species can have cascading effects on their endangered predators (e.g. European wild rabbit, *Oryctolagus cuniculus* – variant Rabbit Haemorrhagic Disease Virus – Iberian lynx, *Lynx pardinus*; Delibes-Mateos et al. 2014). Other diseases do affect game production, and hence rural economy. These probably include all "minor" diseases, if their consequences in terms of losses in game productivity and quality were measured. Some more obvious examples include keratoconjunctivitis and mange in mountain ungulates and parasitic and bacterial infections of gamebirds, among many others (e.g. Turchetto et al. 2014).

One peculiarity of the wildlife disease field is that, in many cases, research efforts must also be focused on the development of diagnostic tools appropriate for wildlife species (Simpson 2002). Fortunately, the progressive switch towards increased antigen-detection instead of serum antibody screening has facilitated wildlife disease outbreak diagnostics/detection. However, antibody detection techniques are still a key component of wildlife disease research because they yield complementary information and are often the only applicable tool in non-outbreak settings, for instance in low-prevalence RNA-virus infections such as flavirirus infections (Boadella et al. 2012). Another peculiarity of wildlife diseases is that the assessment and quantification of the impact of a disease in wildlife is much more difficult than in livestock, depending more frequently on indirect indexes such as yearly census variation rather than on known morbidity and mortality figures. However, long-term studies and intense radio- or GPS-collar surveys have contributed to reveal the real impact of some of the most studied wildlife diseases on host population dynamics even if this field still deserves much research.

Yet another peculiarity of wildlife diseases is that, in some instances, disease-mediated conservation conflicts can be even worse than the direct effect of the disease itself (e.g. carrion removal for BSE control, in BSE-free countries). Such situations may benefit from applied research to develop management options to mitigate these situations and resolve the conflicts surrounding those (Gortazar et al. 2010).

A few general wildlife-related risk factors can be identified in most of the relevant shared infections (updated from Gortazar et al. 2007):

- Introduction of diseases through movements or translocations of wild or domestic animals
- Overabundance of wildlife
- Farming, including wildlife farming and open air farming of domestic animals, as well as the wildlife-livestock interactions

– The expansion or introduction of vectors
– The expansion or introduction of hosts
– Habitat or ecosystem modifications related to global change

All abovementioned factors deserve further research, particularly regarding risk mitigation.

Wildlife as Reservoirs for Livestock

Most pathogens relevant to livestock are able to cross-infect multiple host species, including wildlife, and therefore in areas where wildlife and livestock co-occur (i.e. interface areas), pathogens can emerge and establish themselves in these sympatric host populations. From the livestock health perspective, multi-host diseases that are notifiable, eradicated or almost under control in domestics are the most concerning ones, because a single spill over from wildlife to livestock may have severe consequences not only on animal health, but also on economy (Gortazar et al. 2007). Examples of multi-host diseases with a known wildlife reservoir include viral diseases such as bluetongue or West Nile, bacterial diseases such as animal tuberculosis (TB) and brucellosis, and many parasitic diseases such as for instance neosporosis. Also single-host diseases, such as classical and African swine fever, deserve attention given their severe economic consequences for the pig industry. Focusing on these diseases from the One Health perspective, where livestock and wildlife are considering as one in the disease control plan (obviously attending the particularities in management for each host group) is mandatory to overcome that transmission threat in the future.

However, and despite much recent progress, in many cases the available information is still not sufficient to decide if a given "disease–wildlife species–livestock" triangle is of concern, or not (Simpson 2002). In consequence, these many unclear "disease–wildlife species–livestock" triangles are targets for future research, and those with a higher socio-economic relevance should be prioritized among them. In this respect, birds merit special attention due to their high local mobility that can connect very different and spatially separated habitats. Even more importantly, migration, a feature intrinsic to the great majority of avian species, has proven a key feature for the epidemiology of diseases of worldwide importance such as Low and Highly pathogenic avian influenza (Wallensten et al. 2007).

Wildlife and Zoonoses

Since prehistory, the human-animal interface has evolved and expanded, ever allowing new pathogens to access the human host and cross species barriers (Reperant et al. 2013). Consequently, the number of pathogens known to infect humans is increasing. Whether such increase reflects improved surveillance and detection or actual emergence of novel pathogens is unclear. Nonetheless, infectious diseases

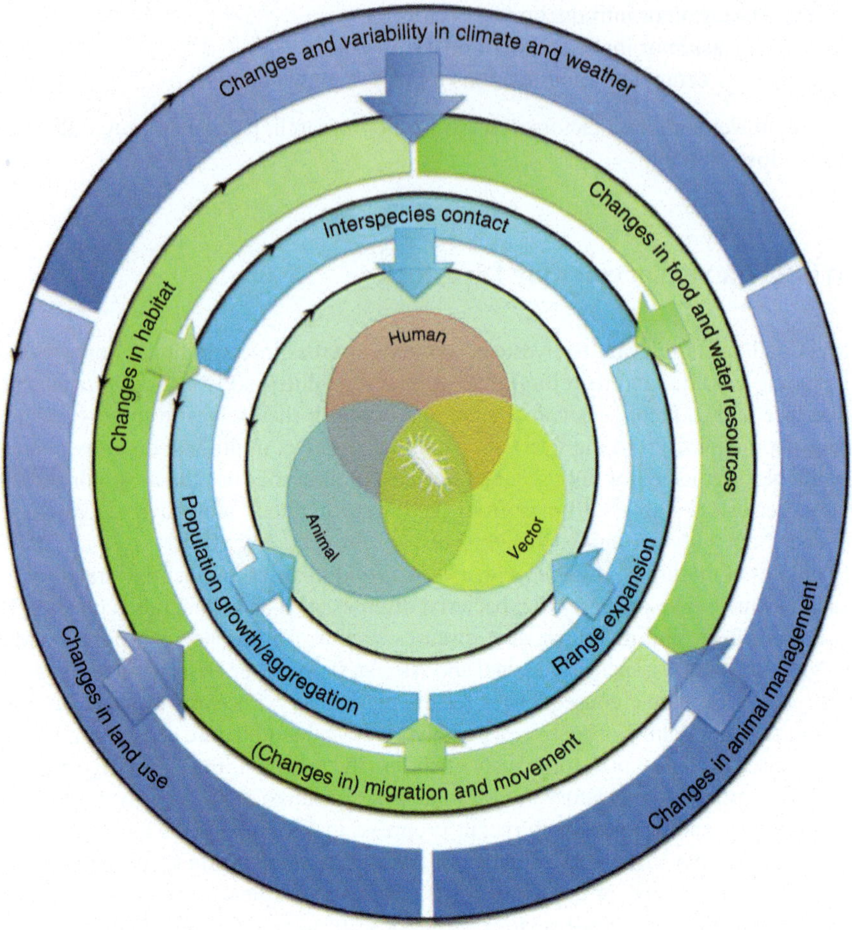

Fig. 1 Proximate and ultimate drivers of emergence of zoonotic pathogens. Reproduced from Gortazar et al. (2014)

are the second leading cause of human mortality worldwide (Fauci 2001). Most of these emerging pathogens originate from nonhuman animal species. The over 1000 known zoonotic pathogens represent approximately 60 % of all pathogens able to infect humans (Taylor et al. 2001). Revently, a limited number of drivers of zoonotic pathogen emergence have been defined, including several which are linked to wildlife such as land use change, climate change, and changes in animal (including wildlife) management (Fig. 1; Gortazar et al. 2014).

Currently, there is still a number of knowledge gaps regarding zoonotic disease emergence (Gortazar et al. 2014). From a wildlife perspective, these gaps include (a) the characterization of the relationships between environmental conditions and host (vector) population dynamics, as well as exploration of pathogen survival in the environment; (b) the need to switch from single host-pathogen systems to a multihost and multipathogen perspective; and (c) the need to consider all kind of interactions between hosts, vectors, and infectious agents (Gortazar et al. 2014).

Current Trends in Wildlife Disease Research

This section presents four aspects where both the methods and tools, as well as the insights into wildlife disease epidemiology, are evolving at a high speed, namely (1) the applications of novel molecular technologies ("omics") to wildlife disease research; (2) the growing relevance of vectors and vector-borne diseases in the wildlife disease field; (3) the incorporation of modern tools from animal ecology; and (4) the special characteristics of wild birds as pathogen hosts.

Omics Go Wild: Characterization of Host Response to Pathogen Infection and Vector Infestation and Possibilities for Disease Control

Recently, genome-scale or omics technologies produced nucleotide and protein databases from different hosts, pathogens and vectors that provide new possibilities for characterizing molecular events at the host-pathogen, host-vector and vector-pathogen interfaces (Fig. 2). As exemplified here using domestic and wild hosts, vector-borne and non-vector-borne pathogens, ticks and mosquitoes, recent publications illustrate the application of omics technologies to the study of host-pathogen, host-vector and vector-pathogen interactions using genomics, transcriptomics or proteomics (Chouard et al. 2002; Naranjo et al. 2006, 2007; de la Fuente et al. 2007, 2008; Kocan et al. 2008; Fernández de Mera et al. 2008; Galindo et al. 2008, 2009a, b, 2010a, b, 2012a, b; Galindo and de la Fuente 2012; Mastronunzio et al. 2012; Troese et al. 2011; Marcelino et al. 2012; Zivkovic et al. 2010; Villar et al. 2012; Ricci et al. 2012; Antunes et al. 2012; Dunning Hotopp et al. 2006; Lin et al. 2011; Ramabu et al. 2010; Sonenshine et al. 2011; Ciota and Kramer 2013; Ayllón et al. 2013; Huang et al. 2014; Crompton et al. 2014; Popara et al. 2013; Liu et al. 2014). Recent publications have also reported on the application of omics technologies to the study of host/vector response to pollutants and other stress conditions (Baillon et al. 2015; Fang et al. 2014; Villar et al. 2014a), evolution (Mans et al. 2008; Villar et al. 2014b; Genomic Resources Development et al. 2014) and microorganism communities in both hosts and vectors (Nakao et al. 2013; Blaser 2014; Zhuang et al. 2014; Mediannikov and Fenollar 2014).

Because almost all cellular processes rely on protein–protein interactions that can be easily disturbed by biological stimuli or during disease, the understanding of the dynamics of protein-protein interactions is essential for the control of important biological processes, such as those involved in the host immune response to

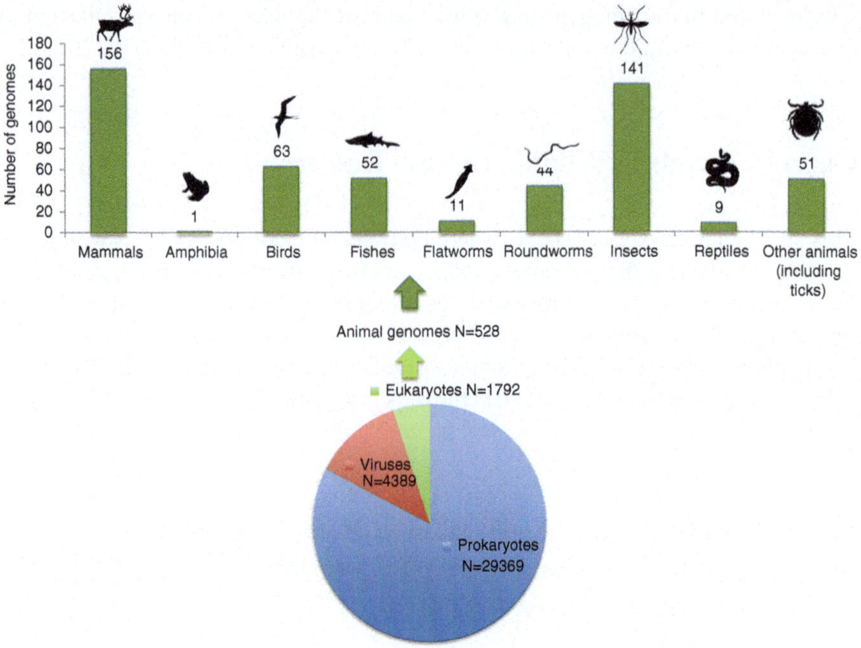

Fig. 2 Genome information by organisms. Information on available complete and partial genomes was obtained from NCBI (http://www.ncbi.nlm.nih.gov/genome/) on December 1st, 2014

infection. The characterization of these interactions by a systems biology perspective that include high throughput data-collection approaches will reveal the dynamics of infectious processes and facilitate the development of new drugs and vaccines for the control of these diseases. Therefore, interactomics is a key step to understand development, transmission and control of infectious diseases and is a very important scientist point to focus next research.

The analysis and integration of data derived from different omics technologies or systems biology will increase our understanding of the molecular mechanisms at the host-pathogen, host-vector and vector-pathogen interfaces, but omics data analysis and integration using bioinformatics tools is complex are requires the development of improved algorithms to integrate multi-level omics data with physiological or ecological drivers (Williams et al. 2011; Veldhoen et al. 2012; He 2012; Evans et al. 2012; Tieri et al. 2011; Kim et al. 2010; Joyce and Palsson 2006; Villar et al. 2014a; Popara et al. 2015; Domingos et al. 2015). One of these tools developed for biomedical research is VANTED, which allows the analysis of extensive experimental data from different omics platforms to structure, filter, evaluate and visualize results in the context of biological networks and signaling pathways (Klukas and Schreiber 2010; Rohn et al. 2012). However, the limited information available in genomics, transcriptomics, proteomics, metabolomics and other omics databases for many host, pathogen and vector species together with limitations in gene annotations due to the lack of functional studies requires validation of the results obtained after data integration and analysis.

The characterization of the molecular mechanisms that mediate host-pathogen, host-vector and vector-pathogen interactions allows the identification of the host factors affected in response to pathogen infection and vector infestation (de la Fuente and Merino 2013). These factors are relevant for the identification of genetic drivers affecting pathogen virulence and transmission that together with other biotic and abiotic factors could influence disease emergence, re-emergence and spread (Gortazar et al. 2014; Estrada-Peña et al. 2014). Furthermore, these interactions affect genetic traits of hosts, pathogens and vectors that will likely provide new targets for the control of vector infestations and pathogen infection and transmission (de la Fuente and Merino 2013). Vaccinomics is a recently developed approach based on the use of omics technologies and bioinformatics for the development of next-generation vaccines with an impact on individualized medicine (Poland et al. 2011a, b; Bernstein et al. 2011; de la Fuente and Merino 2013). These next-generation vaccines could be used to target pathogen or vector control as well as the transmission of vector-borne pathogens (Poland et al. 2011a, b; Bernstein et al. 2011; de la Fuente and Merino 2013).

The control of vector-borne diseases is particularly challenging due to the complexity of host-vector-pathogen interactions that condition pathogen infection and transmission (de la Fuente et al. 2008; Kocan et al. 2008). Therefore, the control of vector-borne diseases could be achieved using vaccines containing arthropod vector antigens alone or in combination with pathogen-derived antigens to control both arthropod populations and the infection and transmission of vector-borne pathogens (de la Fuente 2012). An important advantage of vector-targeted vaccines will likely be the ability to reduce or prevent transmission of several pathogens through immunization of reservoir hosts and human and animal populations at risk (Parizi et al. 2012; de la Fuente 2012). These vaccines could be obtained using proteins such as Subolesin/Akirin that are highly immunogenic antigens conserved across multiple vector species and targeting functionally relevant proteins and pathways for both vector infestations and pathogen infection and transmission (de la Fuente et al. 2013; Moreno-Cid et al. 2013; Merino et al. 2013; da Costa et al. 2014). The efficacy of antigen combinations and chimeras containing protective epitopes from different antigens is difficult to predict and requires vaccine trials, but should also be considered for the development of next-generation vaccines for the control of vector-borne diseases. A vaccinomics approach could be used for the identification of both vector-derived and pathogen-derived antigens, but in this case proteomics data is likely to be more relevant (Fig. 3), provided efficient algorithms can be applied for the selection of candidate protective antigens before entering into vaccine development and animal trails (de la Fuente and Merino 2013).

Indirect Links of Between-Host Pathogen Transmission: The Implication of Arthropod Blood-Feeding Vectors

From a 'One Health' perspective, blood-feeding arthropods constitute an important link for the indirect transmission of multi-host pathogens between wildlife, domestic animals and humans. The number of microbes – potential pathogens – that

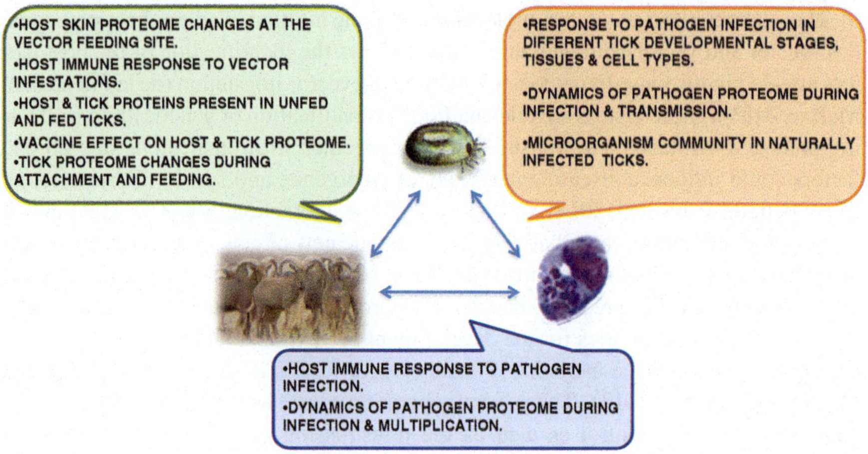

Fig. 3 Characterization of host-vector, vector-pathogen and host-pathogen interactions using proteomics. The characterization of the molecular mechanisms at the host-vector-pathogen interface identifies host, vector and pathogen-derived factors that will likely impact on the development of vaccines for the control of vector infestations and pathogen infection and transmission

circulate in undetected endemic cycles with limited geographic distribution should be high. The chances for low-prevalent, local endemic pathogens to reach new ecosystems favoring high transmission rates and geographic expansion – as pre-requisites for emerging pandemics – have been, perhaps until recently, very limited. However, increasing globalization of trade and travel, in liaison with changes in habitat structure, land use, host population dynamics, climatic conditions and, even, socioeconomic conditions and human behavioral patterns, have made vector-borne pathogens to be among the most concerning health threats for humans and animals (Jones et al. 2008; Semenza and Zeller 2014).

A high number of vector-borne zoonotic diseases have emerged/re-emerged or expanded their distribution range in humans (e.g. West Nile fever, Rift Valley fever, Japanese encephalitis, tick-borne encephalitis, Kyasanur forest disease or Lyme borreliosis) and in animals (e.g. bluetongue, Schmallenberg and leishmaniasis in Europe or epizootic hemorrhagic disease of deer in the USA) in the 20th and 21st centuries. Many shared vector-borne pathogens infect one or more wild species that may act as reservoirs for domestic animals and humans (Vorou et al. 2007; Baneth 2014). However, the role of wildlife in the ecology of many vector-borne pathogens has been largely dismissed with few exceptions, e.g. Lyme borreliosis in Europe and North America (reference needed), tick-borne encephalitis in central Europe (reference needed) or West Nile in the USA (reference needed). Wildlife plays a double role in vector-borne pathogen transmission: (i) influencing vector population dynamics; and (ii) influencing vector-borne pathogen replication (Ruiz-Fons et al. 2014a). Therefore, vector hosts may modulate vector-borne pathogen dynamics even though having no role in the replication of the pathogen, e.g. red deer (*Cervus elaphus*) and mountain hares (*Lepus timidus*) and louping ill virus in Scotland

(Gilbert et al. 2001). Changes in wildlife population dynamics – geographic range expansion and population density increase – in Europe in the last decades (Saez-Royuela and Telleria 1986; Csányi 1995; Lemel et al. 2003; DEFRA 2008; Hartley 2010; Putman et al. 2011) have affected vector population dynamics, e.g. red and roe deer and the tick *Ixodes ricinus* (Ruiz-Fons and Gilbert 2010) or wild ungulates and *Culicoides imicola* (Acevedo et al. 2010), as well as vector-borne pathogen epidemiology, e.g. roe deer and tick-borne encephalitis (Rizzoli et al. 2009). Inadequate wildlife disease surveillance within a scenario of changing wildlife population dynamics may lead to unexpected outbreaks of vector-borne diseases. The outbreak of human leishmaniasis that started in 2010 in the metropolitan area of Madrid was linked to increasing densities of Iberian hares (*Lepus granatensis*), a previously dismissed reservoir for *Leishmania infantum* (Molina et al. 2012; Ruiz-Fons et al. 2013). This example, among many others, shows the implication of wildlife in the maintenance and transmission of vector-borne pathogens causing human and livestock diseases.

Current research efforts directed to understand the role of wildlife population and individual traits in vector-borne pathogen dynamics and transmission are scarce for many threatening vector-borne pathogens such as Crimean-Congo hemorrhagic fever virus (Estrada-Peña et al. 2012), Schmallenberg virus (Fernández-Aguilar et al. 2014), Usutu and Bagaza viruses (Vázquez et al. 2011; Gamino et al. 2012), new strains of louping ill virus (Balseiro et al. 2012; Ruiz-Fons et al. 2014b), new piroplasms (Jouglin et al. 2014), or avian malaria (Merino et al. 2000), among others . A good example is bluetongue virus (BTV; see Falconi et al. 2011 and Ruiz-Fons et al. 2014b). Since its emergence in southern and central Europe in early 2000s, different BTV serotypes have become endemic in southern Europe and North Africa. As soon as BTV4 emerged in Spain, surveys on the epidemiological status of BTV in wild ungulates were carried out (Ruiz-Fons et al. 2008; García et al. 2009; Rodríguez-Sánchez et al. 2010). Later on, experimental infection studies demonstrated that red deer may constitute an important source of BTV infection for the vectors (López-Olvera et al. 2010). However, although it is currently suspected that BTV4 – and perhaps other serotypes – is maintained by red deer in an endemic cycle from which it may emerge and cause new outbreaks in livestock (Falconi et al. 2012), we currently lack of: (i) complete spatial and temporal BTV monitoring in wild ruminant populations; (ii) knowledge on the role of wild hosts in the dynamics of *Culicoides* spp.; (iii) knowledge on the role of vector community assemblages in BTV maintenance and transmission; and (iv) identification of the links that modulate vector-mediated BTV transmission between wild and domestic hosts. Conversely, in most of the European areas where BTV outbreaks were registered in livestock, vaccination campaigns for domestic hosts only demonstrated to be efficient to stop the epidemics, further supporting the importance of including ecosystem variables in the One Health approach to deal with diseases where wildlife may be relevant.

If information on the role of wildlife in the life-cycle of many vector-borne pathogens is scarce in developed societies, there is practically no information in developing areas of the World. The high biodiversity of both vertebrate hosts and vectors in tropical regions suggests a complex network of endemic, unknown, pathogens that may – if they are given the opportunity – jump from one host-vector system to others and become potential threats to human and animal health and wildlife conservation (Gortazar et al. 2014; Estrada-Peña et al. 2014). Moreover, the global change-related rise in population in such biodiversity rich developing countries causes alterations in the ecosystem, increasing the risk of human and livestock contact with vectors that had previously only contact with wildlife (Estrada-Peña et al. 2014). Current globalization evidences the need that developed countries invest in research in vector-borne diseases in developing countries to: (i) improve public and animal health and wildlife conservation programs in developing countries; (ii) contribute to research, development and innovation in those countries; and (iii) counteract the negative effects of possible future global health threats through prevention. This already constitutes a One Health approach.

Future research should also focus on disentangling the effect of environmental factors for particular vector-borne pathogens at different geographical scales – both for locally and widely geographically distributed pathogens – since many of these pathogens exploit, or have the potential to exploit, very different ecosystems. Changes in vector distribution may allow the pathogen to enter in contact with new competent vectors (as occurred with BTV, reference needed), therefore increasing pathogen distribution area and allowing the pathogen to enter in contact also with new potential naïve hosts. Massive pathogen screening techniques should be applied to vectors in order to identify the global diversity of microbes and viruses – both pathogenic and non-pathogenic. The application of new OMICS tools and next generation sequencing techniques may be useful to cope with the diversity of microbes that may potentially constitute future health threats for humans, domestic animals and wildlife.

Impact of Host Demography and Behavior on Disease: Learning to Use New Tools in Ecology

The demography and behavior of host populations can play an important role in intra- and interspecific pathogen transmission by determining contact rates and environmental exposure (Vicente et al. 2007). The spatial or social structure of a population influence the rate of disease spread and disease persistence (Keeling 1999, 2000; Hagenaars et al. 2004), although few empirical studies have described the maintenance and spread of infectious diseases according to host behavior and social structure (Vicente et al. 2007). Such information is, however, crucial to understanding how population structure affects disease transmission (Anderson and May 1991) and the implications for management.

If resources that are commonly used by both domestic and wild species are aggregated, this can result in high spatial and/or temporal overlap and concentration between two or more species (e. g. Proffitt et al. 2011; Nunn et al. 2014; Cooper et al. 2008; Kukielka et al. 2013; Barasona et al. 2014a, b), further increasing the probability of disease transmission. Interactions between wildlife and livestock create significant risks for disease transmission (Phillips et al. 2003; Kuiken et al. 2005), which increase when animals congregate and share resources (e.g. Miller et al. 2003). For instance, Tuberculosis (TB) is endemic in Eurasian wild boar (*Sus scrofa*) and red deer (*Cervus elaphus*) in south-central Spain, and evidence suggests transmission to domestic cattle (Vicente et al. 2005). Under dry Mediterranean conditions, known risk factors for TB (and other diseases) at the livestock/wild ungulates interface include spatial aggregation around waterholes during summer (Kukielka et al. 2013). Some studies remark the possibility of indirect TB transmission between wildlife and cattle (Phillips et al. 2003; VerCauteren et al. 2008). In the UK, a different epidemiological context, cattle could be infected by badger (*Meles meles*) contact with farm buildings (feed stores and cattle sheds) and grazing on grass contaminated by badger urine, feces, sputum or wound exudates (e.g. Delahay et al. 2005). Usually, wildlife and livestock share the interface (space) at different times, so the disease interface between wildlife and livestock usually is indirect contact (e.g. Palmer et al. 2004). Moreover, direct transmission between wildlife and cattle can occur when wildlife visit farms (Böhm et al. 2009).

Determining the patterns of host distribution and abundance at a finer resolution and for large areas, as demanded for epidemiological studies, requires an impressive sampling effort (e.g. Jachmann et al. 1991). For instance, high-resolution images (which can be now obtained by unmanned aerial systems, UAS) have been used to obtain data on ungulate distribution and abundance (e.g. Vermeulen et al. 2013) and epidemiological relationships between host species (Barasona et al. 2014a, b). The difficulty of estimating first the distribution and then realistic contact frequencies between species, most of which are predicted to be indirect, has been highlighted previously by ecologists and epidemiologists (e.g. Proffitt et al. 2011). However, recent studies have attempted to measure interspecific contact rates in relation to the dynamics of disease transmission (Fig. 4). For example, contact rates have been estimated by direct observation of domestic and wild animals in open habitats where they are easily observed, such as alpine meadows (e.g. Richomme et al. 2006), and by camera trapping in less open habitats (Kukielka et al. 2013). Other recent studies using telemetry data have defined critical time and space windows between pairs of GPS locations, and thus only assumed that interspecific contact had occurred within this critical window (Miguel et al. 2013). Approaches based on proximity loggers potentially have the ability to estimate contact rates between individuals often to within a few meters; however, the performance of these devices is sometimes poor, providing data that is only indicative of contact rates rather than actual contact rates where interactions occur (Drewe et al. 2012). Furthermore, within an epidemiological context, their utility is constrained to direct rather than indirect disease transmission. Habitat selection and overlap modeling can be used as a reliable tool to indirectly estimate annual and seasonal similarities in the use of shared resources, which is valuable for the study of diseases for which direct as well as indirect interactions among sympatric species are of importance in transmission dynamics (Barasona et al. 2014a, b).

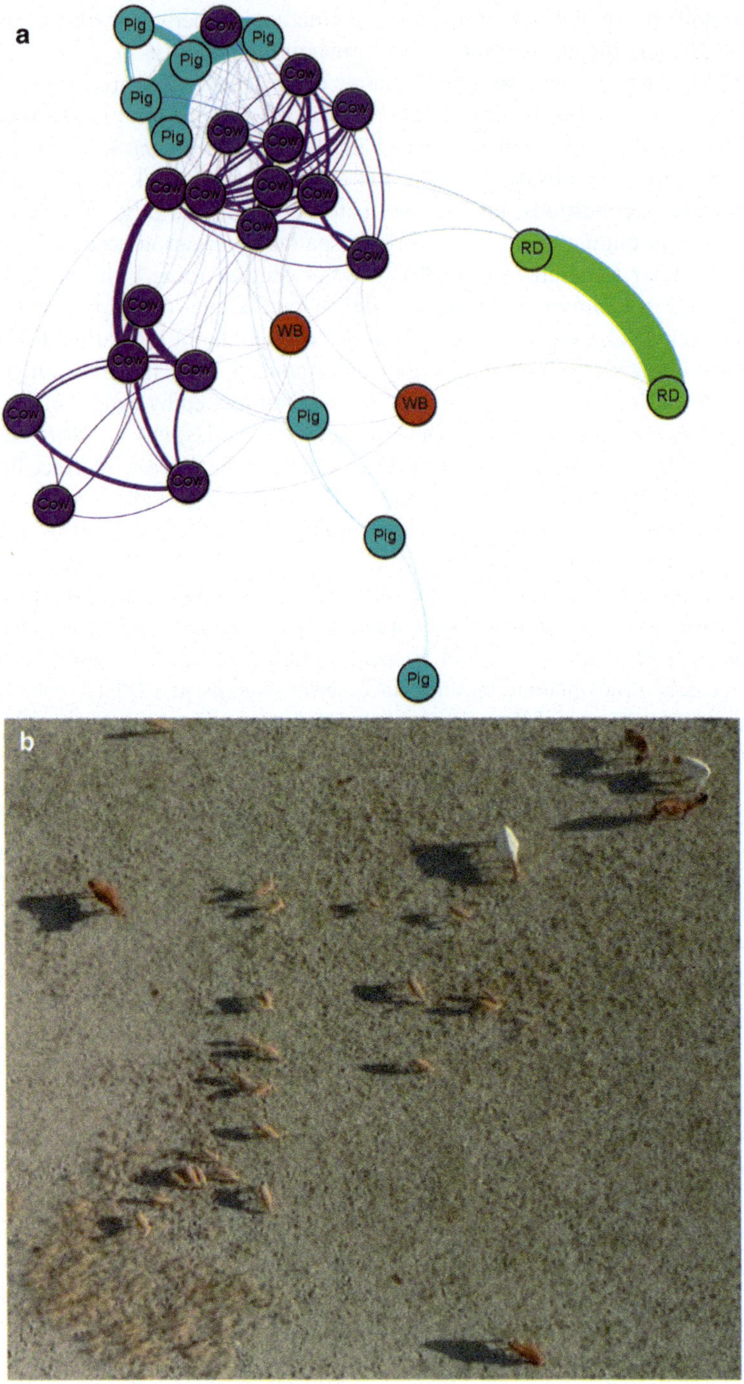
a
Pig
Cow
Pig
Pig
Cow
Pig
Cow
Pig
Cow
Cow
Cow
Cow
Cow
Cow
Cow
RD
Cow
WB
Cow
Cow
WB
RD
Cow
Pig
Cow
Cow
Pig
Pig
b

Integral approaches to control disease at the wildlife/livestock interface require mitigation of contact to reduce disease transmission between wildlife and cattle, including adequate biosecurity, husbandry measures and pasture management (e. g. Judge et al. 2011). Ideally, we must improve biosecurity by using reliable long-term nonlethal barriers in problematic areas, especially in certain spatially limited risk situations. For instance, preventing aggregation and subsequent contact between domestic and wild animals at water points (WP) may be valuable for disease transmission control in Southern Iberia (VerCauteren et al. 2006, 2009, 2010; Phillips et al. 2012; Barasona et al. 2014a, b). However, it is important to stress that the current focus on just one or two actors should progressively switch into more holistic, One Health approaches where all animal hosts, and humans, are considered.

Host Ecology and Disease Transmission in the Flight Division: Avian Mobility and Migration

While mammalian disease transmission systems are to a certain degree spatially defined, avian mobility and especially avian migration add new factors, that include both the introduction of pathogens but also the temporal influx of hosts naïve to local pathogens (Reed et al. 2003; Verhagen et al. 2014). Migration also adds substantial behavioral (aggregation, social stress) and physiologic (energy budget, endocrine household, immune capacity) changes in the individual hosts and host populations that significantly affect infection by pathogens, and pathogen excretion in carriers (Reperant et al. 2011). Especially during and since the H5N1 avian influenza pandemic, avian migration ecology is receiving special attention with view to wildlife disease research reducing a number of existing knowledge gaps while opening others (Reeds et al. 2003). In parallel, significant advances in satellite and GPS tracking technology and development of other fields such as stable isotope analysis have provided new tools for the study of the ecology of avian migratory

Fig. 4 Examples of new ecology tools in wildlife epidemiology. (**a**) Example of proximity data loggers for contact network analysis. The interaction network between host populations defined by nodes and edges, and characterized by the frequency and intensity of the interactions (Figure by Cat Cowie). Species involved in this real example are pigs, cows, red deer and wild boar from an extensive farm in central Spain. Proximity data-logging collars consist of an Ultra High Frequency (UHF) transceivers that broadcast a unique ID code, whilst simultaneously 'listening' for those of others. It is possible to measure the connectedness between individuals and/or spatial features (e. g. aggregation points) and perform a network analysis. (**b**) High resolution image obtained by Unmanned Aircraft Systems (UAS) to model spatial risk for disease transmission in ungulates (tuberculosis in a complex multi-host system: fallow deer and cattle). Picture by Mara Mulero and Juan Jose Negro (Aeromab-EBD, see details in Barasona et al. 2014a, b)

hosts. Research trends in this respect are on one hand integrative approaches using tracking devices and/or stable isotope analysis in combination with documentation of exposure (serosurveillance) or pathogen genome detection and phylogenetic analysis (Gunnarsson et al. 2012; Wille et al. 2013; Verhagen et al. 2014), but also field studies that combine analysis of environmental conditions with host and pathogen diversity (Pérez-Ramírez et al. 2012; McKenzie and Goulet 2010) .

Experimental studies using wild avian hosts that try to integrate physiological changes and the effect of environmental conditions on host susceptibility and pathogen excretion as well as on the impact of infection on the host and thus host capacity for further spread of the disease, are becoming ever more common (Hall et al. 2013; Jourdain et al. 2010; Pérez-Ramírez et al. 2014; Reperant et al. 2011). Experimental and field study settings also try to assess the effects of host pathogen interactions and disease pathogenesis using combinations of classical methods in pathology and virology and more modern and molecular technology such as sequencing, immunohistochemistry or ex-vivo cultures (Costa et al. 2012; Bertran et al. 2013; Höfle et al. 2012). A field of paramount importance for experimental and field studies in the future is the combined study of these pathogenic mechanisms of the host-pathogen interaction, environmental pathogen persistence, host ecology, competition and synergism between strains of the same pathogen and pathogen interactions (e.g. Handel et al. 2014).

Wildlife Disease Control

The control of infections shared with wildlife requires the development of strategies that will reduce pathogen transmission between wildlife and both domestic animals and humans. The first requirement is establishing a proper surveillance and monitoring scheme (disease and population wise; see Boadella et al. 2011). Thereafter, disease control can be achieved by different means such as preventive actions, host population control and vaccination. A review discusses the pros and cons of the currently available options (Gortazar et al. 2015).

As stated in this review, the success of disease control in wildlife depends on many factors, including disease ecology, natural history of the pathogen, availability of suitable diagnostic tools, characteristics of the hosts (and eventually vectors), geographical spread of the problem and scale of the control effort, and stakeholders' attitudes. This highlights two important facts to consider, regarding research on wildlife disease control: first, the need for cross-disciplinarity, and second, the need for many more studies on the effects of intervention on disease prevalence and on wildlife populations (i.e., experimental disease and/or ecosystem management studies), particularly using single-tool approaches, since integrated interventions (using more than one tool at a time) might be more efficient, but limit the assessment of the efficacy of each single intervention (Gortazar et al. 2015).

Research Priorities

Priorities in wildlife disease research can be seen by pathogen, by host, or by knowledge gap. Priority diseases are those where wildlife has a high probability of substantially interfering with existing (and successful) control schemes in domestic animals, such as TB, and those cases where the disease has a strong impact on human health or on conservation (Gortazar et al. 2007). From the host perspective, endangered species and species with a high economic relevance (as game species, as pest, or otherwise) are of particular relevance. Among the abundant species, migratory birds, the Eurasian wild boar and the feral pig (*Sus scrofa*) and deer (subfamily cervinae) are probably the ones deserving more attention regarding research on shared infections. How habitat use by hosts affects direct and indirect interactions among hosts is fundamental in understanding multi-host disease transmission (Cooper et al. 2010), and is critical for designing scientifically-based disease control strategies. Nonetheless, the role that spatial and temporal interactions between hosts (for instance livestock and wildlife) play in exposure to pathogens and disease transmission remains mostly unknown (e.g. Böhm et al. 2009). To assess this question, the combination of new technology (in the lab and in the field) to validate and model rates of contacts and quantify the potential for indirect disease transmission will be essential. Finally, relevant general knowledge gaps identified in this review are listed in Table 1.

Conclusions

This chapter identified a few general wildlife-related risk factors, and pointed out that there are still many insufficiently known "disease–wildlife species–livestock" triangles. Both the risk factors and the triangles deserve further research. Among them, situations with higher socio-economic relevance should be prioritized. Regarding our understanding of disease emergence, knowledge gaps identified include the characterization of the relationships between environmental conditions and host (vector) population dynamics, as well as exploration of pathogen survival in the environment; the need to switch from single host-pathogen systems to a multihost and multipathogen perspective; and the need to consider all kind of interactions between hosts, vectors, and infectious agents. Vector-borne diseases and diseases with avian hosts deserve increased attention. Regarding disease control, we identified two important needs: first, the need for cross-disciplinarity, and second, the need for studies on the effects of intervention on disease prevalence and on wildlife populations.

Acknowledgements This is a contribution to EU-H2020 grant COMPARE (377/14) and to MINECO Spain Plan Nacional de I + D grant AGL2014-56305 and FEDER.

Table 1 Research priorities in the field "shared infections" identified in this review

Research area	Examples of research needs	New tools involved
Pathogen discovery	Exploration of sampling strategies for emerging diseases that account for host and pathogen diversity and optimize detection	OMICS
Host-vector-pathogen interactions	Studies that combine results of host-pathogen interactions in laboratory settings with results of field studies on host ecology	Improved algorithms to integrate multi-level omics data with physiological or ecological drivers Animal behavior tools combined with OMICS pathology and immunology
Epidemiology	Studies on the less known "disease–wildlife species–livestock" triangles, prioritized by conservation, socio-economic or zoonotic relevance	Several
Vector ecology	Characterization of the relationships between environmental conditions and host (vector) population dynamics at different geographical scales	Modelling
Epidemiology	Exploration of pathogen survival in the environment. Need to consider all kind of interactions between hosts (vectors) and infectious agents	Molecular microbiology
Epidemiology	Need to switch from single host-pathogen systems to a multihost and multipathogen perspective	Modelling
Risk mitigation (shared diseases)	Limit the likelihood of contacts between host species, and the opportunities for inter-species pathogen transmission	Innovative contact-prevention tools
Multi-infections and co-infections	Competition and potentiation between pathogen strains and interactions among pathogens	Molecular epidemiology combined with pathology and immunology
Host-vector-pathogen interactions	Development of genomics, transcriptomics, proteomics, metabolomics and other omics databases for host, pathogen and vector species together with gene annotations based on functional studies to validate the results obtained after data integration and analysis.	OMICS, bioinformatics and functional studies
Disease control	Development of next-generation vaccines to target pathogen or vector control as well as the transmission of vector-borne pathogens	Vaccinomics

References

Acevedo P, Ruiz-Fons F, Estrada R, Márquez AL, Miranda MA, Gortázar C, Lucientes J (2010) A broad assessment of factors determining culicoides imicola abundance: modelling the present and forecasting its future in climate change scenarios. PLoS One 5(12):e14236

Antunes S, Galindo RC, Almazán C, Rudenko N, Golovchenko M, Grubhoffer L, Shkap V, do Rosário V, de la Fuente J, Domingos A (2012) Functional genomics studies of Rhipicephalus (Boophilus) annulatus ticks in response to infection with the cattle protozoan parasite. Babesia bigemina. Int J Parasitol 42:187–195

Anderson RM, May RM (1991) Infectious Diseases of Humans: Dynamics and Control. Oxford UP, Oxford, pp 766

Ayllón N, Villar M, Busby AT, Kocan KM, Blouin EF, Bonzón-Kulichenko E, Galindo RC, Mangold AJ, Alberdi P, Pérez de la Lastra JM, Vázquez J, de la Fuente J (2013) Anaplasma phagocytophilum inhibits apoptosis and promotes cytoskeleton rearrangement for infection of tick cells. Infect Immun 81:2415–2425

Baillon L, Pierron F, Coudret R, Normendeau E, Caron A, Peluhet L, Labadie P, Budzinski H, Durrieu G, Sarraco J, Elie P, Couture P, Baudrimont M, Bernatchez L (2015) Transcriptome profile analysis reveals specific signatures of pollutants in Atlantic eels. Ecotoxicology 24(1):71–84

Balseiro A, Royo LJ, Pérez Martínez C, Fernández de Mera IG, Höfle U, Polledo L, Marreros N, Casais R, García Marín JF (2012) Louping ill in goats, Spain, 2011. Emerg Infect Dis 18:976–978

Baneth G (2014) Tick-borne infections of animals and humans: a common ground. Int J Parasitol 44:591–596

Barasona JA, Latham CM, Acevedo P, Armenteros JA, Latham ADM, Gortazar C, Carro F, Soriguer RC, Vicente J (2014a) Spatiotemporal interactions between wild boar and cattle: implications for cross-species disease transmission. Vet Res 45:122

Barasona JA, Mulero-Pázmány M, Acevedo P, Negro JJ, Torres MJ, Gortázar C, Vicente J (2014b) Unmanned aircraft systems for studying spatial aggregation of ungulates: relevance to spatial epidemiology. PloS One 9(12):e115608. doi:10.1371/journal.pone.0115608

Bernstein A, Pulendran B, Rappuoli R (2011) Systems vaccinomics: the road ahead for vaccinology. OMICS 15:529–531

Bertran K, Dolz R, Busquets N, Gamino V, Vergara-Alert J, Chaves AJ, Ramis A, Abad FX, Höfle U, Majó N (2013) Pathobiology and transmission of highly and low pathogenic avian influenza viruses in European quail (Coturnix c. coturnix). Vet Res 28(44):23. doi:10.1186/1297-9716-44-23

Blaser MJ (2014) The microbiome revolution. J Clin Invest 124:4162–4165

Boadella M, Gortazar C, Acevedo P, Carta T, Martín-Hernando MP, de la Fuente J et al (2011) Six recommendations for improving monitoring of diseases shared with wildlife: examples regarding mycobacterial infections in Spain. Eur J Wildl Res 57(4):697–706. doi:10.1007/s10344-011-0550-x

Boadella M, Díez-Delgado I, Gutiérrez-Guzmán AV, Höfle U, Gortázar C (2012) Do wild ungulates allow improved monitoring of flavivirus circulation in Spain? Vector Borne Zoonotic Dis 12:490–495

Böhm M, Hutchings M, White P (2009) Contact networks in a wildlife-livestock host community: identifying high-risk individuals in the transmission of bovine TB among badgers and cattle. PLoS One 4(4):e5016

Chouard T, Weiss U, Dhand R (2002) Good 'omics' for the poor? Nature 3(419(6906)):489

Ciota AT, Kramer LD (2013) Vector-virus interactions and transmission dynamics of West Nile virus. Viruses 5(12):3021–3047

Cooper SM, Perotto-Baldivieso HL, Owens MK, Meek MG, Figueroa-Pagan M (2008) Distribution and interaction of white-tailed deer and cattle in a semi-arid grazing system. Agr Ecosyst Environ 127(1):85–92

Cooper SM, Scott HM, de la Garza GR, Deck AL, Cathey JC (2010) Distribution and interspecies contact of feral swine and cattle on rangeland in south Texas: implications for disease transmission. J Wildl Dis 46:152–164
Costa T, Chaves AJ, Valle R, Darji A, van Riel D, Kuiken T, Majó N, Ramis A (2012) Distribution patterns of influenza virus receptors and viral attachment patterns in the respiratory and intestinal tracts of seven avian species. Vet Res 10(43):28. doi:10.1186/1297-9716-43-28
Crompton PD, Moebius J, Portugal S, Waisberg M, Hart G, Garver LS, Miller LH, Barillas-Mury C, Pierce SK (2014) Malaria immunity in man and mosquito: insights into unsolved mysteries of a deadly infectious disease. Annu Rev Immunol 32:157–187
Csányi S (1995) Wild boar population dynamics and management in Hungary. J Mt Ecol 3:222–225
da Costa M, Pinheiro-Silva R, Antunes S, Moreno-Cid JA, Villar M, de la Fuente J, Domingos A (2014) Mosquito Akirin as a potential antigen for malaria control. Malar J 13:470
de la Fuente J (2012) Vaccines for vector control: exciting possibilities for the future. Vet J 194:139–140
de la Fuente J, Merino O (2013) Vaccinomics, the new road to tick vaccines. Vaccine 31:5923–5929
de la Fuente J, Blouin EF, Manzano-Roman R, Naranjo V, Almazán C, Perez de la Lastra JM, Zivkovic Z, Jongejan F, Kocan KM (2007) Functional Genomic studies of tick cells in response to infection with the cattle pathogen, anaplasma marginale. Genomics 90:712–722
de la Fuente J, Kocan KM, Almazán C, Blouin EF (2008) Targeting the tick-pathogen interface for novel control strategies. Front Biosci 13:6947–6956
de la Fuente J, Moreno-Cid JA, Galindo RC, Almazán C, Kocan KM, Merino O, Pérez de la Lastra JM, Estrada-Peña A, Blouin EF (2013) Subolesin/Akirin vaccines for the control of arthropod vectors and vector-borne pathogens. Transbound Emerg Dis 60(Suppl 2):172–178
Defra (2008) Feral wild boar in England: An action plan. Report available at http://www.naturalengland.org.uk/Images/feralwildboar_tcm6-4508.pdf
Delahay RJ, Smith GC, Ward AI, Cheeseman CL (2005) Options for the management of bovine tuberculosis transmission from badgers (Meles meles) to cattle: evidence from a long-term study. Mamm Stud 30:73–81
Delibes-Mateos M, Ferreira C, Carro F, Escudero MA, Gortázar C (2014) Ecosystem effects of variant rabbit hemorrhagic disease virus, Iberian Peninsula. Emerg Infect Dis 20:2166–2168, http://dx.doi.org/10.3201/eid2012.140517
Domingos A, Antunes S, Villar M, de la Fuente J (2015) Functional genomics of tick vectors challenged with the cattle parasite Babesia bigemina. Methods Mol Biol 1247:475–489
Drewe JA, Weber N, Carter SP, Bearhop S, Harrison XA, Dall SR, McDonald RA, Delahay RJ (2012) Performance of proximity loggers in recording intra-and inter-species interactions: a laboratory and field-based validation study. PLoS One 7(6):e39068
Dunning Hotopp JC, Lin M, Madupu R, Crabtree J, Angiuoli SV, Eisen JA, Seshadri R, Ren Q, Wu M, Utterback TR, Smith S, Lewis M, Khouri H, Zhang C, Niu H, Lin Q, Ohashi N, Zhi N, Nelson W, Brinkac LM, Dodson RJ, Rosovitz MJ, Sundaram J, Daugherty SC, Davidsen T, Durkin AS, Gwinn M, Haft DH, Selengut JD, Sullivan SA, Zafar N, Zhou L, Benahmed F, Forberger H, Halpin R, Mulligan S, Robinson J, White O, Rikihisa Y, Tettelin H (2006) Comparative genomics of emerging human ehrlichiosis agents. PLoS Genet 2:e21
Estrada-Peña A, Palomar AM, Santibáñez P, Sánchez N, Habela MA, Portillo A, Romero L, Oteo JA (2012) Crimean-Congo hemorrhagic fever virus in ticks, Southwestern Europe, 2010. Emerg Infect Dis 18(1):179–180
Estrada-Peña A, Ostfeld RS, Peterson AT, Poulin R, de la Fuente J (2014) Effects of environmental change on zoonotic disease risk: an ecological primer. Trends Parasitol 30:205–214
Evans VC, Barker G, Heesom KJ, Fan J, Bessant C, Matthews DA (2012) De novo derivation of proteomes from transcriptomes for transcript and protein identification. Nat Methods 9:1207–1211
Falconi C, López-Olvera JR, Gortazar C (2011) BTV infection in wild ruminants, with emphasis on red deer: a review. Vet Microbiol 151(3–4):209–219

Falconi C, López-Olvera JR, Boadella M, Camarena J, Rosell R, Alcaide V, Vicente J, Sánchez-Vizcaíno JM, Pujols J, Gortazar C (2012) Evidence for BTV-4 circulation in free-ranging red deer (Cervus elaphus) in Cabañeros National Park, Spain. Vet Microbiol 159:40–46

Fang Y, Gao X, Zhao F, Zhang H, Zhang W, Yang H, Lin B, Xi Z (2014) Comparative proteomic analysis of ovary for Chinese rare minnow (Gobiocypris rarus) exposed to chlorophenol chemicals. J Proteomics 110C:172–182

Fauci AS (2001) Infectious diseases: considerations for the 21st century. Clin Infect Dis 32:675–685

Fernández de Mera IG, Pérez de la Lastra JM, Ayoubi P, Naranjo V, Kocan KM, Gortazar C, de la Fuente J (2008) Differential expression of inflammatory and immune response genes in mesenteric lymph nodes of Iberian red deer (Cervus elaphus hispanicus) naturally infected with Mycobacterium bovis. Dev Comp Immunol 32:85–91

Fernández-Aguilar X, Pujols J, Velarde R, Rosell R, López-Olvera JR, Marco I, Pumarola M, Segalés J, Lavín S, Cabezón O (2014) Schmallenberg virus circulation in high mountain ecosystem, Spain. Emerg Infect Dis 20:1062–1064

Galindo RC, de la Fuente J (2012) Transcriptomics data integration reveals Jak-STAT pathway as a common pathway affected by pathogenic intracellular bacteria in natural reservoir hosts. J Proteomics Bioinform 5:108–115

Galindo RC, Ayoubi P, García-Pérez AL, Naranjo V, Kocan KM, Gortazar C, de la Fuente J (2008) Differential expression of inflammatory and immune response genes in sheep infected with Anaplasma phagocytophilum. Vet Immunol Immunopathol 126:27–34

Galindo RC, Ayoubi P, Naranjo V, Gortazar C, Kocan KM, de la Fuente J (2009a) Gene expression profiles of European wild boar naturally infected with Mycobacterium bovis. Vet Immunol Immunopathol 129:119–125

Galindo RC, Muñoz PM, de Miguel MJ, Marin CM, Blasco JM, Gortazar C, Kocan KM, de la Fuente J (2009b) Differential expression of inflammatory and immune response genes in rams experimentally infected with a rough virulent strain of Brucella ovis. Vet Immunol Immunopathol 127:295–303

Galindo RC, Ayllón N, Carta T, Vicente J, Kocan KM, Gortazar C, de la Fuente J (2010a) Characterization of pathogen-specific expression of host immune response genes in Anaplasma and Mycobacterium species infected ruminants. Comp Immunol Microbiol Infect Dis 33:e133–e142

Galindo RC, Muñoz PM, de Miguel MJ, Marin CM, Labairu J, Revilla M, Blasco JM, Gortazar C, de la Fuente J (2010b) Gene expression changes in spleens of the wildlife reservoir species, Eurasian wild boar (Sus scrofa), naturally infected with Brucella suis biovar 2. J Genet Genomics 37:1–12

Galindo RC, Ayllón N, Strašek Smrdel K, Boadella M, Beltrán-Beck B, Mazariegos M, García N, Pérez de la Lastra JM, Avsic-Zupanc T, Kocan KM, Gortazar C, de la Fuente J (2012a) Gene expression profile suggests that pigs (Sus scrofa) are susceptible to Anaplasma phagocytophilum but control infection. Parasit Vectors 5:181

Galindo RC, Falconi C, López-Olvera JR, Jiménez-Clavero MA, Fernández-Pacheco P, Fernández-Pinero J, Sánchez-Vizcaíno JM, Gortázar C, de la Fuente J (2012b) Global gene expression analysis in skin biopsies of European red deer experimentally infected with bluetongue virus serotypes 1 and 8. Vet Microbiol 161:26–35

Gamino V, Gutiérrez-Guzmán AV, Fernández-de-Mera IG, Ortíz JA, Durán-Martín M, de la Fuente J, Gortázar C, Höfle U (2012) Natural Bagaza virus infection in game birds in southern Spain. Vet Res 43:65

García I, Napp S, Casal J, Perea A, Allepuz A, Alba A, Carbonero A, Arenas A (2009) Bluetongue epidemiology in wild ruminants from Southern Spain. Eur J Wildl Res 55:173–178

Genomic Resources Development Consortium, Contreras M, de la Fuente J, Estrada-Peña A, Grubhoffer L, Tobes R (2014) Transcriptome sequence divergence between Lyme disease tick vectors, Ixodes scapularis and Ixodes ricinus. Genomic Resources Notes accepted 1 April 2014 – 31 May 2014. Mol Ecol Resour 14:1095

Gilbert L, Norman R, Laurenson KM, Reid HW, Hudson PJ (2001) Disease persistence and apparent competition in a three-host community: an empirical and analytical study of large-scale, wild populations. J Anim Ecol 70:1053–1061
Gortazar C, Ferroglio E, Höfle U, Frölich K, Vicente J (2007) Diseases shared between wildlife and livestock: a European perspective. Eur J Wildl Res 53:241–256. doi:10.1007/s10344-007-0098-y
Gortazar C, Ferroglio E, Lutton CE, Acevedo P (2010) Disease-related conflicts in mammal conservation. Wildl Res 37:668–675
Gortazar C, Reperant LA, Kuiken T, de la Fuente J, Boadella M, Martínez-Lopez B, Ruiz-Fons F, Estrada-Peña A, Drosten C, Medley G, Ostfeld R, Peterson T, VerCauteren KC, Menge C, Artois M, Schultsz C, Delahay R, Serra-Cobo J, Poulin R, Keck F, Aguirre AA, Henttonen H, Dobson AP, Kutz S, Lubroth J, Mysterud A (2014) Crossing the interspecies barrier: opening the door to zoonotic pathogens. PLoS Pathog 10(6):e1004129. doi:10.1371/journal.ppat.1004129
Gortazar C, Diez-Delgado I, Barasona JA, Vicente J, de la Fuente J, Boadella M (2015) The wild side of disease control at the wildlife-livestock-human interface: a review. Front One Health 1:27. doi:10.3389/fvets.2014.00027
Gunnarsson G, Latorre-Margalef N, Hobson KA, Van Wilgenburg SL, Elmberg J, Olsen B, Fouchier RA, Waldenström J (2012) Disease dynamics and bird migration – linking mallards Anas platyrhynchos and subtype diversity of the influenza A virus in time and space. PLoS One 7(4):e35679. doi:10.1371/journal.pone.0035679
Hagenaars TJ, Donnelly CA, Ferguson NM (2004) Spatial heterogeneity and the persistence of infectious diseases. J Theor Biol 229:349–359
Hall JS, Krauss S, Franson JC, TeSlaa JL, Nashold SW, Stallknecht DE, Webby RJ, Webster RG (2013) Avian influenza in shorebirds: experimental infection of ruddy turnstones (Arenaria interpres) with avian influenza virus. Influenza Other Respir Viruses 7:85–92
Handel A, Lebarbenchon C, Stallknecht D, Rohani P (2014) Trade-offs between and within scales: environmental persistence and within-host fitness of avian influenza viruses. Proc Biol Sci. 281(1787). pii: 20133051. doi:10.1098/rspb.2013.3051
Hartley M (2010) Qualitative risk assessment of the role of the feral wild boar (Sus scrofa) in the likelihood of incursion and the impacts on effective disease control of selected exotic diseases in England. Eur J Wildl Res 56(3):401–410
He Y (2012) Analyses of Brucella pathogenesis, host immunity, and vaccine targets using systems biology and bioinformatics. Front Cell Infect Microbiol 1(2):2
Höfle U, Van de Bildt MW, Leijten LM, Van Amerongen G, Verhagen JH, Fouchier RA, Osterhaus AD, Kuiken T (2012) Tissue tropism and pathology of natural influenza virus infection in black-headed gulls (Chroicocephalus ridibundus). Avian Pathol 41:547–553
Huang YJ, Higgs S, Horne KM, Vanlandingham DL (2014) Flavivirus-mosquito interactions. Viruses 6(11):4703–4730
Jachmann H (1991) Evaluation of four survey methods for estimating elephant densities. African Journal of Ecology 29:188–195
Jones KE, Patel NG, Levy MA, Storeygard A, Balk D, Gittleman JL, Daszak P (2008) Global trends in emerging infectious diseases. Nature 451:990–994
Jouglin M, Fernández-de-Mera GI, de la Cotte N, Ruiz-Fons F, Gortázar C, Moreau E, Bastian S, de la Fuente J, Malandrin L (2014) Isolation and characterization of Babesia pecorum sp. nov. from farmed red deer (Cervus elaphus). Vet Res 45:78
Jourdain E, Gunnarsson G, Wahlgren J, Latorre-Margalef N, Bröjer C, Sahlin S, Svensson L, Waldenström J, Lundkvist A, Olsen B (2010) Influenza virus in a natural host, the mallard: experimental infection data. PLoS One 5(1):e8935. doi:10.1371/journal.pone.0008935
Joyce AR, Palsson BO (2006) The model organism as a system: integrating 'omics' data sets. Nature Rev Mol Cell Biol 7:198–210
Judge J, McDonald RA, Walker N, Delahay RJ (2011) Effectiveness of biosecurity measures in preventing badger visits to farm buildings. PLoS One 6:e28941
Keeling MJ (1999) The effects of local spatial structure on epidemiological invasions. Proc R Soc Lond, B 266:859–867

Keeling MJ (2000) Metapopulation moments: coupling, stochasticity and persistence. J Anim Ecol 69:725–736

Kim TY, Kim HU, Lee SY (2010) Data integration and analysis of biological networks. Curr Opin Biotechnol 2:78–84

Klukas C, Schreiber F (2010) Integration of omics data and networks for biomedical research with VANTED. J Integr Bioinform 7:112

Kocan KM, de la Fuente J, Blouin EF (2008) Advances toward understanding the molecular biology of the Anaplasma-tick interface. Front Biosci 13:7032–7045

Kuiken T, Leighton FA, Fouchier RAM, LeDuc JW, Peiris JSM, Schudel A, Stohr K, Osterhaus ADME (2005) Pathogen surveillance in animals. Science 309:1680–1681

Kukielka E, Barasona JA, Cowie CE, Drewe JA, Gortazar C, Cotarelo I, Vicente J (2013) Spatial and temporal interactions between livestock and wildlife in South Central Spain assessed by camera traps. Prev Vet Med 112(3):213–221

Lemel J, Truvé J, Söderberg B (2003) Variation in ranging and activity behaviour of European wild boar Sus scrofa in Sweden. Wildl Biol 9(Suppl 1):29–36

Lin M, Kikuchi T, Brewer HM, Norbeck AD, Rikihisa Y (2011) Global proteomic analysis of two tick-borne emerging zoonotic agents: Anaplasma phagocytophilum and Ehrlichia chaffeensis. Front Microbiol 2:24

Liu XY, de la Fuente J, Cote M, Galindo RC, Moutailler S, Vayssier-Taussat M, Bonnet S (2014) IrSP1, a tick serine protease inhibitor involved in tick feeding and Bartonella henselae infection. PLoS Negl Trop Dis 8:e2993

López-Olvera JR, Falconi C, Férnandez-Pacheco P, Fernández-Pinero J, Sánchez MA, Palma A, Herruzo I, Vicente J, Jiménez-Clavero MA, Arias M, Sánchez-Vizcaíno JM, Gortázar C (2010) Experimental infection of European red deer (Cervus elaphus) with bluetongue virus serotypes 1 and 8. Vet Microbiol 145:148–152

Mans BJ, Andersen JF, Francischetti IMB et al (2008) Comparative sialomics between hard and soft ticks: implications for the evolution of blood-feeding behavior. Insect Biochem Mol Biol 38:42–58

Marcelino I, de Almeida AM, Ventosa M, Pruneau L, Meyer DF, Martinez D, Lefrançois T, Vachiéry N, Coelho AV (2012) Tick-borne diseases in cattle: applications of proteomics to develop new generation vaccines. J Proteomics 75:4232–4250

Mastronunzio JE, Kurscheid S, Fikrig E (2012) Postgenomic analyses reveal development of infectious Anaplasma phagocytophilum during transmission from ticks to mice. J Bacteriol 194:2238–2247

McKenzie VJ1, Goulet NE (2010) Bird community composition linked to human West Nile virus cases along the Colorado front range. Ecohealth 7(4):439–447. doi:10.1007/s10393-010-0360-8

Mediannikov O, Fenollar F (2014) Looking in ticks for human bacterial pathogens. Microb Pathog. pii: S0882-4010(14)00132-6. doi:10.1016/j.micpath.2014.09.008

Merino S, Moreno J, Sanz JJ, Arriero E (2000) Are avian blood parasites pathogenic in the wild? A medication experiment in blue tits (Parus caeruleus). Proc R Soc Lond B 267:2507–2510

Merino O, Alberdi P, Pérez de la Lastra JM, de la Fuente J (2013) Tick vaccines and the control of tick-borne pathogens. Front Cell Infect Microbiol 3:30

Miguel E, Grosbois V, Caron A, Boulinier T, Fritz H, Cornélis D, de Garine-Wichatitsky M (2013) Contacts and foot and mouth disease transmission from wild to domestic bovines in Africa. Ecosphere 4(4):51

Miller RA, Kaneene JB, Fitzgerald SD, Schmitt SM (2003) Evaluation of the influence of supplemental feeding of white-tailed deer (Odocoileus virginianus) on the prevalence of bovine tuberculosis in the Michigan wild deer population. J Wild Dis 39:84–95

Molina R, Jiménez MI, Cruz I, Iriso A, Martín-Martín I, Sevillano O, Melero S, Bernal J (2012) The hare (Lepus granatensis) as potential sylvatic reservoir of Leishmania infantum in Spain. Vet Parasitol 23(190(1–2)):268–271

Moreno-Cid JA, Pérez de la Lastra JM, Villar M, Jiménez M, Pinal R, Estrada-Peña A, Alarcón P, Delacour S, Oropeza V, Ruiz I, Molina R, Lucientes J, Prudencio CR, Galindo RC, Almazán C, Nijhof AM, Mangold AJ, Gortázar C, de la Fuente J (2013) Control of multiple arthropod vector infestations with subolesin/akirin vaccines. Vaccine 31:1187–1196

Nakao R, Abe T, Nijhof AM, Yamamoto S, Jongejan F, Ikemura T, Sugimoto C (2013) A novel approach, based on BLSOMs (Batch Learning Self-Organizing Maps), to the microbiome analysis of ticks. ISME J 7(5):1003–1015

Naranjo V, Höfle U, Vicente J, Martín MP, Ruiz-Fons F, Gortazar C, Kocan KM, de la Fuente J (2006) Genes differentially expressed in oropharyngeal tonsils and mandibular lymph nodes of tuberculous and non-tuberculous European wild boars naturally exposed to Mycobacterium bovis. FEMS Immunol Med Microbiol 46:298–312

Naranjo V, Villar M, Martín-Hernando MP, Vidal D, Höfle U, Gortazar C, Kocan KM, Vázquez J, de la Fuente J (2007) Proteomic and transcriptomic analyses of differential stress/inflammatory responses in mandibular lymph nodes and oropharyngeal tonsils of European wild boars naturally infected with Mycobacterium bovis. Proteomics 7:220–231

Nunn CL, Thrall PH, Kappeler PM (2014) Shared resources and disease dynamics in spatially structured populations. Ecol Model 272:198–207

Palmer MV, Waters WR, Whipple DL (2004) Shared feed as a means of deer-to-deer transmission of Mycobacterium bovis. J Wild Dis 40:87–91

Parizi LF, Githaka NW, Logullo C, Konnai S, Masuda A, Ohashi K, da Silva Vaz I Jr (2012) The quest for a universal vaccine against ticks: Cross-immunity insights. Vet J 194:158–165

Pérez-Ramírez E, Acevedo P, Allepuz A, Gerrikagoitia X, Alba A, Busquets N, Díaz-Sánchez S, Álvarez V, Abad FX, Barral M, Majó N, Höfle U (2012) Ecological factors driving avian influenza virus dynamics in Spanish wetland ecosystems. PLoS One 7(11):e46418. doi:10.1371/journal.pone.0046418

Pérez-Ramírez E, Llorente F, Jiménez-Clavero MÁ (2014) Experimental infections of wild birds with West Nile virus. Viruses 6:752–781. doi:10.3390/v6020752. Review

Phillips C, Foster C, Morris P, Teverson R (2003) The transmission of Mycobacterium bovis infection to cattle. Res Vet Sci 74:1–15

Phillips GE, Lavelle MJ, Fischer JW, White JJ, Wells SJ, Vercauteren KC (2012) A novel bipolar electric fence for excluding white-tailed deer from stored livestock feed. J Anim Sci 90:4090–4097

Poland GA, Kennedy RB, Ovsyannikova IG (2011a) Vaccinomics and personalized vaccinology: is science leading us toward a new path of directed vaccine development and discovery? PLoS Pathog 7:e1002344

Poland GA, Ovsyannikova IG, Kennedy RB, Haralambieva IH, Jacobson RM (2011b) Vaccinomics and a new paradigm for the development of preventive vaccines against viral infections. OMICS 15:625–636

Popara M, Villar M, Mateos-Hernández L, de Fernández Mera IG, de la Fuente J (2013) Proteomics approach to the study of cattle tick adaptation to White tailed deer. BioMed Res Int 2013:319812

Popara M, Villar M, de la Fuente J (2015) Proteomics characterization of tick-host-pathogen interactions. Methods Mol Biol 1247:513–527

Proffitt KM, Gude JA, Hamlin KL, Garrott RA, Cunningham JA, Grigg JL (2011) Elk distribution and spatial overlap with livestock during the brucellosis transmission risk period. J Appl Ecol 48(2):471–478

Putman R, Apollonio M, Andersen R (2011) Ungulate management in Europe: problems and practices. Cambridge University Press, Cambridge, Uk, p 410. ISBN 978-0-521-76059-1 (hardback)

Ramabu SS, Ueti MW, Brayton KA, Baszler TV, Palmer GH (2010) Identification of Anaplasma marginale proteins specifically upregulated during colonization of the tick vector. Infect Immun 78:3047–3052

Reperant LA, van de Bildt MW, van Amerongen G, Buehler DM, Osterhaus AD, Jenni-Eiermann S, Piersma T, Kuiken T (2011) Highly pathogenic avian influenza virus H5N1 infection in a long-distance migrant shorebird under migratory and non-migratory states. PLoS One 6(11):e27814. doi:10.1371/journal.pone.0027814

Reperant LA, Cornaglia G, Osterhaus ADME (2013) The importance of understanding the human-animal interface. From early hominids to global citizens. Curr Top Microbiol Immunol 365:49–81

Ricci I, Damiani C, Capone A, DeFreece C, Rossi P, Favia G (2012) Mosquito/microbiota interactions: from complex relationships to biotechnological perspectives. Curr Opin Microbiol 15(3):278–284
Richomme C, Gauthier D, Fromont E (2006) Contact rates and exposure to inter-species disease transmission in mountain ungulates. Epidemiol Infect 134(1):21–30
Rizzoli A, Hauffe HC, Tagliapietra V, Neteler M, Rosá R (2009) Forest structure and Roe deer abundance predict tick-borne encephalitis risk in Italy. PLoS One 4(2):e4336
Rodríguez-Sánchez B, Gortázar C, Ruiz-Fons F, Sánchez-Vizcaíno JM (2010) Bluetongue virus serotypes 1 and 4 in red deer, Spain. Emerg Infect Dis 16:518–520
Rohn H, Junker A, Hartmann A, Grafahrend-Belau E, Treutler H, Klapperstück M, Czauderna T, Klukas C, Schreiber F (2012) VANTED v2: a framework for systems biology applications. BMC Syst Biol 6:139
Ruiz-Fons F, Gilbert L (2010) The role of deer as vehicles to move ticks, Ixodes ricinus, between contrasting habitats. Int J Parasitol 40:1013–1020
Ruiz-Fons F, Reyes-García AR, Alcaide V, Gortázar C (2008) Spatial and temporal evolution of bluetongue virus in wild ruminants, Spain. Emerg Infect Dis 14:951–953
Ruiz-Fons F, Ferroglio E, Gortázar C (2013) Leishmania infantum in free-ranging hares, Spain, 2004–2010. Euro Surveill 18(30):pii=20541
Ruiz-Fons F, Balseiro A, Willoughby K, Oleaga A, Dagleish MP, Pérez-Ramírez E, Havlíková S, Klempa B, Llorente F, Martín-Hernando MP (2014a) Clinical infection of Cantabrian chamois (Rupicapra pyrenaica parva) by louping ill virus: new concern for mountain ungulate conservation? Eur J Wildl Res 60:691–694
Ruiz-Fons F, Sánchez-Matamoros A, Gortázar C, Sánchez-Vizcaíno JM (2014b) The role of wildlife in bluetongue virus maintenance in Europe: lessons learned after the natural infection in Spain. Virus Res 182:50–58
Saez-Royuela C, Telleria JL (1986) The increased population of the wild boar (Sus scrofa L.) in Europe. Mamm Rev 16:97–101
Semenza JC, Zeller H (2014) Integrated surveillance for prevention and control of emerging vector-borne diseases in Europe. Eurosurveillance 9(13):20757
Simpson VR (2002) Wild animals as reservoirs of infectious diseases in the UK. Vet J 163:128–146
Sonenshine DE, Bissinger BW, Egekwu N, Donohue KV, Khalil SM, Roe RM (2011) First transcriptome of the testis-vas deferens-male accessory gland and proteome of the spermatophore from Dermacentor variabilis (Acari: Ixodidae). PLoS One 6:e24711
Taylor LH, Latham SM, Woolhouse ME (2001) Risk factors for human disease emergence. Philos Trans R Soc Lond B Biol Sci 356:983–989
Tieri P, de la Fuente A, Termanini A, Franceschi C (2011) Integrating Omics data for signaling pathways, interactome reconstruction, and functional analysis. Methods Mol Biol 719:415–433
Troese MJ, Kahlon A, Ragland SA, Ottens AK, Ojogun N, Nelson KT, Walker NJ, Borjesson DL, Carlyon JA (2011) Proteomic analysis of Anaplasma phagocytophilum during infection of human myeloid cells identifies a protein that is pronouncedly upregulated on the infectious dense-cored cell. Infect Immun 79:4696–4707
Turchetto S, Obber F, Permunian R, Vendrami S, Lorenzetto M, Ferré N, Stancampiano L, Rossi L, Citterio CV (2014) Spatial and temporal explorative analysis of sarcoptic mange in Alpine chamois (Rupicapra r. rupicapra). Hystrix 25:1
Vázquez A, Jiménez-Clavero MA, Franco L, Donoso-Mantke O, Sambri V, Niedrig M, Zeller H, Tenorio A (2011) Usutu virus – potential risk of human disease in Europe. Euro Surveill 16(31):pii=19935
Veldhoen N, Ikonomou MG, Helbing CC (2012) Molecular profiling of marine fauna: integration of omics with environmental assessment of the world's oceans. Ecotoxicol Environ Saf 76(2):23–38
VerCauteren KC, Lavelle MJ, Hygnstrom S (2006) Fences and deer-damage management: a review of designs and efficacy. Wildl Soc Bull 34:191–200

VerCauteren KC, Lavelle MJ, Phillips GE (2008) Livestock protection dogs for deterring deer from cattle and feed. J Wildl Manag 72:1443–1448

VerCauteren KC, Seward NW, Lavelle MJ, Fischer JW, Phillips GE (2009) Deer guards and bump gates for excluding white-tailed deer from fenced resources. Hum Wildl Conflicts 3:145–153

Vercauteren KC, Vandeelen TR, Lavelle MJ, Hall WH (2010) Assessment of abilities of white-tailed deer to jump fences. J Wildl Manag 74:1378–1381

Verhagen JH, Van Dijk JGB, Vuong O, Bestebroer T, Lexmond P, Klaassen M, Fouchier RAM (2014) Migratory birds reinforce local circulation of avian influenza viruses. PLoS ONE 9:e112366

Vicente J, Höfle U, Garrido JM, Fernandez-De-Mera IG, Juste R, Barral M, Gortazar C (2005) Wild boar and red deer display high prevalences of tuberculosis-like lesions in Spain. Vet Res 37:107–119

Vicente J, Delahay RJ, Walker NJ, Cheeseman CL (2007) Social organization and movement influence the incidence of bovine tuberculosis in an undisturbed high-density badger Meles meles population. J Anim Ecol 76:348–360

Villar M, Popara M, Bonzón-Kulichenko E, Ayllón N, Vázquez J, de la Fuente J (2012) Characterization of the tick-pathogen interface by quantitative proteomics. Ticks Tick Borne Dis 3:154–158

Villar M, Popara M, Ayllón N, Fernández de Mera IG, Mateos-Hernández L, Galindo RC, Manrique M, Tobes R, de la Fuente J (2014a) A systems biology approach to the characterization of stress response in Dermacentor reticulatus tick unfed larvae. PLoS One 9:e89564

Villar M, Popara M, Mangold AJ, de la Fuente J (2014b) Comparative proteomics for the characterization of the most relevant Amblyomma tick species as vectors of zoonotic pathogens worldwide. J Proteomics 105:204–216

Vorou RM, Papavassiliou VG, Tsiodras S (2007) Emerging zoonoses and vector-borne infections affecting humans in Europe. Epidemiol Infect 135(8):1231–1247

Wallensten A, Munster VJ, Latorre-Margalef N, Brytting M, Elmberg J, Fouchier RA, Fransson T, Haemig PD, Karlsson M, Lundkvist A, Osterhaus AD, Stervander M, Waldenström J, Björn O (2007) Surveillance of influenza A virus in migratory waterfowl in northern Europe. Emerg Infect Dis 13:404–411

Wille M, Tolf C, Avril A, Latorre-Margalef N, Wallerström S, Olsen B, Waldenström J (2013) Frequency and patterns of reassortment in natural influenza A virus infection in a reservoir host. Virology 443(1):150–160. doi:10.1016/j.virol.2013.05.004

Williams TD, Turan N, Diab AM, Wu H, Mackenzie C, Bartie KL, Hrydziuszko O, Lyons BP, Stentiford GD, Herbert JM, Abraham JK, Katsiadaki I, Leaver MJ, Taggart JB, George SG, Viant MR, Chipman KJ, Falciani F (2011) Towards a system level understanding of non-model organisms sampled from the environment: a network biology approach. PLoS Comput Biol 7(8):e1002126

Zhuang L, Zhang Z, An X, Fan H, Ma M, Anderson BD, Jiang J, Liu W, Cao W, Tong Y (2014) An efficient strategy of screening for pathogens in wild-caught ticks and mosquitoes by reusing small RNA deep sequencing data. PLoS One 9(3):e90831

Zivkovic Z, Esteves E, Almazán C, Daffre S, Nijhof AM, Kocan KM, Jongejan F, de la Fuente J (2010) Differential expression of genes in salivary glands of male Rhipicephalus (Boophilus) microplus in response to infection with Anaplasma marginale. BMC Genomics 11:186

Wildlife Habitat Requirements: Concepts and Research Approaches

José Luis Tellería

Introduction

Habitat loss and degradation is a main driver of worldwide biodiversity loss (Fahrig 2003; Barnosky et al. 2012). In this context, it is important to establish management guidelines that conserve habitats and species based on a sound knowledge of wildlife-habitat relationships (Morrison et al. 2006). However, habitat studies vary considerably in concepts and methodology, making it difficult to link theoretical with applied approaches in the design of management plans (Cook et al. 2009). From a conceptual standpoint, habitat is one of the most widely used and yet imprecise terms in ecology (Block and Brennan 1993; Jones 2001; Mitchell 2005; Kearney 2006; Morrison 2009). According to Hall et al. (1997), this happens because (a) habitat terminology is used ambiguously, (b) the fundamental concepts supporting habitat studies are poorly defined and poorly understood, and (c) wildlife-habitat relationships are highly idiosyncratic and strongly dependent on the different spatial and temporal scales at which species actually operate. From a methodological standpoint, the main problem is the widespread use of traditional, untested practices in wildlife management (Sutherland et al. 2004). As a response to this approach, it is commonly agreed that the best way to generate sound management guidelines is through the use of the scientific method (Morrison et al. 2006). This requires knowledge of species' biology to propose reliable hypotheses on the features affecting habitat requirements, the statement of clear and testable predictions, and their validation/rejection through suitable experiments (Feisinger 2001).

This chapter reviews the habitat concept and the main features affecting wildlife-habitat interactions to provide a theoretical background for the study of the effect of

J.L. Tellería
Departamento de Zoología y Antropología Física, Facultad de Ciencias Biológicas, Universidad Complutense, Madrid 28040, Spain
e-mail: telleria@bio.ucm.es

R. Mateo et al. (eds.), *Current Trends in Wildlife Research*,
Wildlife Research Monographs 1, DOI 10.1007/978-3-319-27912-1_4

habitat changes on wildlife populations. More explicitly, it sets some basic guidelines for the study of species-habitat interaction from a management perspective.

Habitat Concept

A first challenge in determining wildlife habitat requirements and management strategies is to disentangle the term 'habitat' since two distinct approaches have emerged in recent decades (Table 1):

Community-Oriented Approach

Habitat is often viewed as an area of similar vegetation cover, occupied by particular plant communities that are classified according to some dominant species or phytosociological criteria (the "habitat type" of Daubenmire 1968). For instance, the northern Iberian beech woods (*Fagus sylvatica*) have been classified as *Scillo-Fagion* by botanists. In this way, they differentiate this plant assemblage from other European beech woodlands (Dierschke 2011). This typological view of habitat is used by the European Habitat Directive (Council Directive 92/43/EEC of 21 May 1992 on the conservation of natural habitats and of wild fauna and flora) and other large-scale conservation projects designed to locate and preventively protect the most interesting areas in terms of biodiversity (Brooks et al. 2006; Bunce et al. 2013).

There are several reasons that make this approach to habitat useful from a management perspective. Most of these habitat types are discerned in aerial photos or satellite images, making it easier to explore their size and spatial arrangement (Kerr

Table 1 Main differences between community-oriented and species-oriented approaches to the habitat concept

	Community-oriented	Species-oriented
Definition	A place where many species occur	A place where individual species occur
Focus	Species assemblages	Individuals, populations
Conceptual background	Community assembly rules	Niche
Response variables	Species richness, community indexes	Behavioral decisions, body condition, abundance, growth rate
Main predictor variables	Landscape features	Biotic and abiotic interferences
Management approaches	Mainly preventive	Mainly proactive
Conservation guidelines	Easy to transfer	Difficult to transfer

and Ostrovsky 2003). The conservation interest of species assemblages (not only plants) included in these habitats can be ranked in terms of species richness and other community parameters (singularity, rarity, etc.; Margules and Pressey 2000). Finally, they can often constitute the basic units of ecosystem approaches to environmental conservation (Bunce et al. 2013). As a result, this community-oriented approach to habitat is firmly ingrained in the current ecological thinking and praxis of many researchers, managers and politicians.

The community-oriented approach to habitat is often used to explore the distribution of individual species at broad spatial scales. This is a methodological link between the two habitat concepts since the cartographic information generated by the community approach is habitually used to explore species distribution (Aebischer et al. 1993). In fact, several approaches to habitat distribution of individual species are strongly based in the former classification of the space in habitat types differing in quality on which individuals allocate (see Habitat quality). However, habitat composition of actual landscapes is too complex to be reflected in the simplistic approaches based on homogeneous habitat types perceived by man, which may have little to do with how individual species perceive the environment. It has been suggested that a "species-eye" view of habitats be incorporated (Betts et al. 2014), embracing the spatial gradients in habitat suitability to which each species responds in an idiosyncratic way (Manning et al. 2004; Fischer and Lindenmayer 2007). As a result, species-oriented approaches to habitat seem to be necessary for more sophisticated research on wildlife-habitat interactions.

Species-Oriented Approach

The species-oriented approach defines habitat as the place where an organism lives (Odum 1971) with only its physical elements (Block and Brennan 1993), the range of environments and communities over which a species occurs (Whittaker et al. 1973) or the resources and conditions present in an area that produces occupancy by a given organism (Hall et al. 1997). The variation of the habitat concept from the simple "place where one would go to find an organism" (Odum 1971) to more abstract multivariate definitions, to the effect of a set of variables on a species has led to some confusion. This is particularly due to the fact that the definition of habitat as "resources and conditions affecting organisms" resembles the *Grinnelian niche* concept, defined as the suite of biotic and abiotic factors that permit an organism to use part of the environment (Hutchinson 1957; James et al. 1984; Holt 2009).

To establish an unambiguous definition of habitat in the species-oriented approach (the main focus of this chapter), the three-step approach proposed by Kearney (2006; see also Mitchell 2005) is useful. He suggested that *habitat* be defined as "a physical place, at a particular scale of space and time, where an organism either actually or potentially lives". Within the habitat, he described *environment* as the "biotic and abiotic phenomena surrounding and potentially interacting with an organism", and *niche* as "as the subset of those environmental conditions which affect a particular organism where the average absolute fitness of individuals in a population is greater than or equal to one". For instance, the Short-toed treecreeper (*Certhia brachydactyla*) – a trunk-specialist insectivorous passerine (Fig. 1)- and the European green lizard (*Lacerta viridis*) occur together in northern Iberian beech woods. In this *habitat*, these insectivorous vertebrates share a common *environment* (e.g. weather, vegetation cover, invertebrate abundance, preda-

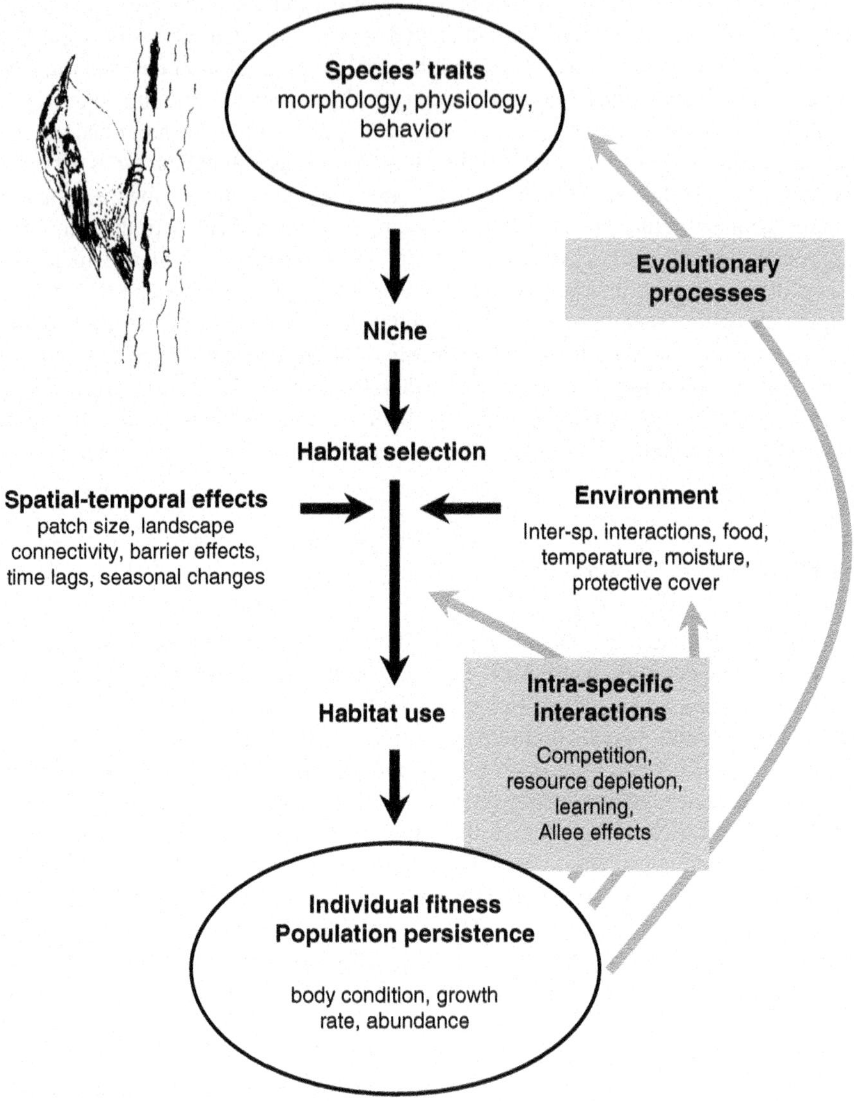

Fig. 1 Relationships between habitat selection and habitat use. Habitat selection is the potential use of habitat according to the species niche. Habitat use is the actual use of habitat after the constraints imposed by environment, spatial-temporal effects and intra-specific interactions affecting the species. See text for details

tors, management practices) but show different *niches*: treecreepers depend on trunks to feed on invertebrates and to nest in tree holes, and lizards depend on protective shrub cover and the availability of sunny patches for thermoregulation. They react to the same environment according to the particular morphological or behavioral traits making up their niche, which is a species-specific trait resulting from independent evolutionary histories.

Consistent with this view, the basic work of wildlife managers will be to discern which of the environmental variables making up the species niche should be managed to drive the population in a given place (the habitat) according to a programmed target (e.g. to increase trunk density to recover tree-creeper populations in a beech wood). This requires the design of studies whose conceptual foundation and practical limitations must be clarified.

Habitat Selection and Use

A first challenge in the study of wildlife-habitat interactions is to recognize the differences between *habitat selection* and *habitat use* (Fig. 1). *Habitat selection* refers to a hierarchical process of animal responses resulting in a disproportionate use of some places or substrates to meet life requisites (foraging, breeding, roosting, etc.; Hall et al. 1997). However, rarely if ever do individuals of a given species use the full range of environmental conditions within their capabilities due to the effect of intra- and inter-specific interactions, or difficulty in accessing suitable locations (Fig. 1). *Habitat use* represents the actual pattern of space occupation resulting from those processes that constrain the free manifestation of *habitat selection* in a given place (Udvardy 1959; Block and Brennan 1993; Jones 2001).

These differences between habitat selection and habitat use have three practical consequences in the study of wildlife-habitat relationships: (a) field studies of animal-habitat interactions will report information on *habitat use*; (b) habitat use will change if the constraints to innate choices change among places, meaning that the patterns reported in one place may be unsuitable to predict habitat use in other places; and (c) *habitat selection* must be considered a species-specific template derived from the niche and will only be known indirectly through suitable experiments (Wiens 1989; Johnson 2007). More explicitly, habitat selection is the species-eye view of habitat that researchers attempt to understand through the study of habitat use by individuals or populations in different places.

Habitat Quality

Environmental changes influence reproduction and survival of individuals (fitness), contribute to the regulation and persistence of populations and shape the spatial-temporal abundance patterning of species (Brown et al. 1995). In this context, *habitat quality* refers to the ability of a place to provide appropriate conditions for individual fitness and population persistence. It should be considered a continuous variable ranging from low to medium to high according to resources available for survival, reproduction and population persistence (Hall et al. 1997). It is not surprising, then, that wildlife managers try to disentangle *habitat selection* of species by exploring the way in which they react to environmental conditions (Johnson 2007). However, this approach must address two challenges: (a) to assess habitat quality and (b) to relate changes in habitat quality with environmental changes.

Measures of Habitat Quality

Wildlife managers must differentiate habitat quality from the perspective of individuals, who seek to maximize their own fitness, and from the perspective of populations, usually assessed by size and numerical trends (Johnson 2007). This is very important for a full understanding of the meaning of habitat quality because, despite the fact that these traits are often related (e.g. use of high quality habitats will enhance individual fitness and will increase population size and persistence), individuals and populations may show non-coincident patterns (e.g. very dense populations can be composed of starving individuals; Van Horne 1983). From a population perspective, habitat quality is usually assessed with the per capita rate of population increase (*r*; MacArthur 1972; Southwood 1977) that can be expressed by the theta-logistic model:

$$r = (1/N)(dN/dt) = r_m (1 - N/K)^{\theta}$$

where N is the population size, r_m is the maximum growth rate (the rate at which a population grows in a place when no resource is in short supply), K is the carrying capacity, or the maximum number of individuals one habitat can maintain indefinitely, and Θ is an exponent that adjusts $\mathrm{r_m}$ changes to different relationships with N (Gilpin et al. 1976; see however Clark et al. 2010). Accordingly, the theta-logistic model is a simple model for describing how the growth rate of a population slows as abundance increases (and it becomes increasingly difficulty for individuals to thrive in a given habitat), where the growth response decreases from r_m in a linear, convex or concave way (according to the shape parameter θ) to zero when the population reaches carrying capacity (K). This model can also be used to obtain a regional perspective of habitat quality (Morris 2011). For instance, the mean habitat quality (r') of habitat types 1, 2 and 3 in a region where they host a proportion p, q and z of individuals ($p + q + z = 1$) and a per capita rate of increase of r_1, r_2 and r_3 will produce a mean score of

$$r' = p.r_1 + q.r_2 + z.r_3$$

$$r' = p.r_{m1}(1 - N_1/K_1)^{\theta_1} + q.r_{m2}(1 - N_2/K_2)^{\theta_2} + z.r_{m3}(1 - N_3/K_3)^{\theta_3}$$

This means that regional habitat quality for individuals will ultimately be influenced by the composition of habitat types over the study region in which researchers will inform land management decisions (Dunning et al. 1992; Johnson 2007). However, from a population perspective, it is important to note that high quality habitats for individuals (e.g. $r_1 < r_2 < r_3$) could contribute to a minor part of managed populations if compared to lower quality but broadly extended habitats containing most individuals; (e.g. $p > q > z$; Johnson 2007). This will occur, for instance, if low rates of increase (r_i) are related to situations of high numerical saturation (N_i/K_i) with the concomitant low resource availability inherent to this model. As habitats can vary in r_{mi} and K_i, these parameters are often used to assess differences in habitat quality independent of the numerical saturation of a population. This permits us, for instance, to speculate from a management perspective on the potential quality of habitats at spatial (e.g. *core habitats,* where r_{mi} is the highest; *source* vs. *sink habitats,* where source habitats with $r_{mi} > 0$ emit individuals to sink habitats with $r_{mi} = 0$; Pulliam 1988; Pulliam and Danielson 1991) and temporal scales (e.g. seasonal or inter-annual changes in habitat quality; Lawton 1993).

However, despite the conceptual usefulness of these demographic approaches to habitat quality, they are difficult to achieve empirically under most circumstances (Holt 2009). Alternative approaches include the use of some surrogates of habitat quality, such as body condition (body mass, fluctuating asymmetry, hematocrit, stress hormones, etc.), reproduction-mortality rates or abundance trends of populations (Manly et al. 2002; Wikelski and Cooke 2006; Johnson 2007; Nichols and Williams 2006).

Wildlife-Habitat Studies

Habitat studies are usually carried out using the Hutchinsonian approach to Grinnellian niche, viewed as a n-dimensional hyper-volume in which a suite of biotic and abiotic variables act on species synergistically (Hutchinson 1957). More explicitly, studies on wildlife-habitat interactions explore some hypotheses on the causal relationships between habitat quality scores reported by individuals and populations (response variables) and a set of predictor variables (weather, vegetation structure, food availability, predator abundance, etc.; Johnson 2007). This is carried out by checking the changes in habitat quality over habitat gradients (mensurative experiments) or through the *ad hoc* manipulation of some predictor variables (manipulative experiment; *sensu* Hurlbert 1984). This last approach can be challenging to carry out in those cases in which habitat manipulation is difficult, expensive or illegal. However, some methodological approaches can be used to test the predictions. For instance, it is possible to check the positive effect of trunk density on treecreeper abundance by comparing the two parameters in different forest plots (mensurative experiment) or by testing the effect of changes in trunk density resulting from commercial logging (manipulative experiment). The inclusion of scientific approaches within management practices is known as *adaptive management* (Holling 1978), a way of testing hypotheses on wildlife-habitat interactions (Runge 2011). Finally, the construction of reliable hypotheses on the features affecting wildlife-habitat interactions, and the experimental design to test the predictions, will require the consideration of three additional effects: the species' traits, the spatial-temporal scale of the approach, and intra- and inter-specific interactions.

Species Traits

The evolutionary process leads to a set of morphological, physiological and behavioral traits that conform to the niche (Fig. 1). This produces a habitat selection template that, after environmental restrictions, results in a given habitat use (Wiens 1989). In this way, the traits of the study species will inevitably affect hypotheses on the features affecting habitat use. This is obvious in most organisms (e.g. treecreepers vs. lizards), but similar species may show subtle differences in habitat use. For instance, slight differences in body mass or in the functional complexes related to feeding and locomotion have often been used to explain differences in habitat use of birds (Leisler and Winkler 1985) and minor changes in physiological tolerance can play a role in the distribution of some species (Kearney and Porter 2009). For instance, the first mention in this chapter to habitat use by the Short-toed treecreeper omitted its active search of sunny patches to attenuate the costs of thermoregulation during winter, a common strategy in many other small passerines (Carrascal et al. 2001; Fig. 1). In fact, it has been reported that, as large birds show less thermal conductance, they are able to cope with thermal restrictions and thus occur in colder habitats during winter (Meehan *et al.* 2004). Similarly, innate behavioral differences can also affect habitat use through the effect or neophobic reactions or learning processes in some species (Harris and Reed 2002).

Nonetheless, species niche can change as a reaction to environmental novelties shifting innate habitat selection. The tendency to retain niche despite environmental changes has been called *niche conservationism* (Harvey and Pagel 1991) and some empirical evidence supports an uneven distribution of this trait among species (Holt 2009; Wiens et al. 2010). In the case of birds, for instance, Tingley et al. (2009) reported niche conservationism in 48 out of 53 California bird species that were forced to shift their range as a reaction to climate warming. However, the influential study of Grant and Grant (1995) reported short-term changes in bill morphology of Darwin finches to cope with changes in seed availability resulting from drought; and Desrochers (2010) described recent morphological changes in forest birds to improve flight effectiveness in the increasingly cleared woodlands of North America. In this way, the extent and intensity of these changes in niche is today an emerging focal point for wildlife researchers and managers in a world under the effect of accelerating human impacts (Wiens and Graham 2005). From a practical standpoint, niche conservationism makes a species less resilient to changes and, alternatively, niche evolution maintains species in changing environments through evolutionary change. This means that, in managed habitats, changes in habitat use could result from evolutionary changes. As a result, the need for evolution-enlightened management to identify, understand and influence (where possible or necessary) environmental changes that affect species evolution has been emphasized (Stockwell et al. 2003; Ashley et al. 2003). This dynamic approach to wildlife management is driven by Frankel's idea that "wild species, increasingly endangered by loss of habitats, will depend on organized protection for their survival. On a long term basis this is feasible only within natural communities in a state of continuing evolution…" (1974).

Spatial and Temporal Effects

Species-habitat interactions occur along a continuum of spatial scales (Johnson 1980). Hutto (1985) visualized this process describing the hierarchical chain of decisions of a hypothetical migratory forest bird arriving to wintering grounds: after reaching the region, it selects woodlands vs. grasslands, then selecting broadleaved trees vs. conifer patches and, finally, perches in oak foliage to look for invertebrates avoiding trunks or gross branches. This top-down spatial approach to habitat distribution from macro to micro-scales can also be reversed and has been used in some mechanistic approaches to species distribution intended to explain large-scale distribution from local processes (e.g. thermal restrictions at local scales could predict the geographical range of species; Kearney and Porter 2009). However, this can be a difficult task in many cases because constraints affecting habitat use are scale-dependent (Morrison 2009).

Geographical Scale

Large-scale approaches to habitat use can be critical in explaining the absence of species in suitable habitats beyond the range limits. This occurs due to the effect of barriers affecting historical processes, such as continental drift or Pleistocene changes, that have arranged the current distribution of most species. The ability of species to reach remote areas has often been explored in the conceptual framework of the *island biogeography theory*, in which the distance to source areas (the continent) is considered a main determinant of species occurrence in suitable habitat islands (MacArthur and Wilson 1967). However, pure geographical effects can also affect species occurrences in habitat patches of marginal areas within the limits of the species range (Pulliam 2000). Despite that, this has frequently been explained by the fact that peripheral areas -and their habitats- are near the species' limits of environmental tolerance (Brown et al. 1995), evidence does not always support this view (Sagarin and Gaines 2002). Alternative hypotheses suggest that is may be increasingly difficulty for peripheral populations to recover from local extinctions if the rescue effect produced by the input of individuals is constrained by the distance from source areas in the main range of species (Sexton et al. 2009). As a result, suitable habitats may be unoccupied, or occupied under their actual carrying capacity, due to pure geographical effects.

Landscape

Similar geographical approaches have been used to interpret habitat occupancy at landscape scales. Landscapes are traditionally represented as fragmented mosaics of suitable habitat patches and corridors within hostile matrixes in which species move according to their dispersive abilities. In this scenario, species occurrence is usually predicted by size and isolation of suitable habitat patches within the conceptual framework of the *island biogeography theory* (see above) and *meta-population dynamics* (Hanski 1998). These approaches, based on a typological classification of habitat patches (see the *community-oriented approach* to habitat) have often shown a poor ability to predict species distribution (Connor et al. 2000; Prugh et al. 2008; Pardini et al. 2010). As a result, two main ideas have been advanced to improve current hypotheses on wildlife-habitat distribution at landscape scales. First, the need of a "species-eye" approach to habitat quality to avoid overly simplistic interpretations of habitat quality has been suggested (Betts et al. 2014). This includes a re-evaluation of connectivity, a concept related to the dispersive ability of species that is loosely defined in most cases. To avoid this, some authors (e.g. Uezu et al. 2005) differentiate two connectivity concepts. *Structural connectivity* refers to landscape patterns from a human perspective, which are measured with the usual landscape metrics (patch distance, matrix structure, corridors, etc.). *Functional connectivity* is the extent to which an individual species can move throughout a landscape (a "species-eye" approach), and must be measured on focal individuals. Second, the assumption has been challenged that habitat patches are the natural units within which to explore habitat use at landscape scales in terms of reduced patch size, increasing inter-patch distance, border effects, etc. typical of former approaches to habitat fragmentation (Wiens 1989). Included in this debate is the *fragmentation threshold hypothesis*, which suggests that the regional abundance of species should increase with the regional cover of suitable habitats and that only under a given *threshold* of habitat cover will the particular effect of fragmentation emerge (Bascompte and Solé 1996; Fahrig 2013). Several authors have suggested that the effect of habitat fragmentation on habitat use by vertebrates is only important when the regional cover of the reduced habitat is under 10–30 % (Andrén 1994; Fahrig 1997; Pardini et al. 2010). In this context, both linear and non-linear responses of species to habitat loss have been reported in the literature and it is commonly agreed that the presence and value of critical thresholds is strongly affected by characteristics of the species (e.g. dispersal ability, ecological density, edge sensitivity, etc.; Swift and Hannon 2010; Hanski 2011).

Local Scales

At the smallest spatial scales, most vertebrates occur in home ranges where individuals move routinely in the search for food, shelter, nesting places, etc. However, it is unlike that animals use all the area with equal frequency (Morrison 2009). The uneven utilization distribution of space within home ranges provides an opportunity to explore the use of habitat by individuals at micro-scales (e.g. treecreepers select oak vs. beech patches within the home ranges of beech woods of northern Spain, where they search for invertebrates in lichens vs. bare barks within trees; Carrascal and Tellería 1989). This small spatial scale approach is particularly suitable to interpret habitat use within the conceptual framework of some behavioral hypotheses. While optimal foraging theory predicts the use of food patches (Pyke 1984), it has been argued that habitat selection theory is a subset of optimal foraging theory extended to a broader set of resources (Rosenzweig 1985; Block and Brennan 1993). For instance, the habitat matching rule (Pulliam and Caraco 1984) predicts that if both resource and population abundance are measured along a habitat gradient, any spatial change in resource abundance will be followed by a concomitant change in population density (Fagen 1987). However, at larger spatial scales (see above), this approach is strongly constrained by the difficulty of individuals to perceive resource distribution and the cost associated with displacement among habitat patches (Kennedy and Gray 1993). Only in some highly mobile, free-ranging species are these behavioral models successfully applied to explore habitat use at large geographical scales (e.g. fruit-tracking migratory birds; Tellería et al. 2014).

Temporal Effects

Finally, all spatial patterns described above can change over time. Seasonal changes (winter vs. summer conditions, dry vs. rainy seasons, etc.) usually correspond to different life-history requirements (e.g. breeding season is usually adjusted to the most productive period). These affect food preferences and the selection of feeding-nesting substrata within home ranges. In the case of migratory species, individuals show seasonal home ranges distributed over different geographical regions. The itinerancy of migratory species may complicate the study of the causal relationships between the fate of individuals and populations and the traits of seasonal habitats, since changes in one habitat may have carry-over effects in the subsequent one (Norris 2005). Temporal differences in habitat use also have practical consequences in the study of wildlife-habitat relationships: (a) inter-annual differences in habitat use can only reflect inter-annual changes in the distribution of some particular resources such that a comprehensive approach to habitat use requires the design of long-term studies; (b) time lags are frequent between environmental changes and the reaction of species impeding the immediate assessment of the causal relationships between environment and habitat use. This is clear in those processes of habitat loss and fragmentation that leave large numbers of species to gradually decline and go extinct (the "extinction debt"; Tilman et al. 1994).

Intra- and Inter-Specific Interactions

The popular model of Fretwell and Lucas (1970) hypothesizes that in an ideal situation, a species uses the habitat of highest quality first, and then expands its distribution to use secondary habitats with increasing population size. This "ideal free distribution" rests on three basic assumptions: (a) individuals are free to occur in any habitat, (b) individuals select the best habitats in terms of quality, and (c) habitat quality decreases with increasing population density because the shared use of resources decreases availability. Once population levels reach a given threshold, the decreasing quality of the best habitat is reduced to the level of the "second-best" habitat in terms of suitability, a continuous process that ultimately distributes individuals with equal fitness across different habitats. This density-dependent approach to habitat use has proven valuable to predict habitat use in some cases but has been overly simplistic in others for a number of reasons: (a) dominant individuals exclude subordinates from high-quality habitats, which results in unequal fitness benefits (Brown 1969; Fretwell 1972); (b) individuals have imperfect information on habitat quality distribution misusing the extant availability of resources (Kennedy and Gray 1993); (c) individuals at familiar sites (e.g. home ranges) may exploit a known low-quality habitat more effectively than an unknown high quality habitat (Piper 2011); (d) individuals can learn habitat use from relatives so that inter-habitat distribution can also be flawed by a "cultural" imprint (Nielsen et al. 2013); and, finally, (e) increasing densities may improve habitat quality if, over a given density threshold and prior to the negative effect of resource competition, Allee effects occur (improved anti-depredatory strategies, increasing availability of mates, shared information on food patches, etc.; Courchamp et al. 1999).

In addition, animals are interconnected to other species in a complex, dynamic and frequently difficult to define cause-effect relationship that affects habitat use (resource partitioning, food webs, etc.; Wiens 1989). For instance, ecologically similar species may preclude potential competitors from using resources either by using them more efficiently (exploitation competition) or by impeding access (interference competition; Maurer 1984). The short-toed treecreeper is included in the guild of pariforms, a set of insectivorous forest birds (*Regulus*, *Parus*, *Cyanistes*, *Sitta*, *Aegithalos*, etc.) that forms multispecific flocks in winter and offers a good example of inter-specific interaction to habitat use. These species distribute within trees according to a given pattern (e.g. treecreepers and nuthatches exploit trunks and gross branches while tits and firecrests occur in different parts of the canopy). It has been observed that the lack of one of these species in a flock enlarges the use of feeding substrata of nearby birds (Alatalo et al. 1987). However, competition is not the only biotic interaction affecting habitat use since predators, parasites and diseases can affect distribution. The ecological networks representing the functional interactions of individual species in ecosystems are particularly suitable to demonstrate that habitat use is strongly affected by a synergetic interaction with other species (Sole and Montoya 2001; Bascompte 2007). These interactions are far from stable since species assemblages are rarely at a stable equilibrium and, ultimately, may be related to unpredictable, stochastic processes (Hubbell 2001). More explicitly, the traits affecting the composition of communities occurring in a given habitat strongly affect the habitat use of individual species. This means that the species approach to habitat is not independent of the community approach to habitat.

Conclusions

This review suggests that the habitat concept can be easily dealt with if we agree that habitat is "a place" in which one or several species actually or potentially occur (Table 1). For instance, northern Iberian beech woods are the habitat of the short-toed treecreeper, but are also the habitat of pariforms, of a plant community called *Scillo-Fagion* by some botanists, and of many other organisms. From a management perspective, the community-oriented approach to habitat is useful to delimit and protect the best places in terms of species richness and other community indexes. The results are easily transferred to managers and politicians and, in a world of restricted resources for conservation, this approach is suitable for conserving the most species. However, the species-oriented approach to habitat is required to manage individual species (e.g. threatened and commercial species). This approach must cope with several conceptual and methodological challenges: (a) habitat selection is an idiosyncratic species trait resulting from particular evolutionary processes making up the niche; (b) habitat use is a contingent version of innate habitat selection (the species-eye view of habitat) resulting from intra- and inter-specific interactions in a given place and time; and (c) habitat use is not always related to extant habitat quality since it is strongly affected by the ability of species to track the environment at different spatial and temporal scales. All these constraints make habitat use of individual species a very context-dependent process.

Acknowledgements I am indebted to Dr. M.L. Morrison (Oregon State University, USA) and Dr. J.T. García (Instituto de Investigación de Recursos Cinegéticos, CSIC, Spain) for valuable comments on a previous version of this review. This paper is a contribution to the project CGL2011-22953/BOS of the Spanish Ministry of Economy and Competitiveness.

References

Aebischer NJ, Robertson PA, Kenward RE (1993) Compositional analysis of habitat use from animal radiotracking data. Ecology 74:1313–1325

Alatalo RV, Eriksson DAG, Gustafsson L, Larsson K (1987) Exploitation competition influences the use of foraging sites by tits: experimental evidence. Ecology 68:284–290

Andrén H (1994) Effects of habitat fragmentation on birds and mammals in landscapes with different proportions of suitable habitat: a review. Oikos 71:355–366

Ashley MV, Willson MF, Pergams ORW, O'Dowd DJ, Gende SM, Brown JS (2003) Evolutionarily enlightened management. Biol Conserv 111:115–123

Barnosky AD et al (2012) Approaching a state shift in Earth's biosphere. Nature 486:52–58

Bascompte J (2007) Networks in ecology. Basic Appl Ecol 8:485–490

Bascompte J, Solé RV (1996) Habitat fragmentation and extinction thresholds in spatially explicit models. J Anim Ecol 65:465–473

Betts MG et al (2014) A species-centered approach for uncovering generalities in organism responses to habitat loss and fragmentation. Ecography 37:517–527

Block WM, Brennan A (1993) The habitat concept in ornithology: theory and applications. In: Power DM (ed) Current ornithology, vol 11. Plenum Press, New York, pp 35–91

Brooks TM, Mittermeier RA, da Fonseca GAB et al (2006) Global biodiversity conservation priorities. Science 313:58–61

Brown JL (1969) Territorial behavior and population regulation in birds. Wilson Bull 81:293–329

Brown JH, Mehlman DW, Stevens GC (1995) Spatial variation in abundance. Ecology 76:2028–2043

Bunce RGH, Bogers MMB, Evans D, Halada L, Jongman RHG, Mucher CA, Bauch B, de Blust G, Parr TW, Olsvig-Wjittaker L (2013) Field identification of habitats directive Annex I habitats as a major European biodiversity indicator. Ecol Indic 33:105–110

Carrascal LM, Tellería JL (1989) Comportamiento de búsqueda del alimento y selección de especies arbóreas: análisis con el Agateador Común (*Certhia brachydactyla*). Ardeola 36:149–160

Carrascal LM, Díaz JA, Huertas DL, Mozetich I (2001) Behavioral thermoregulation by treecreepers: trade-off between saving energy and reduced crypsis. Ecology 82:1642–1654

Clark F, Brook BW, Delean S, Reşit Akçakaya H, Bradshaw CJ (2010) The theta-logistic is unreliable for modeling most census data. Methods Ecol Evol 1:253–262

Connor EF, Courtney AC, Yoder JM (2000) Individuals-area relationships: the relationship between animal population density and area. Ecology 81:734–748

Cook CN, Hockings M, Carter RW (2009) Conservation in the dark? The information used to support management decisions. Front Ecol Environ 8:181–186

Courchamp F, Clutton-Brock T, Grenfell B (1999) Inverse density dependence and the Allee effect. Trends Ecol Evol 14:405–410

Daubenmire R (1968) Plant communities: a textbook of plant synecology. Harper and Row, New York

Desrochers A (2010) Morphological response of songbirds to 100 years of landscape change in North America. Ecology 91:1577–1582

Dierschke H (2011) Syntaxomomical survey of European beech forests: some general conclusions. Annali di Botanica 55:17–23

Dunning JB, Danielson BJ, Pulliam PR (1992) Ecological processes that affect populations in complex landscapes. Oikos 65:169–175

Fagen R (1987) A generalized habitat matching rule. Evol Ecol 1:5–10

Fahrig L (1997) Relative effects of habitat loss and fragmentation on population extinction. J Wildl Manag 61:603–610

Fahrig L (2003) Effects of habitat fragmentation on biodiversity. Ann Rev Ecol Evol Syst 34:487–515

Fahrig L (2013) Rethinking patch size and isolation effects: the habitat amount hypothesis. J Biogeogr 40:1649–1663

Feisinger P (2001) Designing field studies for biodiversity conservation. The Nature Conservancy, Washington

Fischer J, Lindenmayer DB (2007) Landscape modification and habitat fragmentation: a synthesis. Global Ecol Biogeogr 16:265–280

Frankel OH (1974) Genetic-conservation: our evolutionary responsibility. Genetics 78:53–65
Fretwell SD (1972) Populations in seasonal environment. Princeton University Press, Princeton
Fretwell SD, Lucas HL (1970) On territorial behavior and other factors influencing habitat distribution in birds. I. Theoretical development. Acta Biotheor 19:16–36
Gilpin ME, Case TJ, Ayala FJ (1976) θ-selection. Math Biosci 32:131–139
Grant PR, Grant BR (1995) Prediciting microevolutionary responses to directional selection on heritable variation. Evolution 49:241–251
Hall LS, Krausman PR, Morrison ML (1997) The habitat concept and a plea for standard terminology. Wildl Soc Bull 25:173–182
Hanski I (1998) Metapopulation dynamics. Nature 396:41–49
Hanski I (2011) Habitat loss, the dynamics of biodiversity, and a perspective on conservation. Ambio 40:248–255
Harris RJ, Reed JM (2002) Behavioral barriers to non-migratory movements of birds. Ann Zoo Fennici 39:275–290
Harvey PH, Pagel MD (1991) The comparative method in evolutionary biology. Oxford University Press, Oxford
Holling CS (1978) Adaptive environment assessment and management. John Willey, Nueva York
Holt RD (2009) Bringing the Hutchinsonian niche into the 21st century: ecological and evolutionary perspectives. Proc Natl Acad Sci U S A 17:19659–19665
Hubbell SP (2001) The unified neutral theory of biodiversity and biogeography. Princeton University Press, Princeton
Hurlbert SH (1984) Pseudoreplication and the design of ecological field experiments. Ecol monogr 54:187–211
Hutchinson GE (1957) Concluding remarks. Cold Spring Harbor Symp Quant Biol 22:415–427
Hutto R (1985) Habitat selection by nonbreeding migratory land birds. In: Cody M (ed) Habitat selection in birds. Academic, London, pp 455–476
James FC, Johnson RF, Wamer NO, Neimi GJ, Boecklen WJ (1984) The Grinnelian niche of the Wood Thrush. Am Nat 124:17–30
Johnson DH (1980) The comparison of usage and availability measurements for evaluating resource preference. Ecology 61:65–71
Johnson MD (2007) Measuring habitat quality: a review. Condor 109:489–504
Jones J (2001) Habitat selection studies in avian ecology: a critical review. Auk 118:557–562
Kearney MR (2006) Habitat, environment and niche: what are we modeling? Oikos 115:186–191
Kearney MR, Porter W (2009) Mechanistic niche modeling: combining physiological and spatial data to predict species' ranges. Ecol Lett 12:334–350
Kennedy M, Gray RD (1993) Can ecological theory predict the distribution of foraging animals? A critical analysis of experiments on the Ideal Free Distribution. Oikos 68:158–166
Kerr JT, Ostrovsky M (2003) From space to species: ecological applications for remote sensing. Trends Ecol Evol 18:299–305
Lawton JH (1993) Range, population abundance and conservation. Trends Ecol Evol 8:409–413
Leisler B, Winkler H (1985) Ecomorphology. Curr Ornithol 2:155–186
MacArthur RH (1972) Geographical ecology. Harper and Row, New York
MacArthur RH, Wilson EO (1967) The theory of island biogeography. Princeton University Press, Princeton
Manly BFJ, McDonald LL, Thomas DL, McDonald TL, Erikson WP (2002) Resource selection by animals, 2nd edn. Kluwer Academic Publishers, Dordrecht
Manning AD et al (2004) Continua and Umwelt: novel perspectives on viewing landscapes. Oikos 104:621–628
Margules CR, Pressey RL (2000) Systematic conservation planning. Nature 405:243–253
Maurer BA (1984) Interference and exploitation in bird communities. Wilson Bull 96:380–395
Meehan TD, Jetz W, Brown JH (2004) Energetic determinants of abundance in winter land bird communities. Ecol Lett 7:532–537
Mitchell SC (2005) How useful is the concept of habitat?- a critique. Oikos 110:634–638

Morris DW (2011) Adaptation and habitat selection in the eco-evolutionary process. Proc R Soc B 278:2401–2411
Morrison ML (2009) Restoring Wildlife. Ecological concepts and practical applications. Island Press, Washington, DC
Morrison ML, Marcot BG, Mannan RW (2006) Wildlife-habitat relationships: concepts and applications, 3rd edn. Island Press, Washington, DC
Nichols JD, Williams BK (2006) Monitoring for conservation. Trends Ecol Evol 21:668–673
Nielsen SE, Shafer AB, Boyce MS, Stenhouse GB (2013) Does learning or instinct shape habitat selection? PLoS One 8(1):e53721
Norris RD (2005) Carry-over effects and habitat quality in migratory populations. Oikos 109:179–186
Odum E (1971) Fundamentals of ecology, 3rd edn. Saunders, Philadelphia, 574
Pardini R, Bueno AA, Gardner TA, Prado PI, Metzger JP (2010) Beyond the fragmentation threshold hypothesis: regime shifts in biodiversity across fragmented landscapes. PLoS One 5(10):e13666
Piper WH (2011) Making habitat selection more "familiar": a review. Behav Ecol Sociobiol 65:1329–1351
Prugh LR, Hodges KE, Sinclair AR, Brashares JS (2008) Effect of habitat area and isolation on fragmented animal populations. Proc Natl Acad Sci 105:20770–20775
Pulliam HR (1988) Sources, sinks, and population regulation. Am Nat 132:652–661
Pulliam HR (2000) On the relationships between niche and distribution. Ecol Lett 3:349–361
Pulliam HR, Caraco T (1984) Living in groups: is there an optimal group size? In: Krebs JR, Davies NB (eds) Behavioural ecology: an evolutionary approach, 2nd edn. Sinauer, Sunderland, pp 122–147
Pulliam HR, Danielson BJ (1991) Source, sinks, and habitat selection – a landscape perspective on population dynamics. Am Nat 137:50–66
Pyke GH (1984) Optimal foraging theory: a critical review. Ann Rev Ecol Syst 15:523–575
Rosenzweig ML (1985) Some theoretical aspects of habitat selection. In: Cody M (ed) Habitat selection in birds. Academic, London, pp 517–540
Runge MC (2011) An introduction to adaptive management for threatened and endangered species. J Fish Wildl Manag 2:220–233
Sagarin RD, Gaines SD (2002) The 'abundant center' distribution: to what extent is it a biogeographical rule? Ecol Lett 5:137–147
Sexton JP, McIntyre PJ, Angert AM, Rice JK (2009) Evolution and ecology of species range limits. Annu Rev Ecol Evol Syst 40:415–436
Sole RV, Montoya JM (2001) Complexity and fragility in ecological networks. Proc R Soc Lond B 268:2039–2045
Southwood TRE (1977) Habitat, the templet for ecolgical strategies? J Anim Ecol 46:337–365
Stockwell CA, Hendry AP, Kinnison MT (2003) Contemporary evolution meets conservation biology. Trends Ecol Evol 18:94–101
Sutherland WJ, Pullin AS, Dolman PM, Knight TM (2004) The need for evidence-based conservation. Trends Ecol Evol 19:305–308
Swift TL, Hannon SJ (2010) Critical thresholds associated with habitat loss: a review of the concepts, evidence, and applications. Biol Rev 85:35–53
Tellería JL, Carrascal LM, Santos T (2014) Species abundance and migratory status affects large-scale fruit tracking in thrushes (Turdus spp.). J Ornithol 155(1):157–164
Tilman D, May RM, Lehman CL, Nowak MA (1994) Habitat destruction and the extinction debt. Nature 371:65–66
Tingley MW, Monahan WB, Beissinger SR, Moritz C (2009) Birds track their Grinnellian niche through a century of climate change. Proc Natl Acad Sci 106:19637–19643
Udvardy MFD (1959) Notes on the ecological concepts of habitat, biotope and niche. Ecology 40:725–728

Uezu A, Metzger JP, Vielliard JM (2005) Effects of structural and functional connectivity and patch size on the abundance of seven Atlantic Forest bird species. Biological Conservation 30: 507–519

Van Horne B (1983) Density as a misleading indicator of habitat quality. J Wildl Manag 47:893–901

Whittaker RH, Levin SA, Root RB (1973) Niche, habitat, and ecotope. Am Nat 107:321–338

Wiens JA (1989) The ecology of bird communities. Cambridge University Press, Cambridge

Wiens JJ, Graham CH (2005) Niche conservatism: integrating evolution, ecology, and conservation biology. Ann Rev Ecol Evol Syst 36:519–539

Wiens JJ, Ackerly DD, Allen AP, Anacker BL, Buckley LB, Cornell HV, Stephens PR (2010) Niche conservatism as an emerging principle in ecology and conservation biology. Ecol Lett 13:1310–1324

Wikelski M, Cooke SJ (2006) Conservation physiology. Trends Ecol Evol 21:38–46

Effects of Renewable Energy Production and Infrastructure on Wildlife

José Antonio Sánchez-Zapata, Miguel Clavero, Martina Carrete, Travis L. DeVault, Virgilio Hermoso, Miguel Angel Losada, María José Polo, Sonia Sánchez-Navarro, Juan Manuel Pérez-García, Francisco Botella, Carlos Ibáñez, and José Antonio Donázar

Introduction

The high levels of human demands of resources—from food to space and energy—are one of the main drivers of global change and are causing large negative impacts on ecosystems functioning worldwide (Vitousek et al. 1997). Global change components range from climate change to habitat destruction, species invasions, pollution and eutrophication. Although these factors can produce ecosystem changes independently, the final

J.A. Sánchez-Zapata (✉) • J.M. Pérez-García • F. Botella
Department Applied Biology, Universidad Miguel Hernández, Elche, Spain
e-mail: toni.sanchezzapata@gmail.com

M. Clavero • J.A. Donázar
Departament of Conservation Biology, Estación Biológica de Doñana-CSIC, Américo Vespucio s.n, Sevilla 41092, Spain

M. Carrete
Department of Physical, Chemical and Natural Systems, University Pablo de Olavide, Ctra. Utrera km 1, Sevilla 41013, Spain

T.L. DeVault
Department of Agriculture, Wildlife Services, National Wildlife Research Center, Sandusky, OH, USA

V. Hermoso
CEMFOR-CTFC, Forest Sciences Center of Catalonia, Solsona, Lleida 25280, Spain

M.A. Losada
Departament Structural Mechanics and Hydraulic engineering, University of Granada, Granada, Spain

M.J. Polo
Departament of Hydraulic Engineering, University of Córdoba, Cordoba, Spain

S. Sánchez-Navarro • C. Ibáñez
Departament of Evolutionary Ecology, Estación Biológica de Doñana-CSIC, Américo Vespucio s.n, Sevilla 41092, Spain

R. Mateo et al. (eds.), *Current Trends in Wildlife Research*,
Wildlife Research Monographs 1, DOI 10.1007/978-3-319-27912-1_5

descent is often driven by synergistic processes. The resulting amplifying feedbacks can be disconnected from the original driver of change, leading to a state shift in the biosphere with unexpected consequences (Barnosky et al. 2012). Climate change seems to have taken prominence over other drivers of global change, leading to larger funding and attention with respect to other major components of global change (Veríssimo et al. 2014).

Human population growth and growing per capita consumption are causing an increasing energy demand to support industrial and domestic activities. Reducing the emissions of greenhouse gases responsible for current climate change is the main goal of renewable energy production. Major efforts have been devoted to the development of this kind of energy, including longstanding sources such as hydroelectric infrastructure along with more recent technologies such as wind farms and solar plants (Johansson et al. 1993). Renewable energy aims to provide humans with sustainable resources (Dincer 2000), although the development of infrastructure aimed to produce and distribute it may also have detrimental effects on ecosystems. Paradoxically, the development of renewable energy might also jeopardize biodiversity by increasing extinction rates of endangered species (Hooper et al. 2012). Thus, it is important to reconcile the production of renewable energy with the conservation of biodiversity to meet the primary objective of sustainable development.

No energy source is entirely 'biodiversity-neutral': nuclear power has obvious waste-disposal problems, the manufacture of solar cells can create hazardous waste, hydropower alters freshwater ecosystems, and wastewater from geothermal plants affects aquatic ecology. The transmission lines, roads, and other infrastructure associated with all 'clean' power projects can also have extensive impacts on ecosystems. Given that energy consumption is set to double by 2030, the environmental and business communities must find ways to assess and manage the tradeoffs between energy generation and use and biodiversity conservation. Conservationists are racing to gather the necessary information about the impacts and make sure it feeds into the decision-making process. In this chapter, we review previously described as well as less explored effects of renewable energy production and infrastructure on biodiversity, from species to populations and communities, and discuss the main guidelines to mitigate their impacts.

Hydroelectric Production Infrastructures: Overview

Hydroelectricity refers to the electric energy generated by the movement of water though turbines. Hydroelectric production most often requires damming a river to simultaneously ensure the availability of water and provide the water jump necessary to move the turbines. Worldwide, there are almost 10,000 large dams (height >15 m) devoted to hydroelectric production, constituting almost 20 % of all built dams (ICOLD 2014). The number of smaller dams (<15 m) is not precisely known, but they are thought to outnumber larger ones by some tenfold (Carpenter et al. 2011). China alone has almost 50,000 hydroelectric facilities (Kosnik 2008). Hydropower is used to produce over 16 % of total electricity globally, constituting over 75 % of the overall renewable electric production (REN21 2013). The main advantage of hydroelectric production over other renewables is the possibility of responding quickly to demand peaks. During periods of low demand (e.g. at night) electricity can be used to pump water to reservoirs at higher altitudes, thus ensuring water availability for production during high demand periods.

Hydroelectricity has been presented as a "green" energy source, based on the lack of greenhouse gas emissions directly related to electricity production (Kosnik 2008). However, this view has been challenged in recent times due to the accounting of emissions from reservoirs and downstream water reaches (Fearnside 2004; Kemenes et al. 2007). Reservoir-linked emissions can be related to the deforestation of the area occupied by the water mass and the decay of submerged organic matter (Fearnside 2000), and by the production of methane due to microbial metabolism in anoxic environments (Bastviken et al 2011). Hydroelectric development also has socioeconomic impacts on local communities, being cited as one of the main drivers of forced displacement and resettlement (Scudder 2005). Resettlement usually involves a change in living means due to environmental constraints, e.g. when people are moved from fertile rural areas to cities (Wilmsen et al. 2011), and most often results in the impoverishment of affected households (Scudder 2005). Flow regulation for hydroelectric production can negatively affect downstream agricultural lands (Kuenzer et al. 2013) and fisheries (Silvano et al. 2009), while the creation of hydroelectric reservoirs can have negative health implications in some areas (Yewhalaw et al. 2009). Nevertheless, we will not further discuss these global climate and social issues here and will henceforth focus on the ecological impacts of hydroelectric production on aquatic systems and their biota. Many of these impacts are related to the presence of dams and are thus not exclusive of hydroelectric infrastructures, being shared with those of dams devoted to irrigation, urban and industrial water supply, or other uses.

Barrier Effect

Dams constitute impassable barriers for the movement of most strictly aquatic animals. The fragmentation of river networks by dams has led to the collapse of migratory fish populations in many areas of the world (e.g. Holmquist et al. 1998; Limburg and Waldman 2009; Hall et al. 2012). A search in the IUCN red list (www.iucnredlist.org) provides a list of 452 threatened fish species that are negatively affected by dams (threats 7.2.9, 7.2.10 or 7.2.11 in the IUCN threat classification), including 85 critically endangered and 8 extinct taxa. These numbers underestimate the real number of fish species affected by dams, because fish species have not been globally assessed and the red list covers less than 50 % of existing freshwater fish species.

The impacts of river fragmentation can be extreme for anadromous migratory fish, such as migratory salmonids (Fam. Salmonidae) or clupeids (Fam. Clupeidae), because the establishment of a dam can completely impede reproduction. Sturgeons (Fam. Acipenseridae) are perhaps the clearest example of highly threatened anadromous fish affected by river fragmentation, with all 25 sturgeon species assessed by the IUCN negatively affected by dams. For example, the construction of the Gezhouba (in 1981) and Three Gorges (2003) dams in the Yangtze river has led to large reductions in the ranges of the critically endangered Chinese (*Acipenser nudiventris*) and Yangtze (*A. dabryanus*) sturgeons (Zhang et al. 2015).

Catadromous fishes, i.e. those that reproduce in the sea but spend an important part of their life in freshwater systems, are often less severely impacted by dams than anadromous ones. However, the barriers can impede the occupancy of an important proportion of the original habitat of a species, which can in turn be translated into significant population declines. For example, dam construction and associated loss of freshwater habitats is one of the main drivers of the recent collapse of the European eel (*Anguilla anguilla*) (Kettle et al. 2011). Dams also act as effective barriers for tropical catadromous fish assemblages, even precluding the upstream migration of climb-adapted gobies (Fam. Gobiidae; Cooney and Kwak 2013). Dams can also constitute important or absolute barriers for species that move along river systems, using different habitats within the year and/or during different life-history stages. This is the case for several fish species in the Amazon basin, which have been severely affected by hydropower development in Brazil (Godinho and Kynard 2009), and also of different river dolphins in Asia (Dudgeon et al. 2006).

The impediment of downstream movement of the river biota is also a relevant impact of hydroelectric infrastructures. A key element of the downstream barrier is not the dam itself but the reservoir behind it. Reservoirs are stagnant water masses that differ radically from river systems, and thus can eliminate most of the ecological cues followed by migrating fishes, causing disruptions in their migrations (Pelicice et al. 2014).

Most, if not all, of the outflow of hydroelectric dams passes through a turbine system to produce electricity. Fish mortality in these turbines can be high, even approaching 100 %, although it is dependent on the type of turbines, their operation, and the different species and life history stages involved (Larinier and Travade 2002). Because the risk posed by turbines increases with body length, long-bodied species such as freshwater eels (Fam. Anguillidae) often are more severely affected by the downstream barrier effect of hydroelectric infrastructures (Calles et al. 2010).

For more than a century there have been important technological developments to facilitate the movement of fish across dam barriers (Katopodis and Williams 2012). These structures have proven useful in several situations for good swimmer (e.g. salmonids) or good climber (e.g. eels) species (Laine et al. 2002; Feunteun 2002), especially for relatively small barriers. However, passing facilities designed for northern hemisphere fish (most commonly salmonids) often fail to be effective in other environments (e.g. Roberts 2001; Mallen-Cooper and Brand 2007). On the other hand, although technical solutions have been relatively successful in facilitating upstream fish migration, the elimination of the barrier effect for downstream movement is often much more complicated (Feunteun 2002). Proposed actions include the design of modified turbines (Čada 2001) or the implementation of lateral bypasses (Gosset et al. 2005), but their effectiveness and widespread applicability has not been proven. Some successful restorations of fish populations by eliminating barrier effects exist, but failures have also been reported, even in cases where enormous efforts had been invested.

An analysis of anadromous fish restoration programs in large North American rivers fragmented by hydroelectric dams concluded that strategies based on fish passes had failed and that dam removal was the only viable and realistic option (Brown et al. 2013).

Flow Regulation and Ecosystem-Level Impacts

The alterations of natural flow regimes (Poff et al. 2007) are one of the most widespread human impacts on aquatic systems, with the overall potential for water retention in existing dams being more than five times larger than the total volume of water of all the rivers in the world (Dudgeon et al. 2006). Flow regulation affects more than 60 % of the world's large river systems, and there are regions (e.g. Europe) where unregulated large rivers do not exist anymore (Nilsson et al. 2005). In natural conditions, flow regimes differ among regions and among rivers within regions (Poff et al. 1997). This diversity in the functioning of river systems drives the high intersystem variability (or β-diversity) of the river biota. Flow regulation has homogenized this originally diverse scenario so that river systems are nowadays much more alike in their functioning than they were before dam construction (Poff et al. 2007).

Reservoirs retain water during flood periods and often increase baseline flows during naturally dry seasons, thus smoothing the temporal variability in flow conditions (Poff et al. 1997). Of particular relevance is the buffering or elimination of peak floods due to their importance for river ecosystems, including their floodplains, riparian ecosystems, estuaries and deltas, as well as for their biota (FitzHugh and Vogel 2011). After the closure of the High Aswan dam in the mid-1960s, the discharge of the Nile River during the annual flood period decreased by around tenfold (Tockner and Stanford 2002) with severe consequences for floodplain and estuary areas, even affecting marine fisheries in the Eastern Mediterranean (White 1988). Elimination of floods reduces the lateral connectivity of river systems, often precluding the inundation of floodplains that are critical habitats for the reproduction of several aquatic and semiaquatic organisms (Nilsson and Dynesius 1994; FitzHugh and Vogel 2011). The impact of flow regulation extends beyond quantity of water and also affects water quality. Dams often release hypolimnetic water, which is cold and has little dissolved oxygen, with severe consequences on biological communities in downstream areas close to the dam (Ward and Stanford 1982; Jurajda et al. 1995; Zhong and Power 1996).

In water-abundant regions throughout the world, hydroelectric dams, especially small ones, are run-of-the-river facilities (Richter and Thomas 2007), in which the discharge running through the turbines is similar to that entering the reservoir. Dams operated in this way may have small effects, if any, for the flow regimes of river systems. However, hydroelectric facilities in drier areas and (especially) large hydroelectric dams do store water and disrupt hydrological regimes. Hydroelectric dams may also have specific, short-term impacts on river flow related to the variability in electricity demand (Bevelhimer et al. 2015). The daily demand fluctuations can be translated into large, completely unnatural daily variations in flow below hydroelectric dams, with strong negative impacts on the aquatic biota (Cushman 1985).

Upstream Habitat Modifications

World's reservoirs occupy a combined area of about 500,000 km^2, similar to that of France (Nilsson 2009). The loss of terrestrial habitat affects valley bottoms especially, which are the most productive environments in many areas (e.g. mountains, high latitudes) and often host diverse and abundant biological communities (Nilsson and Dynesius 1994; Nilsson and Berggren 2000). The new water masses can also constitute effective barriers precluding the movement of terrestrial species and disrupting migration behaviors (Nellemann et al. 2003). The loss of generally diverse riparian ecosystems is almost never compensated by the establishment of riparian vegetation around the reservoir shores, due to the high water-level fluctuations. Whenever there is a chance for the formation of a riparian vegetation fringe, it is most often dominated by invasive plants and/or those typical of eutrophic systems (Hill et al. 1998; Nilsson 2009). However, reservoirs (including some hydroelectric facilities) can also constitute relevant feeding and breeding habitat for threatened species (McCartney 2009) and become refuge habitat in semi-arid territories during water shortages (Prenda et al. 2001).

The impoundment of a river generates a radically different and simplified stagnant water mass that results in a hostile environment for an important part of the fluvial biota (Baxter 1977). The lack of water currents and turbulence, and the loss of complexity in the substrate due to the deposition of fine sediments, produce a radical homogenization of the water landscape. As a consequence, species linked to running water or those needing complex habitat structures at some stage of their life history tend to be absent from reservoirs (Gido et al. 2009). The fluctuation in water levels, which in hydroelectric dams can be large within each day (Nilsson 2009), hinders the occupation of the littoral zone by many slow-moving animals, including many invertebrate taxa (Baxter 1977). In general, the aquatic biota occupying reservoirs is simplified and impoverished when compared to rivers (Gido et al. 2009; Clavero and Hermoso 2011), and results in low quality habitats for aquatic top predators (Palmeirim et al. 2014).

Reservoirs, Flows and Invasive Species

A relevant ecological impact of reservoirs is their relationship with invasive aquatic species (Havel et al. 2005). Reservoirs host more invasive species than rivers (Clavero and Hermoso 2011) and natural lakes (Johnson et al. 2008). The invasibility of reservoirs can be related to two phenomena. First, reservoirs are stable environments and very homogeneous when compared to rivers (Clavero et al. 2013). These characteristics may facilitate the establishment of many widely introduced species such as the common carp (*Cyprinus carpio*) or the zebra mussel (*Dreissena polymorpha*). Second, reservoirs are associated with many of the activities related to the release of non-native aquatic organisms, such as sport-fishing and recreational navigation (Havel et al. 2005; Johnson et al. 2008).

Reservoirs are not only easily invaded, but also act as facilitators for the invasion of associated river systems (Havel et al. 2005). Reservoirs may function as a source of individuals of invasive species to upstream river stretches (Rincón et al. 1990; MacIsaac et al. 2007). Naturally flowing rivers are suboptimal habitats for many invasive species, especially in areas with high environmental fluctuations for which invasive species may lack adaptations. However, reservoirs, being more stable and thus less hostile systems for invasive species, may compensate the environmental harshness of rivers by being a constant source of colonizing individuals. Reservoirs can also be a source of individuals of non-native species for downstream reaches (MacIsaac et al. 2007), the invasion process being in this case reinforced by the alteration of flow regimes. Most frequently, non-native species are not adapted to cope with the specific natural flow regimes of the river systems where they are released. The softening of discharge extremes due to flow regulation eases their establishment in river sectors downstream from reservoirs (Hermoso et al. 2011). Thus, apart from the barrier effects and habitat changes, reservoirs may exert an indirect negative impact on native species (both within the reservoir and in associated river systems) by enhancing populations of harmful invasive species (Hermoso et al. 2011; Clavero et al. 2013)

Solar and Wind-Energy Infrastructures: Overview

During the last decades, the number of wind and solar energy developments has increased substantially all over the world, receiving support as alternative energy sources that can achieve substantial avoidance of greenhouse gas emissions. By the end of 2013, worldwide installed capacity for wind and solar energy were 318.1 and 138.9 GW, respectively. China, USA, Germany and Spain accounted for more than 65 % of the worldwide installed capacity of wind energy (GWEC 2014), whereas solar energy installation has been led by Europe, followed by China, Japan, and the United States.

In this section, we review the impacts of solar and wind energy production and infrastructure. We consider direct impacts on wildlife populations through habitat transformation and demographic unbalances (e.g., Langston and Pullan 2003; Baerwald et al. 2008; Garvin et al. 2011) as well as changes in ecosystem functioning at different scales, which can be a source of further concern for communities and populations of both plant and animals. We include not only terrestrial infrastructure but also the increasingly implemented and less studied offshore developments.

Effects on Landscapes and Ecosystems

The occupation of the territory by solar and wind energy systems (SWES) involves a change in land use, and their operational characteristics can locally modify mass and energy fluxes, with potential effects on soil-plant processes and ecosystem services such as soil formation and nutrient cycles (supporting services), climate and hydrology (regulating services), water and food supply (provisioning services), and recreational and aesthetic activities (cultural services). Some of these changes have been extensively reported, whereas others, mainly related to indirect impacts, still require further assessment (Tsousos et al 2005; Saidur et al. 2011; Leung and Yang 2012; Aman et al. 2015).

The main environmental impact of SWES is land use change. On average, the footprint area associated to SWES is estimated at 13–20 m^2 for a 5-MW wind turbine, 1.9 m^2 for a 160-W solar photovoltaic system, and 1.9–2.4 km^2 for a 100-MW concentrated solar power system (Jacobson 2009), which results in 1 m^2 of land being required to produce 0.38–0.25 MW (wind energy) or 83–42 W (solar energy). Vegetation is usually removed and the soil is graded, which enhances soil erosion or aeolian sediment transport, as well as loss of organic carbon and nitrogen, especially in semiarid/arid conditions. Nonetheless, SWES impacts in terms of landscape (use and cover) disturbances are much lower than those due to conventional energy systems (Tsousos et al. 2005). Photovoltaic systems, for example, have higher land use efficiency than other renewable energy sources and surface coal mining (70 % of global United States coal extraction). These numbers highlight the need for addressing solar energy environmental impacts (and energy issues, in general) in the global framework of substitution, and not in their absolute values (Jacobson and Delucchi 2011).

Microclimatic conditions can be influenced by SWES (Baidya 2011; Hernandez et al. 2014). Ground-mounted photovoltaic arrays directly intercept precipitation and atmospheric deposition, change surface albedo and increase shading on a local basis, greatly altering the radiation budget on the ground surface. Moreover, their presence affects the wind speed and enhances turbulence at the near-surface atmospheric boundary layer. All these factors may significantly modify energy fluxes over the affected surface and cause changes in local temperature, air and soil moisture, thus affecting evaporation dynamics. Wind farms have little influence on the radiation components, but they strongly change the wind profile distribution and magnitude, and affect turbulence and mixing within the atmospheric boundary layer, which also modifies temperature and air moisture profiles, and again the final energy and water budgets (Baidya et al. 2004).

These effects are also dependent on the diurnal cycle of solar radiation. Photovoltaic panels reduce the shortwave reflected solar radiation due to their lower albedo when compared to ground surface, but they increase the ratio of diffuse/direct shortwave radiation below the panel (Scherba et al. 2011). Changes of soil conductance and air warming below panels are not clear, because the effects on wind must be taken into account and different results can be obtained depending on different additional factors (Taha 2013). Finally, the rainfall interception results in an enhanced draining along the points from the lower end of the panel surface, which significantly alters the infiltration/runoff ratio over the ground beneath, especially for light and intermittent rainfall events. Drop erosion processes may also occur due to this redistribution of rainfall over the ground. At larger scales, the different importance of each hydrological component in the water cycle would be altered, but the significance of this change is dependent on the time distribution pattern of precipitation at the area and the local arid/wet regime (Pisinaras et al. 2014). Since solar energy facilities are efficient in regions with high insolation rates, arid and semiarid areas would be more affected by these impacts, not all of them being negative (Turney and Fthenakis 2011).

As for wind-farms, the impacts on temperature have been observed in both directions. Night-time surface temperature may be increased downwind from turbines, since warmer air eddies would mix into the cooler air; the inverse effect would be observed during daytime. Under stable atmosphere conditions, an increase in evapotranspiration would be expected. Direct effects on microclimate may also induce indirect impacts on the physical environment. Besides water in air and soil, other gases like CO2, methane, and nitrous oxide could modify their concentration profiles over the area affected by SWES, not only due to mixing condition alterations but also to the forcing action of changes in temperature. Changes of 0.7–3.5 °C have been measured within the influence area of wind farms (Baidya and Traiteur 2010; Zhou et al. 2012), and increases of 2.5–26.0 °C in the area surrounding solar facilities in the developed environment (Scherba et al. 2011). Plant and soil processes involving water and energy exchanges, microbial activity, C and N cycling, and other biogeochemical cycles could be significantly altered depending on the initial conditions, and the extent and location of the SWES, with further consequences on the ecosystem dynamics (Armstrong et al. 2014).

Potential effects on climate may work at larger scales. For example, Wang and Prinn (2010) concluded from modelling that the induced changes by wind farms in surface heat fluxes and temperature could result in modified cloud cover distributions, while Fiedler and Bukovsky (2011) obtained an increase of rainfall after modelling on a large scale. However, global circulation models and their downscaling to surface models still have a significant uncertainty related to the adequate inclusion of interactions between physical and biological processes (Cramer et al. 2001; Hawkins and Sutton 2011).

Effects of Solar Plants and Wind Farms on Wildlife

Solar plants are relatively new and their effects on biodiversity have been scarcely documented (but see DeVault et al. 2014). On a local scale, impacts are associated with the above-mentioned habitat transformation and wildlife mortality (Lovich and Ennen 2011; Hernandez et al. 2014). For example, avian mortalities at a 10 MW concentrating solar thermal power plant in California, USA averaged 1.9–2.2

individuals per week, and were mainly caused by collision with site infrastructure (81 %), particularly with heliostats, and to a lesser degree, burning when heliostats were oriented towards standby points (19 %), especially for aerial foraging species (McCrary et al. 1986). Moreover, mortality rates can be increased because of the soil degradation and the creation of roads, and exotic species invasions could even occur.

One potential solution proposed to reduce the negative impacts of solar energy production on wildlife is to locate some solar facilities on unused portions of airport lands (DeVault et al. 2012, 2014). Airports represent one of the only land use types where reduction in wildlife occurrence is justified and socially acceptable, due to the risk of wildlife-aircraft collisions (DeVault et al. 2013). Thus, wildlife conservation is largely discouraged at airports because of safety concerns (Blackwell et al. 2013). Further, airports often occupy extensive areas and much of the land is undeveloped; for example, airports in the USA collectively contain well over 3300 km^2 of idle grasslands (DeVault et al. 2012). Also, photovoltaic solar facilities are largely compatible with airport operation from a safety perspective (Barrett and DeVita 2011), and solar energy production using photovoltaic arrays is generally economically advantageous for airports (DeVault et al. 2012).

Concern about wildlife mortality at wind farms began to surface in the late 1980s and early 1990s, and it was mostly focused on the Altamont Pass Wind Resource Area (APWRA), a 165 km^2 wind farm near San Francisco, California. Hundreds to thousands of birds, including more than 40 species, some of them endangered, died there every year (Asmus 2005; Thelander and Rugge 2000; Smallwood and Thelander 2005; Thelander 2004; Smallwood and Thelander 2008). Such fatalities are not limited to California or the USA (Erickson et al. 2001; US GAO 2005). European countries such as Spain and Belgium also have reported hundreds of birds killed by turbines (Lowther 1998; Everaert and Stienen 2007). Thus, much wind farm-wildlife research has been devoted to investigating how wind farm developments impact bird populations (e.g., Langston and Pullan 2003; Baerwald et al. 2008; Garvin et al. 2011), in particular collision rates of birds with turbines as well as factors influencing interspecific and local variability (reviewed in Drewitt and Langston 2006, 2008; Kuvlesky et al. 2007; Stewart et al. 2007; Loss et al. 2013; Marqués et al. 2014). These studies show that the effects of wind farms on birds are highly variable (while most wind turbines actually kill none or very few individuals, some turbines kill many), and depend on a wide range of factors, including the development type, the topography of the surrounding land, the habitats affected and the number and species of birds present (Barrios and Rodriguez 2004). In general, risks are higher when turbines are placed on ridges and upwind slopes, built close to migration routes, or operated during periods of poor visibility such as fog, rain, and at night (Sovacool 2009). Also, modern, monopole wind turbines might kill fewer birds than older, lattice turbines, such as those that were located at APWRA (Loss et al. 2013; but see Barrios and Rodriguez 2004).

There is a general acceptance of the idea that collision mortality would increases with bird abundance (Musters et al. 1996; Osborn et al. 2000; Drewitt and Langston 2006; Tellería 2009a, b). Although linearity in this relationship could be, a priori, a simplistic assumption because of interspecific differences in susceptibility to this infrastructure, higher abundance of individuals of species sensitive to collision at wind farms would increase fatality rates. In particular, raptors, grouse, gulls and terns tend to collide more often than expected from their occurrence and numbers (Carrete et al. 2009, 2012). Thus, the most effective measures to minimize negative effects on birds is to identify the dangerous locations and avoid locating wind turbines there. Most accidents with birds occur in places where the more vulnerable species concentrate, so the use of reliable estimates of abundances (such as those derived from the location and size of breeding or roosting sites; Carrete et al. 2012) should be preferred over highly variable punctual, year estimates (Barrios and Rodriguez 2004; de Lucas et al. 2008). Otherwise, as is case for most current risk assessments studies in countries such as Spain, it is not surprising to find weak relationships between the predicted risk and the recorded mortality at wind farms (Ferrer et al. 2011).

A major difficulty in assessing the impact of wind farms on bird populations is the scarcity of long-term studies at operational wind farms. Thus, there is a widespread belief that wind farms have, at most, a low impact on animal populations (Marris and Fairless 2004). However, the few studies evaluating the long-lasting effects of wind farms on wildlife advise caution. A recent study carried out in the UK using long-term data of breeding bird abundances show that wind farm developments may result in significant reductions in habitat usage (from 100 to 800 m from the turbines after construction) by some species. This result in the decline in the abundance of some breeding birds such as red grouses *Lagopus lagopus scoticus*, snipes *Gallinago gallinago* and curlews *Numenius arquata* (in some cases by up to 50 % within 500 m of the turbines during construction), without general recoveries after the first year of operation (Pearce-Higgins et al. 2009, 2012). Another study focused on the impact of wind farm mortality on the demography of an endangered species, the Egyptian vulture *Neophron percnopterus*, show that even very low levels of additional mortality can be significant for this long-lived species with low productivity and slow maturation rate (Carrete et al. 2009; Sanz-Aguilar et al. 2015), as is the case with many endangered or rare long-lived species (Saether and Bakke 2000). Very low reductions in survival rates of territorial and non-territorial birds (−0.015 and −0.008, respectively) associated with wind-farms can have significant population impacts (Carrete et al. 2009; Bellebaum et al. 2012; Sanz-Aguilar et al. 2015). Regrettably, this a widespread scenario affecting many other long-lived endangered species also killed at wind farms in different European countries (e.g., at least 10 white-tailed sea eagle *Haliaeetus albacilla* per year in Norway; http://www.statkraft.com/pub/wind_power/feature_articles), in the USA (e.g., 65 golden eagles *Aquila chrysaetos* in California per year; Smallwood and Thelander 2008) and in Australia (e.g., at least 12 Tasmanian wedge-tailed eagles *Aquila audax fleayi* in 4 years; http://www.windaction.org/news/17683).

The other taxonomic group highly affected by wind farm mortality are bats. The first reports of bat fatalities at wind farms occurred in North America and Europe during the 1990s (Kunz et al. 2007; Rydell et al. 2010). However, the occurrence of bat fatalities at wind farms is a global phenomenon with cases described from numerous countries including Australia (Hull and Cawthen 2013) and South Africa (Doty and Martin 2013). The lack of information on the impact of wind farms on bat populations reflects the severity and poor understanding of the problem. This is concerning because in a country such as the United States, which produced 51,630 MW of wind-energy in 2012, some studies have estimated that wind farms kill between 600,000 and 888,000 bats per year (Hayes 2013; Smallwood 2013). Spain is one of the world leaders in wind farm technology and in 2010 produced 19,148 MW of wind-energy. Despite deficiencies in post-construction monitoring of the impact of wind farms on wildlife (e.g. biases in searcher efficiency and carcass scavenging), the estimated number of wind-energy related bat fatalities in Spain can be comparable to the highest estimates available from North America (Camina 2012). In fact, bat mortality at wind farms tends to be higher than that of birds (Barclay et al. 2007). For example, in USA, estimates are 888,000 bats killed per year *vs* 573,000 birds (Smallwood 2013). This is of great concern because bats are extremely long-lived for their size and they have a low reproductive potential (Barclay and Harder 2003). Thus, the cumulative impacts of this new hazard could result in long-term population declines or even extinctions in certain areas (Kunz et al. 2007), especially given that many bat populations are already under severe stress due to mortality from white-nose syndrome (Foley et al. 2011).

Bats may be attracted to wind turbines for many different reasons, including curiosity, searching for food or potential roosts, or social interactions (Cryan and Barclay 2009). The presence of aviation warning lights on the turbines does not increase mortality rate (Johnson et al. 2004; Bennett and Hale 2014); however a recent study has shown that tree roosting bat species from North America were attracted to the turbines (Cryan et al. 2014). This attraction could be due to the visual confusion of the turbines silhouettes with trees, reinforced by other cues such as similar downwind airflow patterns. Attraction of tree bats to other tree-like structures such as tall communication towers only appears to occur in the late summer and autumn, possibly because of social rather than foraging behavior, as they emit fewer feeding buzzes around these sites (Jameson and Willis 2014).

About half of dead bats examined around wind turbines do not show any external injury caused by direct collisions with turning blades. Instead, most of them have pulmonary lesions and internal hemorrhages compatible with barotrauma caused by rapid air-pressure reduction near fast moving turbine blades (Baerwald et al. 2008). Barotrauma has not been suggested as a cause of bird mortality because of differences in their respiratory anatomy. Although the ability to echolocate allows bats to detect and avoid turbines blades, it seems they cannot detect rapid pressure reductions and may die even if they do not come in contact with the blades.

Out of the 23 species of bats reported to be affected by wind turbines in North America (http://www.batcon.org/), about 80 % of fatalities involve migratory tree-roosting bats (mainly *Lasiurus cinereus, L. borealis* and *Lasionycteris noctivagans*; Arnett et al. 2008). In Europe, 27 bat species have been reported as killed by turbines (http://www.eurobats.org/), and 98 % of these fatalities belong to a "high-risk" species group formed by bats included in the aerial-hawking guild (i.e., species that catch their prey in open spaces; Rydell et al. 2010. Although it could seem that different bat species groups are susceptible to wind turbines in America and Europe, it should be taken into account that all migratory tree-roosting bats from America are aerial-hawking, while mortality of migratory species also occurs in Europe but in smaller proportions (64 % in NW Europe, Rydell et al. 2010; 56 % in Greece, Georgiakakis et al. 2012; 40 % in Portugal, Amorim et al. 2012). However, a significant proportion of migratory killed bats in Europe belong to resident populations (Lehnert et al. 2014).

In both North America and Europe most bat fatalities (90 %) occur during the late summer and early autumn with another minor peak occasionally seen in the spring (Kunz et al. 2007; Rydell et al. 2010). There are no consistent sex or age biases in bat mortality during these periods (Arnett *et al.* 2008; Rydell et al. 2010; Hull and Cawthen 2013). Young bats are not more vulnerable despite their typical dispersal behavior and expected lack of experience. In North America, the mortality peak coincides with the migration period of tree-roosting bats.

As among birds, some studies have attempted to generalize bat mortality patterns associated with wind farms. Highest bat mortality has been observed along forested ridge tops in the Appalachian Mountains and forested hilltops in southern Germany, with lowest records in flat agricultural landscapes. Although local concentrations of mortality at specific turbines has been occasionally described (Piorkowski and O'Connell 2010; Georgiakakis et al. 2012), most bat fatalities are randomly distributed across turbines (Arnett et al. 2008), making it difficult to draw clear guidelines for conservation planning (Kunz et al. 2007; Rydell et al. 2010). However, most fatalities occur during low wind nights (<6 m/s) in late summer and the first half of autumn, thus increasing the cut-in speed of the turbines on nights with high risk of bat collision would be an effective management tool to reduce mortality. Indeed, these methods have achieved reductions of bat fatalities from 50 to 90 % with marginal power loss ($\leq$ 1 % of total annual output) (Baerwald et al. 2009; Arnett et al. 2011). Conversely, other management measures such as acoustic deterrents are less effective, in part due to rapid atmospheric attenuation of ultrasounds (Arnett et al. 2013).

Offshore wind energy is developing rapidly and is rapidly occupying marine areas to produce low carbon energy. Whilst acknowledging that research into the impacts of the offshore renewable industry is still in its infancy, it is widely regarded that the risk for impacts on the marine environment may not be negligible and must be taken seriously. Noise disturbance, electromagnetic fields, and migration barriers have had some negative effects on fish, marine mammals, birds and seabed communities (Wilhelmsson et al. 2010). On the other hand, these installations create 'no-take zones' around them and their underwater vicinity that can function as artificial reefs, which leads to a greater abundance of many species (Wilhelmsson et al. 2010). But if not properly planned and managed, these installations can adversely affect marine biodiversity through habitat loss, collisions with turbines, deviation of the migratory routes, noise and electromagnetic disturbance and navigational hazards for ships (Desholm and Kahlert 2005; Larsen and Guillemette 2007; Wilhelmsson et al. 2010).

In spite of the previous information, the understanding of the potential implications of large-scale renewable energy developments has not kept pace with the recent rise in the number of development proposals. The risk that an animal could be killed at a wind turbine is probably small compared to the risks faced from other human activities (Calvert et al. 2013; DeVault 2015), and some not very successful attempts have been made to compare wind turbine mortality with fatality rates associated with energy sources (for example, by calculating a number of birds killed per kWh generated for wind electricity, fossil-fuel, and nuclear power systems; Sovacool 2009). However, the point here is not to ascertain that turbines are or are not the leading cause of bird deaths, something that can change when considering the current pace and scale of wind energy development (Loss et al. 2013). The point is to recognize that renewable energy can help to reduce greenhouse gas emissions and slow-down climate change, but we should develop them in ways that account for and minimize their impacts on wildlife. Unlike fossil-fuel and nuclear power plants, which spread their wildlife-related impacts across large scales, most of the impact from wind farms occurs locally, so solutions are relatively straightforward.

Energy Infrastructure: Power Lines and Wildlife

Renewable energy produced by wind, solar and hydroelectric facilities not only impact biodiversity during the production stage; these facilities also need power lines to transport the electricity to final consumers. An extraordinarily dense network of power poles and lines is located around cities and industrial areas, and have impacts on wildlife in various landscapes around the world. Power lines have significant potential impacts on biodiversity, mainly through changes in habitat structure and wildlife mortality.

The presence of poles and wires introduces lineal anthropogenic structures that alter the visual natural quality and create division lines on the landscape (Arriaza et al. 2004). This applies especially to transmission lines, the higher voltage power lines (> 66 kV). Their presence causes severe changes in habitat structure, increasing fragmentation as a consequence of the removal of natural vegetation below the lines (Luken et al. 1992; Forrester et al. 2005). This change does not always have negative consequences; some species could benefit from the new habitats created (Askins et al. 2012), e.g. forest ungulates could benefit from foraging in power line rights-of-way where there is increased availability of pastures compared to adjacent forest (Bartzke et al. 2014). However, transmission power lines can behave as barriers to animal movements by disrupting migratory routes and promoting the development of avoidance strategies, as described for the reindeer (*Rhandifer tarandus*) (Reimers et al. 2007; Vistnes et al. 2004). Moreover, as a consequence of electric transmission, power lines generate strong electromagnetic fields, UV discharges and acoustic pollution which can affect animal health and behavior (Phernie et al. 2000; Tyler et al. 2014) and have also been identified as causes of wildfires (Tenforde 1992; Haas et al. 2005).

Probably the most serious environmental impact of power lines is avian mortality caused by electrocution, entangling, and collision (Bevanger 1998; Gangoso and Palacios 2002). Mortality associated with power lines can accelerate the declines of several species and affect occupation patterns (Sergio et al. 2004) or population dynamics (Schaub et al. 2010). Power lines are currently considered one of the main human-related causes of bird mortality worldwide (Bevanger 1998; Loss et al. 2014).

Electrocutions and Collisions

Electrocution and collision with power lines are among the main causes of population declines for some species, mainly raptors (Lehman et al. 2007). These include the Cape Vulture (*Gyps capensis*) in South Africa (Ledger and Hobbs 1999), the Egyptian vulture (*Neophron pernopterus*) in Canary Islands and East Africa (Donázar et al. 2002; Angelov et al. 2011), the griffon vulture (*Gyps fulvus*) in Israel (Leshem 1985), the eagle owl (*Bubo bubo*) in France (Bayle 1999) and Italy (Rubolini et al. 2001), the golden eagle (*Aquila chrysaetos*) and the bald eagle (*Haliaetus leucocephalus*) in USA (Harness and Wilson 2001) and Canada (Wayland et al. 2003) and the Spanish imperial eagle (*Aquila adalberti*) (González et al. 2007) and Bonelli's eagle (*Aquila fasciata*) (Real et al. 2001) in Spain.

Several studies have found that bird mortality at electric facilities is not randomly distributed, but concentrated in a very small percentage of pylons (Mañosa 2001; Guil et al. 2011). For example, electrocutions mostly occur in distribution power lines (<66 kV), where the dimensions of the supports are conducive to animals simultaneously touching the wires and the support. Other factors affecting electrocution risk are the characteristics of the landscape (topography, vegetation, prey abundance), that of the pylon (cross harm design, material) and weather conditions (external factors), with all of them usually being spatially correlated (Bevanger 1998; Haas 1980; Ferrer et al. 1991; APLIC 1996; Janss and Ferrer 1999, 2001; Mañosa 2001; Lehman et al. 2007). Identifying the most dangerous pylons and correcting or replacing them can significantly reduce the number of fatalities (Tintó et al. 2010; López-López et al. 2011; Guil et al. 2011).

Collisions occur when a flying bird hits any of the wires (conductors or ground wires). Collisions can occur at any type of power line, and even at other lines such as telephone and telegraph wires or railway catenary (Bevanger 1994, 1998). Many studies have reported annual estimates of bird mortality due to collisions with power lines, and extrapolations from these studies produce estimates ranging from hundreds of thousands to millions of dead individuals (Manville 2005; Rioux et al. 2013; Loss et al. 2014). Although estimates may be biased upward due to the lack of random selection of sampling sites (Bevanger 1999; Jenkins et al. 2010), there is a general consensus that this impact is one of the main causes behind the population declines of some endangered species either locally or globally (Bevanger 1998; APLIC 2012). This is the case for the Whooping crane (*Grus americana*) and the

California condor (*Gymnogyps californianus*) in the USA and the Great bustard (*Otis tarda*) and the Little bustard (*Tetrax tetrax*) in the Iberian peninsula (BirdLife International 2004; Silva et al. 2010; APLIC 2012). More than 350 bird species are considered susceptible to collisions with power lines (Manville 2005; Prinsen et al. 2011a), including more than 50 % of the Spanish (Pérez-García and Botella 2012) and 17 % of the Italian breeding bird species (Rubolini et al. 2005).

Mortality rates due to power line collisions depend on biological, environmental, and engineering-related factors (Loss et al. 2014). Collision vulnerability varies between species due to several biological traits such as size, wing loading, flocking and flight behavior, habitat use, maneuverability in flight and vision. Species with high wing loading such as herons, cranes, swans, vultures and condors tend to be more frequently reported in collision casualties (APLIC 2012, and references therein). Flocking species, like waterfowl, and colonial species that move daily between resting or breeding sites to foraging areas are more vulnerable than solitary ones because these individuals have less space to maneuver and limited vision of the obstacle (Bevanger 1998; Janss 2000; Martin and Shaw 2010; Martin 2011). Vision in bird species and its relationship to collisions have been reviewed by Martin (2011), who suggested that bird collisions may be the result of both visual and perceptual constraints. Environmental factors such as weather conditions and visibility interact with biological characteristics (see above), enhancing risk of collisions. Stormy weather, fog, and wind can alter flight patterns (i.e. altitudes) and affect visual detection, increasing the probability of accidents (APLIC 2012). The location and technical design of power lines are among the most studied factors affecting bird collisions. The presence of shield wires above the conductor phases in transmission lines (> 60 kV) and its smaller diameter increases the risk of collisions and ground wires are involved in most bird fatalities (Bevanger and Brøseth 2001; Prinsen et al. 2011b; APLIC 2012). The location of power lines in areas with high bird abundances (i.e. wetlands) and landscape features can affect the path of bird flight directing them to wires (Janss and Ferrer 2001; Martin and Shaw 2010). Lines crossing migration, frequent displacement routes, or mountain ridges also can increase collision risk.

Mitigation Measures

The design and implementation of mitigation measures to avoid or reduce fatalities with power lines peaked in 1990s (Bevanger 1999; Janss and Ferrer 1999; Lehman et al. 2007). Despite the extensive literature during the last 25 years, and that modification of power lines has proven to be an effective method for reducing mortality in dangerous power poles (Janss and Ferrer 1999; Harness and Garrett 1999; Guil et al. 2011; López-López et al. 2011), power lines still remain an important bird mortality source worldwide (Bayle 1999; Rubolini et al. 2005; Lehman et al. 2007). The first step to reduce the negative effect of power line installation should be an efficient planning of electric transmission and distribution, in order to minimize the extension of the actual and the future electric network. This can be achieved through a spatial aggregation of distribution and transmission lines or by bringing power generation closer to users (Prinsen et al. 2011a, b). The burial of the lines is the most effective solution to prevent the majority of the impacts of power lines on biodiversity and is the safest modification for birds. In fact, it is the only measure which eliminates the risk of electrocution and collision (APLIC 2006). But unfortunately, the economic cost is 3–20 higher than traditional overhead lines (APLIC 1994; Prinsen et al. 2011a, b) and can only be performed under certain conditions (e.g., low relief, medium voltage lines). Indeed, only in some countries of Central Europe it has been widely implemented as a common practice (Netherland, Belgium, Germany, and Norway).

The most widely used measure to mitigate avian electrocutions are the use of deterrents and modification of the supports, increasing the distance between conductors, and isolating the supports or spreaders to ensure that there is no contact between birds and wires (Harness and Garrett 1999; Haas et al. 2005; APLIC 2006; Prinsen et al. 2011b). Flight diverters (spirals, plates or spheres) attached to the wires are widely used to increase their visibility and reduce collisions. Effectiveness of flight diverters has been evaluated with diverse results, ranging from no reduction in mortality (e.g. Scott et al. 1972; Janss and Ferrer 1998; Anderson 2001) to slight (9 %, Barrientos et al. 2012) and strong reductions (60–80 %, Alonso et al. 1994; Bevanger and Brøseth 2001).

Conclusions

The production and transportation of renewable energy has several environmental impacts, ranging from the population to the ecosystem level (Table 1). Hydroelectric production is the major source of renewable energy worldwide and probably the most impacting one, not only because of its geographical extent but also because it affects ecosystem processes at the large scale. Migratory species like fishes are the most dramatically impacted taxa by hydroelectric infrastructure. Wind energy production is also an emerging source of environmental impact at both local and regional scales, with strong effects on certain bird and bat populations. Solar facilities impact mostly at the local scale through habitat alteration, although their effects on wildlife have rarely been studied. All these sources of energy share the need for transportation by means of power lines that have significant negative effects, particularly on bird populations at local and regional scales.

Fighting climate change is one of the major challenges of contemporary society and renewable energies are a key instrument to reduce greenhouse emissions. However, the greener energy is the one that it is not consumed, so reducing energy consumption should be the highest priority to minimize the effects of energy production on ecosystems and wildlife. Nevertheless, the increasing demand and even the need to turn from conventional fuel-dependent to renewable energies require the understanding of the potential effects of the latter on the environment. Under this

Table 1 Summary of the main effects of renewable energy on ecosystems and wildlife

Energy source	Ecosystem/habitats	Scale	Processes	Wildlife
Hydroelectric	Freshwater Riverine Coastal	Large Regional Local	Flow regime Barrier Habitat alteration Invasive species Regional climate	Anadromous migratory fish Catadromous fish
Wind	Terrestrial Marine	Regional Local	Habitat alteration Mortality Local climate	Birds (mainly raptors, waders, seabirds, passerines). Bats
Solar	Terrestrial	Local	Habitat alteration Mortality	Birds (mainly passerines)
Transmission	Terrestrial	Regional Local	Habitat alteration Mortality	Birds (mainly raptors, steppe birds, storks)

scenario, we consider that reducing energy consumption, planning infrastructures, and adopting mitigation measures should be, in that order, the key strategies to minimize the effects of renewable energy production and transportation. We would also recommend improving research on the emerging wind and solar facilities through more comprehensive assessments that require large spatio-temporal data sets. The scientific evidence of the long-term effects of hydroelectric production on species, populations, and ecosystems might help to visualize the potential effects of the other emerging renewable energy sources.

Acknowledgements MC and VH had Ramón y Cajal contracts funded by the Spanish Ministry of Science and Innovation. JAD, MAL, MJP, SSN and CI were funded by project PI-57055 Junta de Andalucía. FB, JP & JASZ were funded by Generalitat Valenciana. We are thankful to Dr. Juan Jiménez for his support.

References

Alonso JC, Alonso JA, Muñoz-Pulido R (1994) Mitigation of bird collisions with transmission lines through groundwire marking. Biol Conserv 67:129–134

Aman MM, Solangi KH, Hossain MS et al (2015) A review of Safety, Health and Environmental (SHE) issues of solar energy system. Renew Sustain Energy Rev 41:1190–1204

Amorim F, Rebelo H, Rodrigues L (2012) Factors influencing bat activity and mortality at a wind farm in the Mediterranean region. Acta Chiropterologica 14(2):439–457

Anderson MD (2001) The effectiveness of two different marking devices to reduce large terrestrial bird collisions with overhead electricity cables in the eastern Karoo, South Africa. Draft report to Eskom Resources and Strategy Division. Johannesburg, South Africa

Angelov I, Hashim I, Oppel S (2011) Persistent electrocution mortality of Egyptian Vultures Neophron percnopterus over 28 years in East Africa. Bird Conserv Int 23(1):1–6

APLIC (Avian Power Line Interaction Committee) (1994) Mitigating bird collisions with power lines: The state of the art in 1994. Edison Electric Institute, Washington, D.C

APLIC (Avian Power Line Interaction Committee) (1994) Suggested practices for raptor protection on power lines: the state of the art in 1996. Edison Electric Institute. Washington, DC

APLIC (Avian Power Line Interaction Committee) (2006) Suggested practices for Avian protection on power lines: the state of the art in 2006. Edison Electric Institute, APLIC/California Energy Commission, Washington, DC/Sacramento

APLIC (Avian Power Line Interaction Committee) (2012) Reducing Avian collisions with power lines: the state of the art in 2012. Edison Electric Institute and APLIC, Washington, DC

Armstrong A, Waldron S, Whitaker J, Ostle NJ (2014) Wind farm and solar park effects on plant-soil carbon cycling: uncertain impacts of changes in ground-level microclimate. Glob Chang Biol 20:1699–1706

Arnett EB, Brown WK, Erickson WP et al (2008) Patterns of bat fatalities at wind energy facilities in North America. J Wildl Manag 72:61–78

Arnett EB, Huso MM, Schirmacher MR, Hayes JP (2011) Altering turbine speed reduces bat mortality at wind-energy facilities. Front Ecol Environ 9(4):209–214

Arnett EB, Hein CD, Schirmacher MR et al (2013) Evaluating the effectiveness of an ultrasonic acoustic deterrent for reducing bat fatalities at wind turbines. PLoS One 8(6), e65794

Arriaza M, Cañas-Ortega JF, Cañas-Madueño JA, Ruiz-Aviles P (2004) Assessing the visual quality of rural landscape. Landsc Urban Plan 69(1):115–125

Askins RA, Folsom-O'Keefe CM, Hardy MC (2012) Effects of vegetation, corridor width and regional land use on early successional birds on powerline corridors. PLoS One 7(2), e31520

Asmus P (2005) Wind and wings: the environmental impact of wind power. Electr Perspect 30(3):68–80

Baerwald EF, D'Amours GH, Klug BJ, Barclay RM (2008) Barotrauma is a significant cause of bat fatalities at wind turbines. Curr Biol 18(16):R695–R696

Baerwald EF, Edworthy J, Holder M, Barclay RM (2009) A large-scale mitigation experiment to reduce bat fatalities at wind energy facilities. J Wildl Manag 73(7):1077–1081

Baidya RS (2011) Simulating impacts of wind farms on local hydrometeorology. J Wind Eng Ind Aerodyn 99:491–498

Baidya RS, Traiteur JJ (2010) Impacts of wind farms on surface air temperatures. Proc Natl Acad Sci 107:15679–15684

Baidya RS, Pacala SW, Walko RL (2004) Can large wind farms affect local meteorology? J Geophys Res 109(D19):1–6

Barclay RMR, Harder LD (2003) Life histories of bats: life in the slow lane. In: Kunz TH, Parsons S (eds) Bat ecology. University of Chicago Press, Chicago, pp 209–253

Barclay RMR, Baerwald EL, Gruver JC (2007) Variation in bat and bird fatalities at wind energy facilities: assessing the effects of rotor size and tower height. Can J Zool 85:381–387

Barnosky AD, Hadly EA, Bascompte J et al (2012) Approaching a state shift in Earth's biosphere. Nature 486:56–58

Barrett SB, DeVita PM (2011) Investigating safety impacts of energy technologies on airports and aviation. Airport Cooperative Research Program synthesis 28. Transportation Research Board of the National Academies, Washington, DC

Barrientos R, Ponce C, Palacín C et al (2012) Wire marking results in a small but significant reduction in avian mortality at power lines: a BACI designed study. PLoS One 7(3), e32569

Barrios L, Rodriguez A (2004) Behavioural and environmental correlates of soaring-bird mortality at on-shore wind turbines. J Appl Ecol 41:72–81

Bartzke GS, May R, Bevanger K et al (2014) The effects of power lines on ungulates and implications for power line routing and rights-of-way management. Int J Biodivers Conserv 6(9):647–662

Bastviken D, Tranvik LJ, Downing JA et al (2011) Freshwater methane emissions offset the continental carbon sink. Science 331(6013):50

Baxter RM (1977) Environmental effects of dams and impoundments. Annu Rev Ecol Syst 8:255–283
Bayle P (1999) Preventing birds of prey problems at transmission lines in Western Europe. J Raptor Res 33:43–48
Bellebaum J, Korner-Niegevergelt F, Mammen DU (2012) Wind turbine fatalities approach a level of concern in a raptor population. J Nat Conserv 21:394–400
Bennett VJ, Hale AM (2014) Red aviation lights on wind turbines do not increase bat–turbine collisions. Anim Conserv 17(4):354–358
Bevanger K (1994) Bird interactions with utility structures: collision and electrocution, causes and mitigation measures. Ibis 136:412–425
Bevanger K (1998) Biological and conservation aspects of bird mortality caused by electricity power lines: a review. Biol Conserv 86:67–76
Bevanger K (1999) Estimating bird mortality caused by collision and electrocution with power lines; a review of methodology. In: Ferrer M, Janss GFE (eds) Birds and power lines: collision, electrocution, and breeding. Quercus, Madrid
Bevanger K, Brøseth H (2001) Bird collisions with power lines—an experiment with ptarmigan (lagopus sp.). Biol Conserv 99(3):341–346
Bevelhimer MS, McManamay RA, O'Connor B (2015) Characterizing sub-daily flow regimes: implications of hydrologic resolution on ecohydrology studies. River Res Appl 31:867–879. doi:10.1002/rra.2781
Bird Life International (2004) Birds in the European Union: a status assessment. Bird Life International, Wageningen
Blackwell BF, Seamans TW, Schmidt PM et al (2013) A framework for managing airport grasslands and birds amidst conflicting priorities. Ibis 155:18–193
Brown JJ, Limburg KE, Waldman JR et al (2013) Fish and hydropower on the US Atlantic coast: failed fisheries policies from half-way technologies. Conserv Lett 6(4):280–286
Čada GF (2001) The development of advanced hydroelectric turbines to improve fish passage survival. Fisheries 26(9):14–23
Calles O, Olsson IC, Comoglio C et al (2010) Size-dependent mortality of migratory silver eels at a hydropower plant, and implications for escapement to the sea. Freshw Biol 55(10):2167–2180
Calvert AM, Bishop CA, Elliot RD (2013) A synthesis of human-related avian mortality in Canada. Avian Conserv Ecol 8:11
Camina Á (2012) Bat fatalities at wind farms in northern Spain-lessons to be learned. Acta Chiropterologica 14(1):205–212
Carpenter SR, Stanley EH, Vander-Zanden MJ (2011) State of the world's freshwater ecosystems: physical, chemical, and biological changes. Annu Rev Environ Resour 36:75–99
Carrete M, Sánchez-Zapata JA, Benítez JR et al (2009) Large scale risk-assessment of wind-farms on population viability of a globally endangered long-lived raptor. Biol Conserv 142:2954–2961
Carrete M, Sánchez-Zapata JA, Benítez JR et al (2012) Mortality at wind farms is positively related to bird abundances. Biol Conserv 145:102–108
Clavero M, Hermoso V (2011) Reservoirs promote the taxonomic homogenization of fish communities within river basins. Biodivers Conserv 20(1):41–57
Clavero M, Hermoso V, Aparicio E, Godinho FN (2013) Biodiversity in heavily modified water-bodies: native and introduced fish in Iberian reservoirs. Freshw Biol 58(6):1190–1201
Cooney PB, Kwak TJ (2013) Spatial extent and dynamics of dam impacts on tropical island freshwater fish assemblages. Bioscience 63(3):176–190
Cramer W, Bondeau A, Woodward FI et al (2001) Global response of terrestrial ecosystem structure and function to CO2 and climate change: results from six dynamic global vegetation models. Glob Chang Biol 7:357–373
Cryan PM, Barclay RMR (2009) Causes of bat fatalities at wind turbines: hypotheses and predictions. J Mammal 90:1330–1340
Cryan PM, Gorresen PM, Hein CD et al (2014) Behavior of bats at wind turbines. Proc Natl Acad Sci 111(42):15126–15131

Cushman RM (1985) Review of ecological effects of rapidly varying flows downstream from hydroelectric facilities. N Am J Fish Manag 5(3A):330–339
de Lucas M, Janss GFE, Whitfield DP, Ferrer M (2008) Collision fatality of raptors in wind farms does not depend on raptor abundance. J Appl Ecol 45:1695–1703
Desholm M, Kahlert J (2005) Avian collision risk at an offshore wind farm. Biol Lett 1:296–298
DeVault TL (2015) Reprioritizing avian conservation efforts. Human-Wildlife Interact 9:148–149
DeVault TL, Belant JL, Blackwell BF et al (2012) Airports offer unrealized potential for alternative energy production. Environ Manage 49:517–522
DeVault TL, Blackwell BF, Belant JL (eds) (2013) Wildlife in airport environments: preventing animal–aircraft collisions through science-based management. Johns Hopkins University Press, Baltimore
DeVault TL, Seamans TW, Schmidt JA, Belant JL et al (2014) Bird use of solar photovoltaic installations at US airports: implications for aviation safety. Landsc Urban Plan 122:122–128
Dincer I (2000) Renewable energy and sustainable development: a crucial review. Renew Sustain Energy Rev 4(2):157–175
Donázar JA, Palacios CJ, Gangoso L et al (2002) Conservation status and limiting factors of the endangered population of Egyptian Vulture (Neophron percnopterus) in the Canary Islands. Biol Conserv 107:89–98
Doty AC, Martin AP (2013) Assessment of bat and avian mortality at a pilot wind turbine at Coega, Port Elizabeth, Eastern Cape, South Africa. New Zealand. J Zool 40:75–80
Drewitt AL, Langston RHW (2006) Assessing the impacts of wind farms on birds. Ibis 148:29–42
Drewitt AL, Langston RHW (2008) Collision effects of wind-power generators and other obstacles on birds. Ann N Y Acad Sci 1134:233–266
Dudgeon D, Arthington AH, Gessner MO et al (2006) Freshwater biodiversity: importance, threats, status and conservation challenges. Biol Rev 81(2):163–182
Erickson WP, Johnson GD, Strickland MD et al (2001) Avian collision with wind turbines: a summary of existing studies and comparisons to other sources of avian collision mortality in the United States. National Wind Coordinating Committee (NWCC) Resource Department, Washington, DC
Everaert J, Stienen EWM (2007) Impact of wind turbines on birds in Zeebrugge (Belgium): significant effect on breeding tern colony due to collisions. Biodivers Conserv 16:3345–3359
Fearnside PM (2000) Greenhouse gas emissions from land-use changes in Brazil's Amazon. In: Lal R, Kimble JR, Stewart BA (eds) Global climate change and tropical ecosystems: advances in soil science. CRC Press, Boca Raton, pp 231–249
Fearnside PM (2004) Greenhouse gas emissions from hydroelectric dams: controversies provide a springboard for rethinking a supposedly 'clean' energy source. An editorial comment. Clim Change 66(1):1–8
Ferrer M, de la Riva M, Castroviejo J (1991) Electrocution of raptors on power lines in Southern Spain. J Field Ornithol 62:54–69
Ferrer M, de Lucas M, Janss GFE et al (2011) Weak relationship between risk assessment studies and recorded mortality in wind farms. J Appl Ecol 49:38–46
Feunteun E (2002) Management and restoration of European eel population (Anguilla anguilla): an impossible bargain. Ecol Eng 18(5):575–591
Fiedler BH, Bukovsky MS (2011) The effect of a giant wind farm on precipitation in a regional climate model. Environ Res Lett 6:045101
FitzHugh TW, Vogel RM (2011) The impact of dams on flood flows in the United States. River Res Appl 27(10):1192–1215
Foley JD, Clifford K, Castle P et al (2011) Investigating and managing the rapid emergence of white-nose syndrome, a novel, fatal, infectious disease of hibernating bats. Conserv Biol 25:223–231
Forrester JA, Leopold DJ, Hafner SD (2005) Maintaining critical habitat in a heavily managed landscape: effects of power line corridor management on Karner blue butterfly (lycaeidesmelissasamuelis) habitat. Restor Ecol 13:488–498

Gangoso L, Palacios CJ (2002) Endangered Egyptian Vulture (Neophron percnopterus) entangled in a power line ground-wire stabilizer. J Raptor Res 36(3):238–239
Garvin JC, Jennelle CS, Drake D, Grodsky SM (2011) Response of raptors to a windfarm. J Appl Ecol 48:199–209
Georgiakakis P, Kret E, Cárcamo B et al (2012) Bat fatalities at wind farms in north-eastern Greece. Acta Chiropterologica 14(2):459–468
Gido KB, Schaefer JF, Falke JA (2009) Convergence of fish communities from the littoral zone of reservoirs. Freshw Biol 54(6):1163–1177
Godinho AL, Kynard B (2009) Migratory fishes of Brazil: life history and fish passage needs. River Res Appl 25(6):702–712
González LM, Margalida A, Mañosa S et al (2007) Causes and spatio-temporal variations of non-natural mortality in the vulnerable Spanish Imperial Eagle (Aquila adalberti) during a recovery period. Oryx 41:495–502
Gosset C, Travade F, Durif C et al (2005) Tests of two types of bypass for downstream migration of eels at a small hydroelectric power plant. River Res Appl 21(10):1095–1105
Guil F, Fernández-Olalla M, Moreno-Opo R et al (2011) Minimising mortality in endangered raptors due to power lines: the importance of spatial aggregation to optimize application of mitigation measures. PLoS One 6, e28212
GWEC (2014) Global wind energy outlook. Greenpeace & GWEC, Amsterdam
Haas D (1980) Endangerment of our large birds by electrocution – a documentation. Okölogia der vöger 2:7–57
Haas D, Nipkow M, Fiedler G et al (2005) Protecting birds from powerlines. Council of Europe Publishing, Strasbourg
Hall CJ, Jordaan A, Frisk MG (2012) Centuries of anadromous forage fish loss: consequences for ecosystem connectivity and productivity. Bioscience 62(8):723–731
Harness RE, Garrett M (1999) Effectiveness of perch guards to prevent raptor electrocutions. J Colorado Field Ornithol 33:215–220
Harness RE, Wilson KR (2001) Electric-utility structures associated with raptor electrocutions in rural areas. Wildl Soc Bull 29(2):612–623
Havel JE, Lee CE, Vander Zanden JM (2005) Do reservoirs facilitate invasions into landscapes? Bioscience 55(6):518–525
Hawkins E, Sutton R (2011) The potential to narrow uncertainty in projections of regional precipitation change. Climatic Dynam 37:407–418
Hayes MA (2013) Bats killed in large numbers at United States wind energy facilities. Bioscience 63(12):975–979
Hermoso V, Clavero M, Blanco-Garrido F, Prenda J (2011) Invasive species and habitat degradation in Iberian streams: an analysis of their role in freshwater fish diversity loss. Ecol Appl 21(1):175–188
Hernandez RR, Easter SB, Murphy-Mariscal ML et al (2014) Environmental impacts of utility-scale solar energy. Renew Sustain Energy Rev 29:766–779
Hill NM, Keddy PA, Wisheu IC (1998) A hydrological model for predicting the effects of dams on the shoreline vegetation of lakes and reservoirs. Environ Manage 22(5):723–736
Holmquist JG, Schmidt-Gengenbach JM, Yoshioka BB (1998) High dams and marine-freshwater linkages: effects on native and introduced fauna in the Caribbean. Conserv Biol 12(3):621–630
Hooper DU, Adair EC, Cardinale BJ et al (2012) A global synthesis reveals biodiversity loss as a major driver of ecosystem change. Nature 486:105–108
Hull CL, Cawthen L (2013) Bat fatalities at two wind farms in Tasmania: bat characteristics, and spatial and temporal patterns. New Zealand J Zool 40:5–15
ICOLD, International Commission on Large Dams (2014) Dams general synthesis. http://www.icold-cigb.org/ accessed October 1st 2014
Jacobson MZ (2009) Review of solutions to global warming, air pollution, and energy security. Energy Environ Sci 2:148–173

Jacobson MZ, Delucchi MA (2011) Providing all global energy with wind, water, and solar power, part I: technologies, energy resources, quantities and areas of infrastructure, and materials. Energy Policy 39:1154–1169

Jameson JW, Willis CK (2014) Activity of tree bats at anthropogenic tall structures: implications for mortality of bats at wind turbines. Anim Behav 97:145–152

Janss GFE (2000) Avian mortality from power lines: a morphologic approach of a species-specific mortality. Biol Conserv 95:353–359

Janss GFE, Ferrer M (1998) Rate of bird collision with power lines: effects of conductor-marking and static wire-marking. J Field Ornithol 69:8–17

Janss GFE, Ferrer M (1999) Mitigation of raptor electrocution on steel power poles. Wildl Soc Bull 27:263–273

Janss GFE, Ferrer M (2001) Avian electrocution mortality in relation to pole design and adjacent habitat in Spain. Bird Conserv Int 11:3–12

Jenkins A, Smallie J, Diamond M (2010) Avian Collisions with Power Lines: A Global Review of Causes and Mitigation with a South African Perspective. Bird Cons Int 20: 263–278

Johansson TB, Kelly H, Reddy AKN, Williams RH (1993) Renewable energy. Island Press, Washington

Johnson GD, Perlik MK, Erickson WP, Strickland MD (2004) Bat activity, composition, and collision mortality at a large wind plant in Minnesota. Wildl Soc Bull 32(4):1278–1288

Johnson PT, Olden JD, Vander Zanden MJ (2008) Dam invaders: impoundments facilitate biological invasions into freshwaters. Front Ecol Environ 6(7):357–363

Jurajda P, Roux AL, Olivier JM (1995) 0-fish assemblages in a sector of the Rhôcneriver influenced by the Bregniér–Cordon hydroelectric scheme. Regul Rivers: Res Manage 10(2-4):363–372

Katopodis C, Williams JG (2012) The development of fish passage research in a historical context. Ecol Eng 48:8–18

Kemenes A, Forsberg BR, Melack JM (2007) Methane release below atropical hydroelectric dam. Geophys Res Lett 34, L12809

Kettle AJ, AsbjørnVøllestad L, Wibig J (2011) Where once the eel and the elephant were together: decline of the European eel because of changing hydrology in southwest Europe and northwest Africa? Fish Fish 12(4):380–411

Kosnik L (2008) The potential of water power in the fight against global warming in the US. Energy Policy 36(9):3252–3265

Kuenzer C, Campbell I, Roch M et al (2013) Understanding the impact of hydropower developments in the context of upstream–downstream relations in the Mekong river basin. Sustain Sci 8(4):565–584

Kunz TH, Arnett EB, Erickson WP et al (2007) Ecological impacts of wind energy development on bats: questions, research needs, and hypotheses. Front Ecol Environ 5:315–324

Kuvlesky WP, Brennan L, Morrison M et al (2007) Wind energy development and wildlife conservation: challenges and opportunities. J Wildl Manag 71:2487–2498

Laine A, Jokivirta T, Katopodis C (2002) Atlantic salmon, Salmosalar L, and sea trout, Salmotrutta L, passage in a regulated northern river–fishway efficiency, fish entrance and environmental factors. Fish Manag Ecol 9(2):65–77

Langston RHW, Pullan JD (2003) Windfarms and birds: an analysis of the effects of wind farms on birds, and guidance on environmental assessment criteria and site selection issues. Report T-PVS/Inf 12, by BirdLife International to the Council of Europe, Bern Convention on the Conservation of European Wildlife and Natural Habitats. RSPB/BirdLife in the UK

Larinier M, Travade F (2002) Downstream migration: problems and facilities. Bulletin Francaise de la Pecheet de la Pisciculture 364:181–207

Larsen JK, Guillemette M (2007) Effects of wind turbines on flight behaviour of wintering common eiders: implications for habitat use and collision risk. J Appl Ecol 44:516–522

Ledger JA, Hobbs JCA (1999) Raptor use and abuse of powerlines in Southern Africa. J Raptor Res 33:49–5

Lehman RN, Kennedy PL, Savidge SA (2007) The state of the art in raptor electrocution research: a global review. Biol Conserv 136:159–174
Lehnert LS, Kramer-Schadt S, Schönborn S et al (2014) Wind farm facilities in Germany kill noctule bats from near and far. PLos One 9(8):e103–e106
Leshem Y (1985) Griffon Vultures in Israel: electrocution and other reasons for a declining population. Vulture News 13:14–20
Leung DYC, Yang Y (2012) Wind energy development and its environmental impact: a review. Renew Sustain Energy Rev 16:1031–1039
Limburg KE, Waldman JR (2009) Dramatic declines in North Atlantic diadromous fishes. Bioscience 59(11):955–965
López-López P, Ferrer M, Madero A et al (2011) Solving man-induced large-scale conservation problems: the Spanish imperial eagle and power lines. PLoS One 6:e17196. doi:10.1371/journalpone0017196
Loss SR, Will T, Marra PP (2013) Estimates of bird collision mortality at wind facilities in the contiguous United States. Biol Conserv 168:201–209
Loss SR, Will T, Marra PP (2014) Refining estimates of bird collision and electrocution mortality at power lines in the United States. PLoS One 9(7), e101565. doi:10.1371/journal.pone.0101565
Lovich JE, Ennen JR (2011) Wildlife conservation and solar energy development in the desert southwest, United States. Bioscience 61(2):982–992
Lowther S (1998) The European perspective: some lessons from case studies. Proceedings of the National Avian Wind Power Planning Meeting III, San Diego, May 1998 (LGL Limited, King City, June 2000), pp 115–123
Luken JO, Hinton AC, Baker DG (1992) Response of woody plant communities in power-line corridors to frequent anthropogenic disturbance. Ecol Appl 2(4):356–362
MacIsaac HJ, Herborg LM, Muirhead JR (2007) Modeling biological invasions of inland waters. In: Gherardi F (ed) Biological invaders in inland waters: profiles, distribution, and threats. Springer, Netherlands, pp 347–368
Mallen-Cooper M, Brand DA (2007) Non-salmonids in a salmonidfishway: what do 50 years of data tell us about past and future fish passage? Fish Manag Ecol 14(5):319–332
Mañosa S (2001) Strategies to identify dangerous electricity pylons for birds. Biodivers Conserv 10:1997–2012
Manville AM (2005) Bird strikes and electrocutions at power lines, communication towers, and wind turbines: state of the art and state of the science – next steps toward mitigation. USDA Forest Service General Technical Report PSW-GTR-191, pp 1051–1064
Marqués AT, Batalhaa H, Rodrigues S et al (2014) Understanding bird collisions at wind farms: an updated review on the causes and possible mitigation strategies. Biol Conserv 179:40–52
Marris E, Fairless D (2004) Wind farms' deadly reputation hard to shift. Nature 447:126
Martin GR (2011) Understanding bird collisions with man-made objects: a sensory ecology approach. Ibis 153:239–254
Martin GR, Shaw JM (2010) Bird collisions with power lines: failing to see the way ahead? Biol Conserv 143(11):2695–2702
McCartney M (2009) Living with dams: managing the environmental impacts. Water Policy 11(1):121–139
McCrary MD, McKernan RL, Schreiber RL et al (1986) Avian mortality at a solar energy power plant. J Field Ornithol 57(2):135–141
Musters CJM, Noordervliet MAW, ter Keurs WJ (1996) Bird casualties caused by awind energy project in an estuary. Bird Study 43:124–126
Nellemann C, Vistnes I, Jordhøy P et al (2003) Progressive impact of piecemeal infrastructure development on wild reindeer. Biol Conserv 113(2):307–317
Nilsson C (2009) Reservoirs. In: Gene E Likens (ed) Encyclopedia of inland waters. Elsevier, pp 625–633
Nilsson C, Dynesius M (1994) Ecological effects of river regulation on mammals and birds: a review. Regul Rivers: Res Manage 9(1):45–53

Nilsson C, Berggren K (2000) Alterations of riparian ecosystems caused by river regulation. Bioscience 50(9):783–792
Nilsson C, Reidy CA, Dynesius M, Revenga C (2005) Fragmentation and flow regulation of the world's large river systems. Science 308(5720):405–408
Osborn RG, Higgins K, Usgaard R et al (2000) Bird mortality associated with wind turbines at the buffalo ridge wind resource area, Minnesota. Am Midl Nat 143:41–52
Palmeirim AF, Peres CA, Rosas FC (2014) Giant otter population responses to habitat expansion and degradation induced by a mega hydroelectric dam. Biol Conserv 174:30–38
Pearce-Higgins JW, Stephen L, Langston RH et al (2009) The distribution of breeding birds around upland wind farms. J Appl Ecol 46:1323–1331
Pearce-Higgins JW, Stephen L, Douse A, Langston RH (2012) Greater impacts of wind farms on bird populations during construction than subsequent operation: results of a multi-site and multi-species analysis. J Appl Ecol 49:386–394
Pelicice FM, Pompeu PS, Agostinho AA (2014) Large reservoirs as ecological barriers to downstream movements of Neotropical migratory fish. Fish Fish 16:697–715. doi:10.1111/faf.12089
Pérez-García JM, Botella F (2012) Modelo de zonas de riesgo para la colisión de la avifauna en líneas eléctricas de transporte en la Comunidad Valenciana, Red Eléctrica de España (Inédito)
Phernie K, Bird DM, Dawson RD, Lague PC (2000) Effects of electromagnetic fields on the reproductive success of American kestrels. Physiol Biochem Zool 73:60–65
Piorkowski MD, O'Connell TJ (2010) Spatial pattern of summer bat mortality from collisions with wind turbines in mixed-grass prairie. Am Midl Nat 164(2):260–269
Pisinaras V, Wei Y, Bärring L, Gemitzi A (2014) Conceptualizing and assessing the effects of installation and operation of photovoltaic power plants on major hydrologic budget constituents. Sci Total Environ 493:239–250
Poff NL, Allan JD, Bain MB et al (1997) The natural flow regime. Bioscience 47:769–784
Poff NL, Olden JD, Merritt DM, Pepin DM (2007) Homogenization of regional river dynamics by dams and global biodiversity implications. Proc Natl Acad Sci 104(14):5732–5737
Prenda J, López-Nieves P, Bravo R (2001) Conservation of otter (Lutra lutra) in a Mediterranean area: the importance of habitat quality and temporal variation in water availability. Aquat Conserv Mar Freshwat Ecosyst 11(5):343–355
Prinsen HAM, Boere GC, Píres N, Smallie JJ (Compilers) (2011a) Review of the conflict between migratory birds and electricity power grids in the African-Eurasian region. CMS Technical Series No 20, AEWA, Bonn
Prinsen HAM, Smallie JJ, Boere GC, Píres N (Compilers) (2011b) Guidelines on how to avoid or mitigate impact of electricity power grids on migratory birds in the African-Eurasian region. CMS Technical Series No 20, AEWA Technical Series No 20, Bonn
Real J, Grande JM, Mañosa S, Sánchez-Zapata JA (2001) Causes of death in different areas for Bonelli's Eagle (Hieraaetus fasciatus) in Spain. Bird Study 48:221–228
Reimers E, Dahle B, Eftestøl S et al (2007) Effects of a power line on migration and range use of wild reindeer. Biol Conserv 134(4):484–494
REN21 (2013) Renewables 2013 global status report. Worldwatch Institute, Paris
Richter BD, Thomas GA (2007) Restoring environmental flows by modifying dam operations. Ecol Soc 12(1):12
Rincón PA, Velasco JC, Gonzalez-Sanchez N, Pollo C (1990) Fish assemblages in small streams in western Spain: the influence of an introduced predator. ArchivfürHydrobiologie 118(1):81–91
Rioux S, Savard JPL, Gerick AA (2013) Avian mortalities due to transmission line collisions: a review of current estimates and field methods with an emphasis on applications to the Canadian electric network. Avian Conserv Ecol 8(2): art 7
Roberts TR (2001) On the river of no returns: Thailand's Pak Mun Dam and its fish ladder. Nat Hist Bull Siam Soc 49:189–230
Rubolini D, Basso E, Bogliani G, Galeotti P, Garavaglia R (2001) Eagle Owl Bubo bubo and power line interactions in the Italian Alps. Bird Conservation International 11:319–324

Rubolini D, Gustin M, Bogliani G, Garavaglia R (2005) Birds and powerlines in Italy: an assessment. Bird Conserv Int 15(02):131–145
Rydell J, Bach L, Doubourg-Savage MJ et al (2010) Mortality of bats at wind turbines links to nocturnal insect migration? Eur J Wild Res 56:823–827
Saether BE, Bakke Ø (2000) Avian life history variation and contribution of demographic traits to the population growth rate. Ecology 81:642–653
Saidur R, Rahim NA, Islam MR, Solangi KH (2011) Environmental impact of wind energy. Renew Sustain Energy Rev 15:2423–1430
Sanz-Aguilar A, Sánchez-Zapata JA, Carrete M et al (2015) Action on multiple fronts, illegal poisoning and windfarm planning, is required to reverse the decline of the Egyptian vulture in southern Spain. Biol Conserv 187:10–18
Schaub M, Aebischer A, Gimenez O et al (2010) Massive immigration balances high anthropogenic mortality in a stable eagle owl population: lessons for conservation. Biol Conserv 143(8):1911–1918
Scherba A, Sailor DJ, Rosenstiel TN, Wamser CC (2011) Modeling impacts of roof reflectivity, integrated photovoltaic panels and green roof systems on sensible heat flux into the urban environment. Build Environ 46:2542–2551
Scott RE, Roberts LJ, Cadbury CJ (1972) Bird deaths from power lines at Dungeness. British Birds 65(7):273–286
Scudder T (2005) The future of large dams: dealing with social, environmental, institutional and political costs. Earthscan, London
Sergio F, Marchesi L, Pedrini P et al (2004) Electrocution alters the distribution and density of a top predator, the Eagle owl Bubo bubo. J Appl Ecol 41:836–845
Silva JP, Santos M, Queirós L et al (2010) Estimating the influence of overhead transmission power lines and landscape context on the density of little bustard Tetrax tetrax breeding populations. Ecol Model 221(16):1954–1963
Silvano RAM, Juras AA, Begossi A (2009) Clean energy and poor people: ecological impacts of hydroelectric dam on fish and fishermen in the Amazon rainforest. In: V International conference on energy, environmental, ecosystems and sustainable development and II international conference on landscape architecture, WSEAS, Greece, pp 139–47
Smallwood KS (2013) Comparing bird and bat fatality-rate estimates among North American wind-energy projects. Wildl Soc Bull 37:19–33
Smallwood KS, Thelander C (2005) Bird mortality at the altamont pass wind resource area, March 1998–September 2001 final report. National Renewable Energy Laboratory NREL/SR-500-36973, Golden, Colorado
Smallwood KS, Thelander CG (2008) Bird mortality in the Altamont pass wind resource area, California. J Wildl Manag 72:215–223
Sovacool BK (2009) Contextualizing avian mortality: a preliminary appraisal of bird and bat fatalities from wind, fossil-fuel, and nuclear electricity. Energy Policy 37:2241–2248
Stewart GB, Pullin AS, Coles CF (2007) Poor evidence-base for assessment of windfarm impacts on birds. Environ Conserv 34:1–11
Taha H (2013) The potential for air-temperature impact from large-scale deployment of solar photovoltaic arrays in urban areas. Sol Energy 91:358–367
Tellería JL (2009a) Potential impacts of wind-farms on migratory birds crossing Spain. Bird Conserv Int 19:131–136
Tellería JL (2009b) Overlap between wind power plants and Griffon Vultures Gyps fulvus in Spain. Bird Study 56:268–271
Tenforde TS (1992) Biological interactions and potential health effects of extremely-low-frequency magnetic fields from power lines and other common sources. Annu Rev Public Health 13(1):173–196
Thelander CG (2004) Bird fatalities in the Altamont pass wind resource area: a case study, part 1. In: Savitt Schwartz S (ed) Proceedings of the wind energy and birds/bats workshop: understanding and resolving bird and bat impacts, 18–19 May 2004, RESOLVE, Washington, DC, pp 25–28

Thelander CG, Rugge L (2000) Avian risk behavior and fatalities at the Altamont wind resource area. National Technical Information Service U.S., Department of Commerce, Golden, NREL, Colorado, USA
Tintó A, Real J, Mañosa S (2010) Predicting and correcting electrocution of birds in Mediterranean areas. J Wildl Manag 74(8):1852–1862
Tockner K, Stanford JA (2002) Riverine flood plains: present state and future trends. Environ Conserv 29(03):308–330
Tsousos T, Frantzeskaki N, Gekas V (2005) Environmental impacts from the solar energy technologies. Energy Policy 33:289–296
Turney D, Fthenakis V (2011) Environmental impacts from the installation and operation of large-scale solar power plants. Renew Sustain Energy Rev 15:3261–3270
Tyler N, Stokkan KA, Hogg C et al (2014) Ultraviolet vision and avoidance of power lines in birds and mammals. Conserv Biol 28(3):630–631
US GAO (US Government Accountability Office) (2005) Wind power: impacts on wildlife and government responsibilities for regulating development and protecting wildlife. U.S. Government Accountability Office, Washington, DC
Veríssimo D, MacMillan DC, Smith RJ et al (2014) Has climate change taken prominence over biodiversity conservation? Bioscience 64:625–629
Vistnes I, Nellemann C, Jordhoy P, Strand O (2004) Effects of infrastructure on migration and range use of wild reindeer. Journal of Wildlife Management 68:101–108
Vitousek PM, Mooney HA, Lubchenco J, Melillo JM (1997) Human domination of earth's ecosystems. Science 277:494–499
Wang C, Prinn RG (2010) Potential climati c impacts and reliability of very large-scale wind farms. Atmos Chemis Phys 10:2053–2061
Ward JV, Stanford JA (1982) Thermal responses in the evolutionary ecology of aquatic insects. Annu Rev Entomol 27(1):97–117
Wayland M, Wilson LK, Elliot JE et al (2003) Mortality, morbidity, and lead poisoning of eagles in Western Canada, 1986–98. J Raptor Res 37(1):8–18
White GF (1988) The environmental effects of the high dam at Aswan. Environ Sci Policy Sustain Dev 30(7):4–40
Wilhelmsson D, Malm T, Thompson R et al (eds) (2010) Greening blue energy: identifying and managing the biodiversity risks and opportunitties of off shore renewable energy. IUCN, Gland
Wilmsen B, Webber M, Yuefang D (2011) Development for whom? Rural to urban resettlement at the Three Gorges Dam, China. Asian Stud Rev 35(1):21–42
Yewhalaw D, Legesse W, Van Bortel W et al (2009) Malaria and water resource development: the case of Gilgel-Gibe hydroelectric dam in Ethiopia. Malar J 8:21
Zhang X, Gao X, Wang J, Cao W (2015) Extinction risk and conservation priority analyses for 64 endemic fishes in the Upper Yangtze River, China. Environ Biol Fishes 98:261–272
Zhong Y, Power G (1996) Environmental impacts of hydroelectric projects on fish resources in China. Regul Rivers: Res Manage 12(1):81–98
Zhou L, Tian Y, Baidya RS et al (2012) Impacts of wind farms on land surface temperature. Nat Clim Chang 2:539–543

An Overview of Recent Trends in Wildlife Ecotoxicology

Rafael Mateo, Silvia Lacorte, and Mark A. Taggart

Introduction

Advances in chemistry during the twentieth century have facilitated the rapid development of a myriad of new chemicals with applications as varied as the needs of humans (Chalew and Halden 2009). However, this ability to quickly synthesize new molecules comes with an inherent risk – that evolutionary mechanisms simply cannot adapt to these new (and potentially toxic) chemical substances quickly enough (Bourret et al. 2008; Whitehead et al. 2012). Thus, some xenobiotics (e.g., novel halogenated organic compounds) have been very challenging for life on earth, since biota often lack the necessary mechanisms for their detoxification, which in turn, may lead to bioaccumulation and biomagnification along food chains. However, it is not just newly synthesised chemicals that represent a hazard to life. A range of human activities also favour the redistribution of naturally occurring toxic substances (such as heavy metals); introduce invasive species that produce toxins (Southard et al. 2010; Bodkin et al. 2012); or, promote favourable environmental conditions for toxins produced by large-scale blooms of algae or bacteria (Watanabe et al. 2011; Anza et al. 2014) – all of which can threaten ecosystems/populations.

R. Mateo (✉)
Group of Wildlife Toxicology, Instituto de Investigación en Recursos Cinegéticos (IREC) CSIC-UCLM-JCCM,
Ronda de Toledo s/n, Ciudad Real 13071, Spain
e-mail: Rafael.mateo@uclm.es

S. Lacorte
Department of Environmental Chemistry, IDAEA-CSIC,
Jordi Girona 18-26, Barcelona, Catalonia 08034, Spain

M.A. Taggart
Department of Environmental Contamination and Ecological Health,
Environmental Research Institute, University of the Highlands and Islands,
Castle St, Thurso, Caithness KW14 7JD, UK

R. Mateo et al. (eds.), *Current Trends in Wildlife Research*,
Wildlife Research Monographs 1, DOI 10.1007/978-3-319-27912-1_6

Many such hazards have been well documented during our recent history. Global change driven by humans and rapid economic development on earth currently put the conservation and preservation of biodiversity at significant risk and may even jeopardise long term human survival. As humans, our capacity to predict and prevent risk has (at least in part) made us such a successful species, even when many current risks are essentially created by us. The motivation of researchers to work in wildlife ecotoxicology has, since the time of Silent Spring (Carson 1962), largely related to the conservation of higher vertebrates (initially birds and mammals and more recently amphibians and reptiles). However, by seeking to identify health risks in vertebrates that are closely related to us, potential hazards to humans are also sometimes highlighted. In fact, by studying higher vertebrates and the chemical risks that could jeopardize the conservation of their biodiversity we have also identified many key "*sentinel species*" which are particularly useful for the study of bioaccumulation of contaminants on biota (Custer 2011; Elliott and Elliott 2013). Studies conducted with (for example) raptors, whales or polar bear, have often helped identify and thus reduce risks in many other species, including humans (Fox 2001; Movalli and Duke 2008; Gómez-Ramírez et al. 2014).

This chapter aims to reflect the latest trends in wildlife ecotoxicology based on original studies published in the last 10 years, with a particular emphasis on those papers that have been or are currently trending in this field. Between 2004 and 2013 there have been about 4000 articles published on wildlife ecotoxicology (based on search terms "wildlife" AND "ecotoxicology" in combination with related terms like toxicology, pollution, birds, wild mammals, game, etc.) covering a very wide range of chemical contaminants and/or impacts. Some relevant articles that appeared between 2014 and 2015 have also been included during the writing of this review. Studies vary from those focused on adverse effects in individual organism's right up to whole ecosystems, and have involved all types of vertebrate wildlife – from fish to mammals.

Over the last decade, new and poorly predicted threats to the conservation of biodiversity have emerged (Oaks et al. 2004), principles for safeguarding environmental health have been identified (Elliott et al. 2011), and new conceptual frameworks, such as the "*adverse outcome pathway*" (AOP) approach, have appeared – aimed at improving and facilitating the structured study of the overall effect of a chemical substance on life (Ankley et al. 2010). Moreover, several high profile cases (as highlighted below) have shown that many improvements must still be made within the regulatory systems that are designed to protect all biota from the potentially harmful effects of chemicals released into our environment. No doubt existing science will serve to reduce some of the risks posed by some chemicals in coming decades, however, many long taught lessons still remain unheeded (i.e., the continued use of Pb ammunition), and new hazards are emerging all the time. Despite >50 years of development of ecotoxicology as a science, many complex challenges still remain – and if these are not addressed in a timely fashion, we will certainly see our global biodiversity continue to decline.

From Forensic Ecotoxicology to Life Cycle Assessment (LCA) of Chemical Products

The identification of agents causing negative impacts on wildlife populations is a complicated task due to the large diversity of factors that can have an effect on the health and the ecology of a species. Elliott et al. (2011) recently defined forensic ecotoxicology as *"...the investigation of causal linkages between source(s) and presence of a chemical or mixture, and biological effects, with the goal of reducing impact via regulatory or non-regulatory interventions..."*. A recent and remarkable example within this discipline was the identification by Oaks et al. (2004) of diclofenac (a non-steroidal anti-inflammatory drug – NSAID) as the driver behind rapid and widespread population declines in Asian vultures in the late twentieth/early twenty-first century. This event has been likened (in terms of scale and species affected) with that of organochlorine (OC) pesticides and egg-shell thinning in the second half of the twentieth century (Elliott et al. 2011). In both cases, the adequacy of existing risk assessment processes has (after the event) been drawn into question. For both organochlorine pesticides and diclofenac, limited knowledge and understanding regarding how these toxicants passed through relevant food chains and behaved in the environment existed before their negative population level impacts were discovered. In other words, an incomplete Life Cycle Assessment (LCA) for these chemicals was in place. That meant the bioaccumulation and biomagnification of organochlorine pesticides and the exposure pathway (through animal carcasses) for diclofenac were both largely unforeseen. The better use of full Life Cycle Assessment "should" help us to improve our ability to predict the potential impact on the environment of products or services, including chemical substances. By better characterising their toxicity to target organisms, their entire life cycle and their potential ecological consequences (van Zelm et al. 2014; Souza et al. 2015), measures can be taken to curtail impacts (i.e., by preventing certain exposure scenarios with inadequately tested new compounds).

From Single Biomarkers to the Adverse Outcome Pathways (AOPs) in Risk Assessment

Although some pollutants are not always purposefully synthesized by humans (e.g., dioxins and furans) or are products of geological or biological processes (e.g., heavy metal ores, petroleum hydrocarbons and bio-toxins), in many instances the chemicals that pose a challenge to life on earth are produced by humans. In this case, the prevention of environmental hazards relies on thorough systems of risk assessment for all new synthetic molecules. Risk assessment commonly requires the development and application of toxicity tests that assess possible impacts on life (Hoffman et al. 2003). Current trends are looking to reduce animal testing and replace this with other types of in vitro assays (Hecker et al. 2007) or in silico assessments (Wang et al. 2006), but, care should be taken that a sufficient range of possible exposure scenarios continue to be considered (i.e., a suitable range of species, exposure levels, additive effects, mixture effects, etc.). Recent examples involving chlorofluorocarbons (CFCs) and their effect on the ozone layer, diclofenac and the near extinction of Asian vultures, and the observed bioaccumulation of halogenated organic molecules in biota (e.g., organochlorine (OC) pesticides, polychlorinated biphenyls (PCBs), brominated flame retardants or perfluorinated compounds (PFCs)) all indicate that recent regulatory systems have failed to predict or prevent the effects of some extremely widely used chemicals (Elliott et al. 2011). The swift adoption of corrective measures have on occasion minimized predicted impacts (Chipperfield et al. 2015), however, effective changes certainly do not always occur in good time (Dullinger et al. 2013).

New advances in genetics and in "omics" (genomics, proteomics, lipidomics, metabolomics, etc.) should allow us to improve our knowledge regarding the mechanisms of action and kinetics of toxicants and may help us explain phylogenetic susceptibility differences among wildlife species (Snape et al. 2004; Karchner et al. 2006; Gunnarsson et al. 2008). In order to create links between knowledge at the molecular level and effects at the ecological level we also need to understand processes at the biochemical and physiological level (e.g., the hypothalamic-pituitary-thyroid axis and its response to certain endocrine disruptors) (Zoeller et al. 2007). Creating this link between different levels of effect for certain chemical substances is in fact the overall purpose of an "*Adverse Outcome Pathway*" (AOP). This concept also requires risk assessment to consider higher organizational levels (i.e., population level impacts), based on adverse toxic effects identified at lower organizational levels (Ankley et al. 2010; Watanabe et al. 2011; Scholz et al. 2013). This requires comprehensive knowledge of the biochemical effects of toxicants (i.e., macromolecular interactions); the cellular and organ responses (e.g., gene expression) to them, the effects on fitness and health of the individual, and the implications for population dynamics. An AOP can therefore link new "omics" based information (toxico-genomics and toxico-proteomics) (Snape et al. 2004; Karchner et al. 2006; Benninghoff 2007; Baker et al. 2009) with the more classic forensic and clinical data (Oaks et al. 2004) and broader ecological approaches and modelling (Lovvorn et al. 2013). Therein, the former "*biomarker approach*" becomes integrated into a wider mechanistic and holistic scheme – the aim being to understand the mode of action by which a toxicant may exert its effects on an entire wildlife population (Ankley et al. 2010) (Fig. 1 provides a working example).

A significant challenge when working in wildlife ecotoxicology is the vast diversity of species under study. Nevertheless, evolutionary conservation of genes associated with specific chemical targets can mean that phylogenetically distant organisms may be similarly sensitive to toxicants. This means that certain wildlife species can be used when conducting environmental risk assessments and that the data obtained can often (but importantly, not always) be extrapolated to others (Gunnarsson et al. 2008; Baker et al. 2009). Moreover, a "*species sensitivity distribution*" (SSD) to a particular chemical (based on multiple studies) has been shown

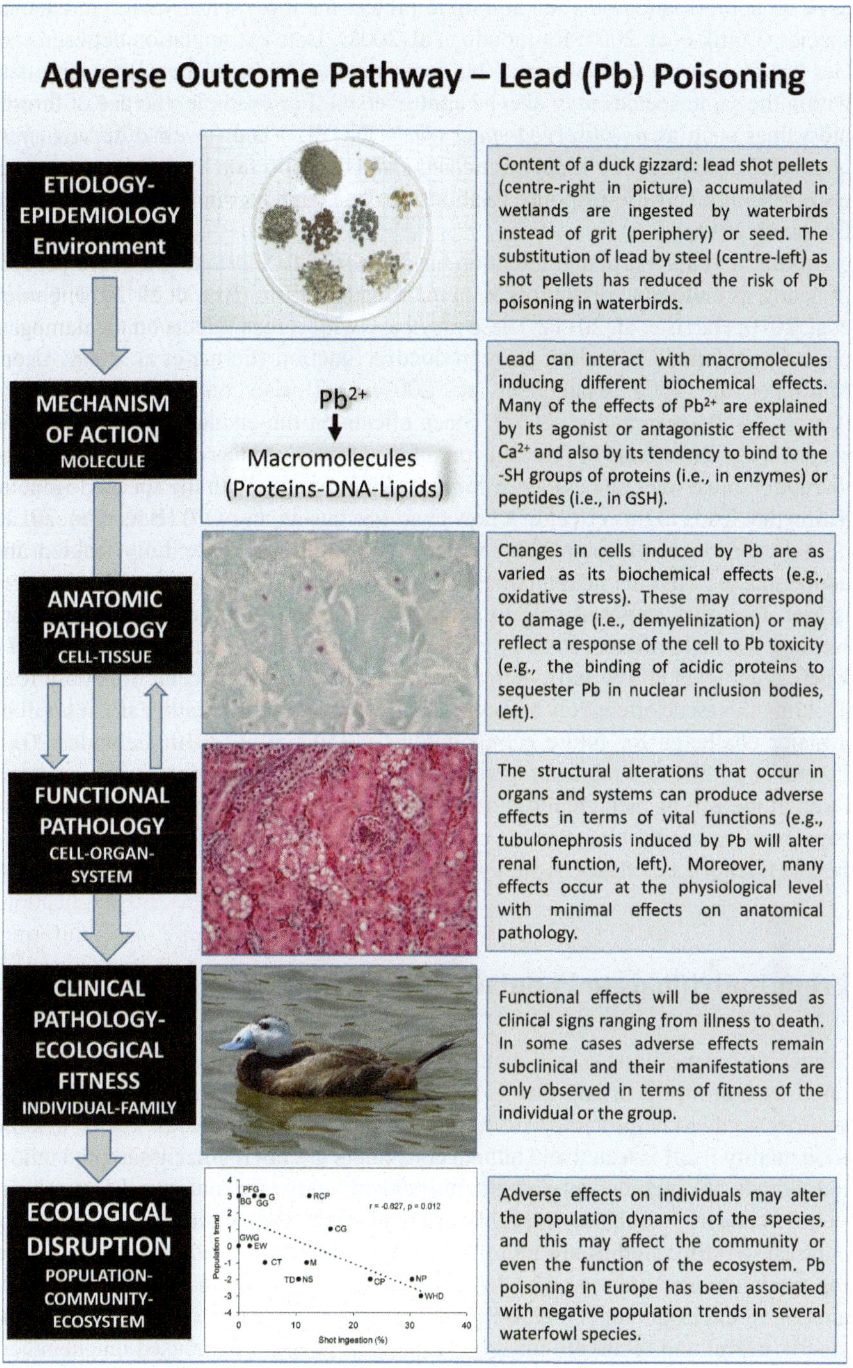

Fig. 1 Adverse Outcome Pathway (AOP) schematic with Pb poisoning in birds as an example. This is only a small part of the potential AOP for this heavy metal, since its potential effects are as diverse as its mechanisms of action (Mateo et al. 2003, 2014; Mateo 2009)

to be an appropriate tool when aiming to protect the most sensitive and threatened species (Luttik et al. 2005; Raimondo et al. 2008). Data extrapolation between species has always been problematic, but interpretation of experimental toxicity data within the same species may also be controversial. For example, the use of threshold values such as *no-observed-effect levels* (NOELs) and *lowest-observed-effect levels* (LOELs) has been widely criticized and its substitution by curve-fitting models regarding exposure-response relationships has been recommended (Landis and Chapman 2011).

In recent years, particular attention has been paid to substances that are capable of acting as endocrine disruptors in humans and wildlife (Brar et al. 2010; Gutleb et al. 2010; Harris et al. 2011). These may have widespread effects on the demography of populations by acting on reproductive function (Fernie et al. 2008; Ucán-Marín et al. 2009; Henny et al. 2009), but also on immune response (Casanova-Nakayama et al. 2011). Since effects on the endocrine system may be secondary to the effects of toxicants on other systems, studies evaluating endocrine disruption must integrate evidence for endocrine toxicity with the specific mode of action that leads to this effect in a flow chart (such as in an AOP) (Bars et al. 2012).

A further level of complexity, which to date has yet to be fully tackled and addressed, regards chemical mixtures present in the environment. No wild organism is ever, in reality, exposed only to a single toxicant (as in an idealistic laboratory setting). As such, mixture toxicity is likely to be highly relevant in the real world – where complex additive and synergistic effects are likely to play an important role. Tackling this issue effectively and consistently in field and laboratory studies will be a major challenge for future ecotoxicologists (Lahr et al. 2010; Celander 2011; Poletta et al. 2011). Progress regarding toxicological interactions can be achieved through the use of meta-analysis of multiple studies involving a variety of organisms exposed to different types of chemicals, as well as to different environmental factors (Hendriks et al. 2005; Laskowski et al. 2010).

From Individuals to Populations in "*Biomonitoring*" Programs

Strategies for monitoring pollutants as they pass along food chains can be structured as bottom-up or top-down. For example, in the human food chain, food security tends to be protected from the bottom-up (Johansen et al. 2004), i.e., the food quality itself is tested and human consumers are not regularly sampled unless individuals are included in an epidemiological study, are monitored for occupational exposure, or, reside in highly contaminated sites (Colles et al. 2008). By contrast, wildlife studies are more frequently based on top-down data – whereby top predators are often used as "*sentinel species*" and "*bioindicators*" of pollutants in an environment (Gómez-Ramírez et al. 2014). This top-down strategy is clearly useful and relatively low cost, especially when it is focused on chemicals that bioaccumulate and biomagnify (i.e., persistent non-polar chemicals, such as halogenated organic compounds). However, this top-down strategy may be of far less use when studying substances that do not bioaccumulate. These may have a higher direct impact on primary producers and consumers (rather than top predators) and an entire food chain may then be destabilized, which in turn affects predators indirectly (Hallmann et al. 2014). For such toxicants (of low persistence), which may also exert acute effects (e.g., anti-cholinesterase carbamate and organophosphate pesticides or the pharmaceutical diclofenac), the concept of "*toxico-* or *ecotoxico-vigilance*" based on long-term opportunistic passive sampling of dead animals (found in the field) can provide a suitable alternative (Shore et al. 2014).

Longer term monitoring of environmental pollutants in wildlife must be sustainable and economically affordable. Good initial planning and the appropriate selection of monitoring species are essential to success (Custer 2011). Bioindicator species with a wide geographic distribution allow the comparison of contaminant levels between different countries or even continents (Eens et al. 2013). However, such cosmopolitan species probably have a high plasticity that may explain their ecological success (Reid et al. 2013). Hence, how the genetic resilience of a species to an environmental pollutant correlates with this plasticity certainly needs further research (Karchner et al. 2006; Vandegehuchte and Janssen 2014). At the other end of the scale, the conservation of endangered species is also a priority, especially where their decline is closely related to their sensitivity to a particular environmental contaminant (Oaks et al. 2004; Finkelstein et al. 2012). Ultimately, it is difficult and probably inappropriate to try to define a single or even a few key biomonitoring species for all possible scenarios – since the aims of any study will inevitably vary with the pollutant in question, its effects and the route of exposure (amongst many other factors).

The development of biomonitoring strategies that utilise non-invasive techniques has been of particular interest in the last decade. Fecal samples allow exposure to some pollutants in wildlife to be measured (Beyer et al. 2007) together with measures of some adverse effects (e.g., metabolism of porphyrins; Mateo et al. 2006; Martinez-Haro et al. 2011). Combining chemical analysis of feces with molecular genotyping can also enable non-intrusive tracking of mammalian meso-carnivores, for example, in time and space in relation to a contaminated source (Guertin et al. 2010), and even allow inference of contaminant effects at the population level (Guertin et al. 2012). Feathers also provide a good matrix for the detection of organic and inorganic pollutants (Jaspers et al. 2009; García-Fernández et al. 2013), but some considerations regarding the possibility of external contamination (Jaspers et al. 2008; Cardiel et al. 2011) or about ptylochronology (Bortolotti 2010) are also necessary to correctly interpret levels detected. Similar to feathers, hair can also be used for biomonitoring contaminants such as mercury that have a high affinity for this protein matrix (Noël et al. 2014). The analysis of stable isotopes of hydrogen, sulfur, nitrogen and carbon in feathers can facilitate the identification of hotspots of mercury exposure in migratory birds (Lavoie et al. 2015). Finally, although also used historically, eggshells remain useful. These may provide information about exposure and effects of pollutants – for example, through studying pigmentation levels (i.e., due to biliverdin and porphyrins) (Jagannath et al. 2008).

Biomonitoring strategies and AOP studies should also consider population dynamics and the concept of community within an ecosystem. Soil conservation is key to ensuring broader biodiversity in terrestrial ecosystems, and considering the whole soil community adds a new scale to the concept of biomonitoring (Decaëns et al. 2006). At this scale chemicals may cause dysfunction that is not classically considered toxicological in nature – in many cases, a toxic effect on one species alters the wider ecological balance within the community by affecting fundamental ecological relationships (e.g., predator-prey). This type of dysfunction is one that is produced by pesticides in agricultural ecosystems, where the artificial simplification of the community (caused by the application of the pesticide) at lower trophic levels drives a loss of ecosystem biodiversity across the system.

Direct Effects of Pesticides on Farmland Biodiversity

The role that pesticides have played in the decline of farmland birds has been widely debated. The effects of pesticides on higher wildlife may often be due to effects at the base of the agricultural ecosystem food chain, i.e., in removing weeds and invertebrates, bird numbers and diversity decline as their food sources become significantly limited. However, direct toxic effects on non-target vertebrates (such as farmland birds) should not be dismissed or overlooked (Goulson 2014; Hallmann et al. 2014). Unlike the organochlorine (OC) insecticides used globally several decades ago (which are still widely used in some developing countries), current products tend to be developed with degradability or lower environmental persistence in mind. Although far less persistent, this does not prevent elevated exposure occurring (to these highly toxic substances) during, for example, aerial application. Topping and Odderskær (2004), using "*Agent Based Simulation Models*", also determined that pesticides can have a negative impact on farmland bird populations in interaction with other environmental/ecological factors, such as landscape diversity, food availability and migratory movement. Concepts such as the "*Toxicity to Exposure Ratio*" (TER) are also now commonly used to perform risk assessments for current pesticides. That is, having defined a pesticide toxicity endpoint in an exposed animal and the dose that can produce such an effect, one can estimate the risk of a pesticide application according to measured or theoretical doses at potential acute to long-term exposures. The risk posed by a pesticide would then be higher when the TER decreases (Prosser and Hart 2005). The key in such a risk assessment is in identifying one or more sensitive endpoints (Mineau 2005), since some adverse effects can be very subtle. This is the case for >100 pesticides known to act as endocrine disruptors (McKinlay et al. 2008) or for neurotoxicants that can alter behaviour in animals (Gill et al. 2012).

At the less subtle end of the effect scale, significant evidence pointing to acute adverse effects caused by commercial formulations (as microgranules) of certain insecticide-nematicide compounds that are applied to soil have also been highlighted. The extremely high non-target toxicity of certain active ingredients in some formulations (e.g., containing compounds such as carbofuran and aldicarb) mean these applications can be very dangerous for terrestrial birds several months after application, especially where they ingest neat/undissolved microgranules or insecticide coated seed products (Elliott et al. 2008; Richards 2012). In addition, the toxicity of such formulations makes them very attractive for use in the deliberate poisoning of many wildlife, especially predators/scavengers (Richards 2012). Despite global evidence regarding the ongoing risk posed by such potent chemicals/formulations when used legally and illegally, sadly, they continue to be widely marketed around the globe (Martínez-Haro et al. 2008; Richards 2012).

Another example where a low TER carries a high risk of poisoning in birds exists when pesticides are used for seed coatings (Prosser and Hart 2005; López-Antia et al. 2015). Products commonly used for seed coatings are fungicides and insecticides. Historically, methylmercury or anti-cholinesterase insecticides were widely applied to seeds and more recently other highly toxic insecticides (like neonicotinoids and fipronil) are being used instead (Gibbons et al. 2015; López-Antia et al. 2015). In this scenario, the risk comes directly from the material that is treated, since coated seeds can be eaten by (and are often attractive to) many farmland species. The repellent effect of some insecticides and fungicides may reduce non-target ingestion and thus dose (López-Antia et al. 2014), but as also occurs with microgranules, the toxicity of some products (i.e., imidacloprid treated seeds) is often so high it can kill wildlife at very minimal exposure levels (López-Antia et al. 2015).

Lastly, the impact of neonicotinoid pesticides in agricultural ecosystems has also been an important and rapidly growing area of concern in recent years. Reports suggest they are behind widespread declines in honeybees and bumblebees because these pesticides adversely affect their feeding behaviour (Mommaerts et al. 2010; Gill et al. 2012; Laycock et al. 2012; Henry et al. 2012). This provides a good example of a subtle effect that may have very wide reaching negative implications (i.e., on plant pollination). Similar examples exist regarding pollutants of a different type (i.e., pharmaceuticals) where behavioural changes have also been described in fish and birds (Almeida et al. 2010; Brodin et al. 2013; Bean et al. 2014). A future research priority is now to develop appropriate assays to detect the often-subtle behavioural effects of neurotoxicants on higher vertebrates, and, to evaluate their implications on population dynamics.

New Bioaccumulative Pesticides

The global impact of OC pesticides is probably hard to match in comparison to the more modern pesticides currently in use. However, some other pesticide families (i.e., pyrethroid insecticides or anticoagulant rodenticides) show some similarities because of their relative persistence in animal tissues. In the case of pyrethroid insecticides, field studies have shown their accumulation and maternal transfer in marine mammals (Alonso et al. 2012), although the effects of this on the health of individuals is still unknown. The adverse effects of anticoagulant rodenticides on wildlife have been studied in more depth (Rattner et al. 2014). For some 2nd generation compounds, developed to overcome the resistance built up by rodents to earlier rodenticides, half-lives in liver of exposed animals can be >100 days (Eason et al. 2002). Because of this persistence, 2nd generation anticoagulant rodenticides (SGARs) have been detected in the livers of predatory animals around the world, and the possible impact on their health has been an emerging research topic. A high prevalence of liver SGAR residues has been found in several bird of prey species (Shore et al. 2006; Albert et al. 2010; Lemarchand et al. 2010; Thomas et al. 2011; Murray 2011; Sánchez-Barbudo et al. 2012) and in mammals (Riley et al. 2007: Dowding et al. 2010; Gabriel et al. 2012). Since SGAR residues now seem ubiquitous in the environment, the question now is whether this affects the health of wildlife. The presence of SGARs above a certain threshold in liver tissue (>0.2 ppm) has been considered diagnostic when it is detected in animals with evidence of altered coagulation and spontaneous bleeding (Berny et al. 1997). However, the presence of macroscopic bleeding in lethally intoxicated animals is not a consistent finding (Sarabia et al. 2008; Rattner et al. 2011). In contrast, other animals may show haemorrhages even with relatively low levels of anticoagulants in their liver. This demonstrates how difficult it can be to define a toxicity threshold for a wide range of species in individuals differing in terms of their existing nutritional and health status. For this reason, a probabilistic approach regarding the interpretation of liver rodenticide residues, based on the variability in sensitivity to pesticides in each species, could be a better approach for AR residues (or in similar cases; Thomas et al. 2011). Such insights, gained through the interpretation of residues, can then be included in the establishment of an AOP for rodenticides or for other pesticides (Rattner et al. 2014). This is just one example of the need within wildlife ecotoxicology to establish reliable residue threshold levels in animal tissues associated with different endpoints; exceedance of which then results in implications for individual health or population demographics (Miller et al. 2007; Baldwin et al. 2009; Beyer and Meador 2011).

Anthropogenic Perturbation and the Impact of "Natural Substances"

Although ecotoxicology largely focuses on environmental contaminants synthesized by humans (intentionally or accidentally), many toxic substances have been present on earth since its formation and since life evolved. We simply need to consider the complex relationship between plants and herbivores to understand the importance of natural phytotoxins in the struggle for survival. However, human action in particular tends to facilitate the emergence of environmental stressors associated with toxic substances. These may be geological (e.g., heavy metals and metalloids released through mining/processing/use) or biological in origin (e.g., toxins produced by bacteria, algae and plants). Ore mining commonly produces focal areas of heavy metal/metalloid contamination in/around the sites of extraction and processing (Taggart et al. 2011). Also, the subsequent use of these toxic elements for multiple applications (throughout history), combined in certain cases with long-range atmospheric transport (i.e., for mercury) has caused widespread increases in "*background levels*" of certain elements (e.g., cadmium, arsenic, mercury and lead) around the world (Scheifler et al. 2006).

Mercury (Hg) is particularly well known for its capacity to bioaccumulate and biomagnify through food chains – in particular in its highly toxic organic forms (which are produced in nature) – this element is now a widespread persistent global pollutant and its increased global presence is closely correlated with the onset of industrialisation. In the last decade, significant advances have been made in research regarding the reproductive effects of Hg in birds (Jackson et al. 2011), including effects on sexual preferences (i.e., increasing copulation between males; Frederick and Jayasena 2011), on sex hormones levels (Heath and Frederick 2005) and on the gonadal hypothalamic-pituitary-axis (Tartu et al. 2013). Neurologic effects (of Hg) have also been highlighted in wild mammals in recent years. In Arctic polar bears, an inverse relationship has been found between the degree of genomic DNA methylation in the brain and levels of Hg, especially in males. This opens up new possibilities for studies in the field of epigenetics, because DNA methylation plays an important role in transcriptional regulation of genes with consequences on the development of different pathologies (Pilsner et al. 2010). Also in polar bear, an inverse relationship has been found between levels of the receptor N-methyl-D-aspartate and Hg (especially organic Hg) in the brain (Basu et al. 2009). Mercury exposure in American mink has also been positively correlated with brain levels of muscarinic acetylcholine receptors and negatively with dopamine receptors (Basu et al. 2005). Interestingly, all these studies suggest new biomarker options for use in the study of reproductive toxicity and neurotoxicity. Finally, interactions between selenium and Hg in different cellular compartments of birds and mammals has also provided new insights regarding mechanisms of action and possible detoxification processes for Hg (Ikemoto et al. 2004).

Lead (Pb) is another heavy metal of significant importance in wildlife ecotoxicology. Although bioaccumulative in bones, Pb does not show the same potential for biomagnification as Hg (Cui et al. 2011). Risks due to Pb in wildlife come mainly from the continued use of Pb-based ammunition in hunting. Lead shot pellets or bullet fragments are commonly ingested by many avian species and may cause lethal poisoning and sublethal effects, with population level implications for the conservation of some species (Fisher et al. 2006; Watson et al. 2009). The possibility of substituting Pb in hunting ammunition has been widely debated over the last decade, from various viewpoints (i.e., regarding its toxicity, safety, effectiveness as ammunition; Watson et al. 2009). The presence of Pb and Pb-free ammunition in the hunting market has facilitated several studies looking at compliance levels in regions where partial bans on Pb ammunition exist (Mateo et al. 2014). Unfortunately, partial bans (rather than a complete global ban) on Pb-based ammunition may be ineffective when attempting to conserve critically endangered species that are vulnerable to lead poisoning, such as the California condor (Finkelstein et al. 2012). Such information shows just how important strict enforcement (alongside good stakeholder engagement and education) of existing environmental regulations can be (Mateo et al. 2014). Also in the last decade, renewed focus has emerged regarding the wider impact of Pb ammunition, especially in terms of the contamination of game meat and subsequent Pb exposure in hunters and game meat consumers (Green and Pain 2012).

Nanoparticles: New Materials and New Challenges

Linked to the challenges posed by heavy metals are those now posed by the suite of new materials known as nanoparticles and nanomaterials. These are being increasingly used in agriculture, electronics, biomedicine, manufacturing and cosmetics, and are also being increasingly detected in the environment (Matranga and Corsi 2012). Some nanoparticles are composed of metals oxides (Ek et al. 2004; Gaiser et al. 2012; Li et al. 2012; Seitz et al. 2013), others are organic compound structures such as microplastics or carbon based fullerene materials (Liu et al. 2009; Mahler et al. 2012). The possible effects of these new materials on biota are now being quite intensively explored; although perhaps not as quickly as they are being developed/put into widespread use. Most studies to date have involved invertebrates and fish, and as such, their potential impact on many other wildlife groups is still unknown. The toxicity of nanoparticles based on metal oxides can (in part) be related to the metal(s) in question (Ek et al. 2004; Gagné et al. 2012), but the particular size and structural characteristics of these nanoparticles can modulate their toxicity significantly – i.e., due to their potential capacity to adhere to or even directly enter different types of cells or tissues (Gaiser et al. 2012; Meyer et al. 2010). Moreover, nanoparticles may also exert physical effects within certain body compartments (e.g., within intestinal microvilli or gills) and cells, and can therefore disrupt normal functional processes (e.g., iron absorption; Gaiser et al. 2012; Mahler et al. 2012).

Ubiquity of Pharmaceuticals and Risks for Wildlife

Another relatively "new" and rapidly shifting field of research in wildlife ecotoxicology currently regards pharmaceuticals. As a diverse suite of potentially very potent bio-active compounds, our knowledge regarding their non-target effects on most wildlife (perhaps excluding some fish) is currently very limited (Arnold et al. 2013). Yet, a multitude of pharmaceuticals for human and veterinary use are now known to be widely present in our environment, albeit commonly at very low levels (i.e., at the high ng/l or low μg/l level in freshwater). Whilst much research has now been published aimed at quantifying the environmental presence of more common pharmaceuticals, predominantly in freshwater systems, increasing attention is now being paid to their actual *impacts* and *effects*.

Research to date has highlighted a relatively limited number of clear cases where pharmaceutical contamination has led to widespread detrimental effects on non-target wildlife. Cases of particular note (and obvious concern) are that of (i) the widely studied synthetic estrogen 17α-ethynylestradiol (EE2; a common ingredient in the human contraceptive pill), which has been linked with the feminisation of wild male fish in many parts of the world (Nash et al. 2004; Jobling et al. 2006; Kidd et al. 2007), and (ii) the non-steroidal anti-inflammatory drug (NSAID) diclofenac, which has now been banned for veterinary use across the Indian subcontinent after being conclusively linked to widespread population declines in three Old World vulture species (Oaks et al. 2004; Cuthbert et al. 2014). In 2013, concern regarding both these compounds (and a third, 17 β-estradiol (E2)) in freshwater systems in Europe also resulted in all three being added to the *Watch List* within the EC Water Framework Directive (which aims to achieve a *'good (quality) status'* for all ground/surface waters in the EU). In terms of NSAIDs specifically, it now seems that diclofenac may be just one of a suite of older generation NSAIDs that have the potential to harm scavenging birds (Cuthbert et al. 2007, 2014; Naidoo et al. 2010; Zorrilla et al. 2015) and *perhaps* other aquatic wild biota (Richards et al. 2011; Simpson et al. 2011; Veldhoen et al. 2014) at environmentally relevant concentrations.

A major challenge with pharmaceutical compounds is that there are simply *so many* differing active ingredients and compound classes now in use within modern society (antibiotics, NSAIDs, antiparasitics, antipsychotics, steroids, immunomodulators, anticancer agents, etc.) – all targeting a very diverse suite of physiological endpoints. To try to address this challenge, several extensive horizon scanning and prioritisation exercises have been undertaken recently to try to focus research attention in the most important areas (Fick et al. 2010; Boxall et al. 2012). In addition to "*legal*" pharmaceuticals, illicit drugs and so-called "*legal-highs*" are also now beginning to receive an increasing level of attention (Zuccato et al. 2008).

Amongst the many important challenges and questions that remain in this particular area of research, several topics have been of particular note recently, i.e., the possible effects of highly polluted effluents related to pharmaceutical manufacturing hubs in developing countries (Larsson et al. 2007; Gunnarsson et al. 2009); the possible effects of contamination related to very large scale, intensive poultry or livestock farming (Burkholder et al. 2007); the global rise in antibiotic resistance and the role that widespread environmental occurrence may be playing in this (Wellington et al. 2013); and, understanding subtle but potentially widespread sub-lethal effects, i.e., on physiology, behaviour or reproduction (Jobling et al. 2006; Bean et al. 2014; Lazarus et al. 2015).

As the global demand for food keeps rising and developing countries continue to rapidly modernise both their health and agricultural practices, the global use (and consequently disposal) of human and veterinary pharmaceuticals is only set to grow – as such, looking ahead, many "*grand challenges*" remain to be addressed in this field (Boxall et al. 2012; Arnold et al. 2013; Margalida et al. 2014).

Presence and Effects of POPs in Wildlife

One of the historic cornerstones of ecotoxicology is the study of persistent organic pollutants (POPs) – this includes polychlorinated biphenyls (PCBs), polychlorinated dibenzo-p-dioxins and furans (PCDD/Fs), polybrominated diphenyl ethers (PBDEs), organochlorinated pesticides (OCs), per- and polyfluorinated alkyl substances (PFASs), and paraffins. POPs represent a serious risk to wildlife, because of their bioaccumulative properties, biomagnification potential, acute toxicity and long-term effects (some of which almost certainly remain unidentified). The effects caused by POPs can vary with species, habitat, biological cycle, dynamics of the ecosystem and climate conditions; whilst the intensity and duration of exposure are also key factors in determining observed effects. Certain POPs are known to cause mortality, reduced fecundity and growth, disrupt development and physiological stress. These effects are often associated with neurotoxicity, endocrine and metabolic disruption and can affect population structure in species ranging from fish to birds to invertebrates (Vasseur and Cossu-Leguille 2006). Endocrine disruption can induce feminisation of males, affect sexual differentiation of the brain, sexual, maternal, play and aggressive behaviour and stress responses in various animals (Weiss 2011). The impact of such effects will depend on life stage and the vulnerability of the exposed species – and impacts may be seen at the individual up to the population and community level (Ross and Birbaum 2003; Letcher et al. 2010). One of the most important concerns about POPs is their high bioaccumulation potential, which then generates long term physiological and biochemical effects (La Merrill et al. 2013) with equally long-term consequences.

Older POPs like PCBs and OCs continue to induce severe effects in wildlife, even in remote ecosystems such as those in the Arctic (Letcher et al. 2010). In birds, examples of a variety of effects continue to be evidenced, i.e., eggshell thinning and effects on reproduction and survival (Elliott et al. 2011); impairment of the immune system, body condition, reproduction and growth in gulls from Bear Island in the Norwegian Arctic (Bustnes 2007); lowered levels of steroid hormones and effects on laying attributed to high concentrations of DDT and PCBs in birds collected near an old DDT factory in Italy (Cortinovis et al. 2008); effects on growth and development in shag (*Phalacrocorax aristotelis*) from Norway associated with PCBs and OCs (and reduced vitamin A in plasma) (Murvoll et al. 2006). PCBs and OCs also impact marine mammals, for example, where PCBs and DDT burdens in blubber of harbour seals (*Phoca vitulina concolor*) from the USA exceed the estimated threshold levels for adverse reproductive and immune system effects (Shaw et al. 2005).

PBDEs, PFASs and other "*emerging*" POPs have also been widely used globally for >30 years and may also pose a risk to wildlife. Fernie et al. (2009) found that captive American kestrels (*Falco sparverius*) exposed to environmentally relevant concentrations of the flame retardant mixture DE-71 exhibited delayed egg laying, laid smaller eggs, and showed reduced fertility and reproductive success. Effects of PBDEs are commonly related to endocrine disruption, as observed in rats exposed to PBDE-99 where circulating sex steroids and sperm production were adversely affected (Lilienthal et al. 2006). In terms of PFAS (used in surfactants, firefighting foams, cleaning products, insecticides, etc.), these can accumulate and biomagnify in aquatic organisms (Houde et al. 2011) and cause an array of effects in wildlife. These include hypertrophy and vacuolization in liver, reduction in serum cholesterol/triglycerides, reduction in body weight gain or body weight, and increased mortality (Stahl et al. 2011). Other emerging POPs, such as chlorinated paraffins, will also bioaccumulate and biomagnify as evidenced in food webs in the Great Lakes (Houde et al. 2008) – however, the effects they may then cause need further investigation.

Whilst many chemicals have provided huge benefits in terms of economic development, POPs have (by definition under the Stockholm Convention) caused significant long-term harm to humans and certain wildlife, especially top predators. Some are so persistent (i.e., DDT) that they will remain within our wider environment for many centuries despite strict limitations on their current use. As such, the benefits of continued use of such compounds (especially in developing countries) must be weighed against the long-term price that will inevitably have to be paid.

Impact of Plastics in Wildlife

Plastics have emerged as global contaminants that can affect wildlife in both terrestrial and aquatic ecosystems. Plastic debris within the aquatic environment, including resin pellets and microscopic and macroscopic plastic fragments, generate several problems. The first regards the migration of plastic monomers and additives into water and the uptake of these contaminants by aquatic organisms and subsequent accumulation along food webs (Moore 2008; Teuten et al. 2009). For example, alkylphenols, phthalates and bisphenol A are monomers and additives used during the manufacture of plastics and these can leach into waters and accumulate in organisms – causing an array of effects from acute toxicity to endocrine disruption (Oehlmann et al. 2009). Second, POPs and other contaminants present in water are readily adsorbed onto the hydrophobic surfaces of plastic material and thus concentrate and become available to organisms (Teuten et al. 2009; Andrady 2011). Third, the physical ingestion of large volumes of indigestible plastic based material is in itself a significant hazard. Macroplastic waste in the environment is typically associated with entanglement, ingestion, suffocation and debilitation of aquatic organisms (Gregory 2009). Many species of seabird, turtle and cetacean are known to commonly ingest plastic debris (Moore 2008). Birds especially often become entangled and die in fishing lines and nets, but in addition, are increasingly consuming an array of plastic debris (Elliott and Elliott 2013) – this can produce gastrointestinal blockages, ulceration, internal perforation and death. Ingestion of plastic that mimics natural food provides no nutritional benefit and use of debris as nest material may pollute otherwise pristine habitats and be detrimental to ecosystem health (Hartwig et al. 2007). In addition, the colonization of plastic marine debris by organisms allows this waste to act as a vector for transport/movement of alien species around the world's oceans, which may again threaten marine biodiversity. Accumulation of plastic debris on the sea floor can also inhibit gas exchange between overlying water and sediment pore water, which may affect benthic organisms. Marine litter can affect coastal species by filling in and damaging nursery habitats.

All such effects are produced due to the immense discharge of plastics to the marine environment because of their use, production and disposal worldwide. At least 267 marine species are now known to have ingested and therefore potentially been negatively affected by plastic debris (Moore 2008).

Conclusions

Wildlife can be exposed to all of the myriad chemicals mobilized, synthesised and released by humans into the environment. In many respects, wildlife can act as better biomonitors and bioindicators of accumulation and adverse effects than humans can. Their role as biomonitors and bioindicators is enhanced by the diversity of wildlife and the important differences that exist between species in terms of sensitivity to toxicants. Many of the most important biomonitoring species sit (like humans) at the top of their respective food chains.

For the wildlife ecotoxicologist, the first question asked often regards identification of the chemical hazard and the exposure route to wildlife (Fig. 2). Here, full LCAs of products is key and these must consider the final fate and behaviour of chemicals when they enter the environment. Detailed LCAs will help identify new chemicals hazards before adverse effects on populations occur. In order to quantify the concentrations of chemicals present in the environment accurately, analytical chemistry must also continue to develop and be kept up to date.

Once potential hazards have been identified, experimental and toxicovigilance studies are essential to investigate if adverse effects are produced at "*environmentally relevant*" field concentrations. Experimental studies permit us to identify the mode of action of toxicants and in-silico assessments may help with data extrapolation – which may in turn reduce the use of experimental animals. Having said this, in many cases, field biomonitoring and toxicovigilance have often been essential in terms of identifying adverse effects in wildlife (e.g., consider the use of NSAIDs in livestock and their subsequent toxicity in scavenging vultures).

By integrating knowledge from both field and experimental studies, AOPs can be established which consider effects from the molecular up to the ecosystem level. This new concept may help create a clearer global picture regarding the impact of a pollutant on wildlife – taking us beyond considering individuals and very specific biomarkers and/or scenarios. Having said this, the mode of action of some types of chemicals can be extremely complex and it will be a significant challenge to integrate all knowledge and establish effect pathways at some of the highest levels of complexity (e.g., for populations). Integrated information compiled by different disciplines (from analytical chemistry to ecology) should ultimately be used to inform detailed environmental risk assessments for chemical substances, in order to predict

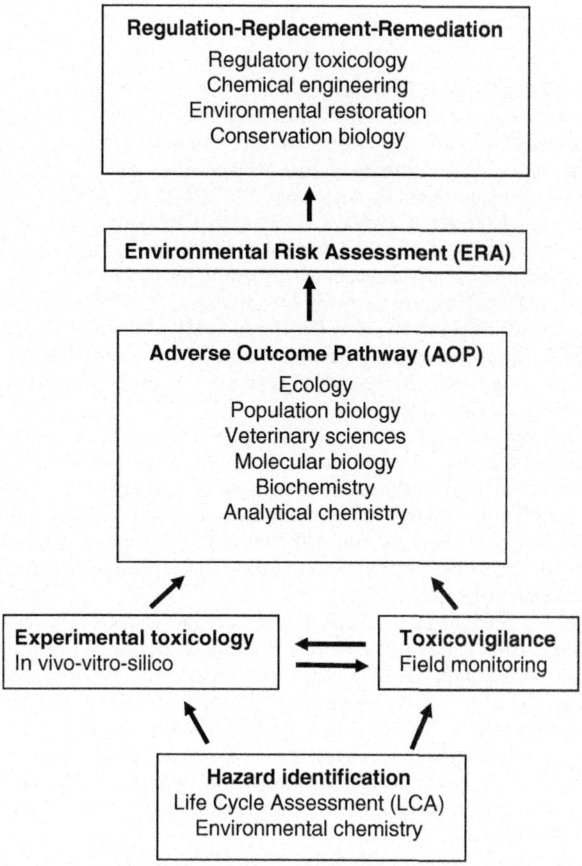

Fig. 2 Schematic of the framework in wildlife ecotoxicology. The identification of new chemical hazards is an essential starting point and leads to the development of new trends in wildlife ecotoxicology. This preliminary step can be followed by compilation of the whole Adverse Outcome Pathway, either by experimental or by field approaches. In fact, forensic toxicology or toxicovigilance may often permit the detection of new hazards in a top – down direction in the absence of full and appropriate Life Cycle Assessment. Information obtained about exposure and effects can be used to perform a corresponding Environmental Risk Assessment – required for decision making by regulatory bodies whose role it is to minimize the adverse effects of chemicals

adverse effects, or in some cases, establish Environmental Policy objectives aimed at reducing impacts within a set timeframe.

In combination, all this research effort would ultimately be meaningless if the information generated were not then publicly available and fully used to facilitate the regulatory actions needed to ensure the sustainable use of chemical substances in the future.

Acknowledgments This study has been supported by the National Parks Projects [755/2012 and 768/2012] from the Spanish Ministry of the Environment.

References

Albert CA, Wilson LK, Mineau P, Trudeau S, Elliott JE (2010) Anticoagulant rodenticides in three owl species from Western Canada, 1988–2003. Arch Environ Contam Toxicol 58:451–459

Almeida JR, Oliveira C, Gravato C, Guilhermino L (2010) Linking behavioural alterations with biomarkers responses in the European seabass *Dicentrarchus labrax* L. exposed to the organophosphate pesticide fenitrothion. Ecotoxicology 19:1369–1381

Alonso MB, Feo ML, Corcellas C, Vidal LG, Bertozzi CP, Marigo J, Secchi ER, Bassoi M, Azevedo AF, Dorneles PR, Torres JP, Lailson-Brito J, Malm O, Eljarrat E, Barceló D (2012) Pyrethroids: a new threat to marine mammals? Environ Int 47:99–106

Andrady AL (2011) Microplastics in the marine environment. Mar Pollut Bull 62:1596–1605

Ankley GT, Bennett RS, Erickson RJ, Hoff DJ, Hornung MW, Johnson RD, Mount DR, Nichols JW, Russom CL, Schmieder PK, Serrrano JA, Tietge JE, Villeneuve DL (2010) Adverse outcome pathways: a conceptual framework to support ecotoxicology research and risk assessment. Environ Toxicol Chem 29:730–741

Anza I, Vidal D, Laguna C, Díaz-Sánchez S, Sánchez S, Chicote A, Florín M, Mateo R (2014) Eutrophication and bacterial pathogens as risk factors for avian botulism outbreaks in wetlands receiving effluents from urban wastewater treatment plants. Appl Environ Microbiol 80:4251–4259

Arnold KE, Boxall ABA, Brown R, Cuthbert RJ, Gaw S, Hutchinson TH, Jobling S, Madden JC, Metcalfe CD, Naidoo V, Shore RF, Smits JE, Taggart MA, Thompson HM (2013) Assessing the exposure risk and impacts of pharmaceuticals in the environment on individuals and ecosystems. Biol Lett 9:20130492

Baker ME, Ruggeri B, Sprague LJ, Eckhardt-Ludka C, Lapira J, Wick I, Soverchia L, Ubaldi M, Polzonetti-Magni AM, Vidal-Dorsch D, Bay S, Gully JR, Reyes JA, Kelley KM, Schlenk D, Breen EC, Šášik R, Hardiman G (2009) Analysis of endocrine disruption in southern California coastal fish using an aquatic multispecies microarray. Environ Health Perspect 117:223–230

Baldwin DH, Spromberg JA, Collier TK, Scholz NL (2009) A fish of many scales: extrapolating sublethal pesticide exposures to the productivity of wild salmon populations. Ecol Appl 19:2004–2015

Bars R, Fegert I, Gross M, Lewis D, Weltje L, Weyers A, Wheeler JR, Galay-Burgos M (2012) Risk assessment of endocrine active chemicals: identifying chemicals of regulatory concern. Regul Toxicol Pharmacol 64:143–154

Basu N, Klenavic K, Gamberg M, O'Brien M, Evans D, Scheuhammer AM, Hing MC (2005) Effects of mercury on neurochemical receptor-binding characteristics in wild mink. Environ Toxicol Chem 24:1444–1450

Basu N, Scheuhammer AM, Sonne C, Letcher RJ, Born EW, Dietz R (2009) Is dietary mercury of neurotoxicological concern to wild polar bears (*Ursus maritimus*)? Environ Toxicol Chem 28:133–140

Bean TG, Boxall ABA, Lane J, Herborn KA, Pietravalle S, Arnold KE (2014) Behavioural and physiological responses of birds to environmentally relevant concentrations of an antidepressant. Philos Trans R Soc B 369:20130575

Benninghoff AD (2007) Toxicoproteomics – The next step in the evolution of environmental biomarkers? Toxicol Sci 95:1–4

Berny PJ, Buronfosse T, Buronfosse F, Lamarque F, Lorgue G (1997) Field evidence of secondary poisoning of foxes (*Vulpes vulpes*) and buzzards (Buteo buteo) by bromadiolone, a 4-year survey. Chemosphere 35:1817–1829

Beyer WN, Meador JP (eds) (2011) Environmental contaminants in biota: interpreting tissue concentrations. CRC Press, Boca Raton, 751 pp

Beyer WN, Gaston G, Brazzle R, O'Connell AF Jr, Audet DJ (2007) Deer exposed to exceptionally high concentrations of lead near the Continental Mine in Idaho, USA. Environ Toxicol Chem 26:1040–1046

Bodkin JL, Ballachey BE, Coletti HA, Esslinger GG, Kloecker KA, Rice SD, Reed JA, Monson DH (2012) Long-term effects of the 'Exxon Valdez' oil spill: sea otter foraging in the intertidal as a pathway of exposure to lingering oil. Mar Ecol Prog Ser 447:273–287

Bortolotti GR (2010) Flaws and pitfalls in the chemical analysis of feathers: bad news-good news for avian chemoecology and toxicology. Ecol Appl 20:1766–1774

Bourret V, Couture P, Campbell PGC, Bernatchez L (2008) Evolutionary ecotoxicology of wild yellow perch (*Perca flavescens*) populations chronically exposed to a polymetallic gradient. Aquat Toxicol 86:76–90

Boxall ABA, Rudd MA, Brooks BW, Caldwell DJ, Choi K, Hickmann S, Innes E, Ostapyk K, Staveley JP, Verslycke T, Ankley GT, Beazley KF, Belanger SE, Berninger JP, Carriquiriborde P, Coors A, DeLeo PC, Dyer SD, Ericson JF, Gagné F, Giesy JP, Gouin T, Hallstrom L, Karlsson MV, Joakim Larsson DG, Lazorchakm JM, Mastrocco F, McLaughlin A, McMaster ME, Meyerhoff RD, Moore R, Parrott JL, Snape JR, Murray-Smith R, Servos MR, Sibley PK, Straub JO, Szabo ND, Topp E, Tetreault GR, Trudeau VL, Van Der Kraak G (2012) Pharmaceuticals and personal care products in the environment: what are the big questions? Environ Health Perspect 120:1221–1229

Brar NK, Waggoner C, Reyes JA, Fairey R, Kelley KM (2010) Evidence for thyroid endocrine disruption in wild fish in San Francisco Bay, California, USA. Relationships to contaminant exposures. Aquat Toxicol 96:203–215

Brodin T, Fick J, Jonsson M, Klaminder J (2013) Dilute concentrations of a psychiatric drug alter behavior of fish from natural populations. Science 339:814–815

Burkholder J, Libra B, Weyer P, Heathcote S, Kolpin D, Thorne PS, Wichman M (2007) Impacts of waste from concentrated animal feeding operations on water quality. Environ Health Perspect 115:308–312

Bustnes JO (2007) Pinpointing potential causative agents in mixtures of persistent organic pollutants in observational field studies: a review of glaucous gull studies. J Toxicol Environ Health A 68:97–108

Cardiel IE, Taggart MA, Mateo R (2011) Use of the ratio Pb-Al to discriminate between internal and external deposition of Pb in feathers. Ecotoxicol Environ Saf 74:911–917

Carson R (1962) Silent spring. Haughton Mifflin, Boston

Casanova-Nakayama A, Wenger M, Burki R, Eppler E, Krasnov A, Segner H (2011) Endocrine disrupting compounds: can they target the immune system of fish? Mar Pollut Bull 63:412–416

Celander MC (2011) Cocktail effects on biomarker responses in fish. Aquat Toxicol 105(3–4 Suppl):72–77

Chalew TEA, Halden RU (2009) Environmental exposure of aquatic and terrestrial biota to triclosan and triclocarban. J Am Water Resour Assoc 45:4–13

Chipperfield MP, Dhomse SS, Feng W, McKenzie RL, Velders GJ, Pyle JA (2015) Quantifying the ozone and ultraviolet benefits already achieved by the Montreal Protocol. Nat Commun 6:7233

Colles A, Koppen G, Hanot V, Nelen V, Dewolf M-C, Noël E, Malisch R, Kotz A, Kypke K, Biot P, Vinkx C, Schoeters G (2008) Fourth WHO-coordinated survey of human milk for persistent organic pollutants (POPs): Belgian results. Chemosphere 73:907–914

Cortinovis S, Galassi S, Melone G, Saino N, Porte C, Bettinetti R (2008) Organochlorine contamination in Crested Grebe (*Podiceps cristatus*). Effects on eggshell thickness and egg steroid levels. Chemosphere 73:320–325

Cui B, Zhang Q, Zhang K, Liu X, Zhang H (2011) Analyzing trophic transfer of heavy metals for food webs in the newly-formed wetlands of the Yellow River Delta, China. Environ Pollut 159:1297–1306

Custer CM (2011) Swallows as sentinel species for contaminant exposure and effect studies. In: Elliott JE, Bishop CA, Morrissey CA (eds) Wildlife ecotoxicology: forensic approaches, emerging topics in ecotoxicology 3. Springer, New York, pp 45–91

Cuthbert R, Parry-Jones J, Green RE, Pain DJ (2007) NSAIDs and scavenging birds: potential impacts beyond Asia's critically endangered vultures. Biol Lett 3:90–93

Cuthbert RJ, Taggart MA, Prakash V, Chakraborty SS, Deori P, Galligan T, Kulkarni M, Ranade S, Saini M, Sharma AK, Shringarpure R, Green RE (2014) Avian scavengers and the threat from veterinary pharmaceuticals. Philos Trans R Soc B 369:20130574

Decaëns T, Jiménez JJ, Gioia C, Measey GJ, Lavelle P (2006) The values of soil animals for conservation biology. Eur J Soil Biol 42:S23–S38

Dowding CV, Shore RF, Worgan A, Baker PJ, Harris S (2010) Accumulation of anticoagulant rodenticides in a non-target insectivore, the European hedgehog (*Erinaceus europaeus*). Environ Pollut 158:161–166

Dullinger S, Essl F, Rabitsch W, Erb KH, Gingrich S, Haberl H, Hülber K, Jarosík V, Krausmann F, Kühn I, Pergl J, Pysek P, Hulme PE (2013) Europe's other debt crisis caused by the long legacy of future extinctions. Proc Natl Acad Sci U S A 110:7342–7347

Eason CT, Murphy EC, Wright GR, Spurr EB (2002) Assessment of risks of brodifacoum to non-target birds and mammals in New Zealand. Ecotoxicology 11:35–48

Eens M, Jaspers VLB, Van den Steen E, Bateson M, Carere C, Clergeau P, Costantini D, Dolenec Z, Elliott JE, Flux J, Gwinner H, Halbrook RS, Heeb P, Mazgajski TD, Moksnes A, Polo V, Soler JJ, Sinclair R, Veiga JP, Williams TD, Covaci A, Pinxten R (2013) Can starling eggs be useful as a biomonitoring tool to study organohalogenated contaminants on a worldwide scale? Environ Int 51:141–149

Ek KH, Rauch S, Morrison GM, Lindberg P (2004) Platinum group elements in raptor eggs, faeces, blood, liver and kidney. Sci Total Environ 334–335:149–159

Elliott JE, Elliott KH (2013) Tracking marine pollution. Science 340:556–558

Elliott JE, Birmingham AL, Wilson LK, McAdie M, Trudeau S, Mineau P (2008) Fonofos poisons raptors and waterfowl several months after granular application. Environ Toxicol Chem 27:452–460

Elliott JE, Bishop CA, Morrissey CA (eds) (2011) Wildlife ecotoxicology: forensic approaches, emerging topics in ecotoxicology 3. Springer, New York

Fernie KJ, Shutt JL, Letcher RJ, Ritchie JI, Sullivan K, Bird DM (2008) Changes in reproductive courtship behaviors of adult American kestrels (*Falco sparverius*) exposed to environmentally relevant levels of the polybrominated diphenyl ether mixture, DE-71. Toxicol Sci 102:171–178

Fernie KJ, Shutt L, Letcher RJ, Ritchie IJ, Bird DM (2009) Environmentally relevant concentration of DE-71 and HBCD alter eggshell thickness and reproductive success of American kestrels. Environ Sci Tech 43:2124–2130

Fick J, Lindberg RH, Tysklind M, Larsson DGJ (2010) Predicted critical environmental concentrations for 500 pharmaceuticals. Regul Toxicol Pharmacol 58:516–523

Finkelstein ME, Doak DF, George D, Burnett J, Brandt J, Church M, Grantham J, Smith DR (2012) Lead poisoning and the deceptive recovery of the critically endangered California condor. Proc Natl Acad Sci U S A 109:11449–11454

Fisher IJ, Pain DJ, Thomas VG (2006) A review of lead poisoning from ammunition sources in terrestrial birds. Biol Conserv 131:421–432

Fox GA (2001) Wildlife as sentinels of human health effects in the Great Lakes--St. Lawrence basin. Environ Health Perspect 109(Suppl 6):853–861

Frederick P, Jayasena N (2011) Altered pairing behaviour and reproductive success in white ibises exposed to environmentally relevant concentrations of methylmercury. Proc R Soc B Biol Sci 278:1851–1857

Gabriel MW, Woods LW, Poppenga R, Sweitzer RA, Thompson C, Matthews SM, Higley JM, Keller SM, Purcell K, Barrett RH, Wengert GM, Sacks BN, Clifford DL (2012) Anticoagulant rodenticides on our public and community lands: spatial distribution of exposure and poisoning of a rare forest carnivore. PLoS One 7, e40163

Gagné F, André C, Skirrow R, Gélinas M, Auclair J, van Aggelen G, Turcotte P, Gagnon C (2012) Toxicity of silver nanoparticles to rainbow trout: a toxicogenomic approach. Chemosphere 89:615–622

Gaiser BK, Fernandes TF, Jepson MA, Lead JR, Tyler CR, Baalousha M, Biswas A, Britton GJ, Cole PA, Johnston BD, Ju-Nam Y, Rosenkranz P, Scown TM, Stone V (2012) Interspecies comparisons on the uptake and toxicity of silver and cerium dioxide nanoparticles. Environ Toxicol Chem 31:144–154

García-Fernández AJ, Espín S, Martínez-López E (2013) Feathers as a biomonitoring tool of polyhalogenated compounds: a review. Environ Sci Tech 47:3028–3043

Gibbons D, Morrissey C, Mineau P (2015) A review of the direct and indirect effects of neonicotinoids and fipronil on vertebrate wildlife. Environ Sci Pollut Res Int 22:103–118

Gill RJ, Ramos-Rodriguez O, Raine NE (2012) Combined pesticide exposure severely affects individual-and colony-level traits in bees. Nature 491:105–108

Gómez-Ramírez P, Shore RF, van den Brink NW, van Hattum B, Bustnes JO, Duke G, Fritsch C, García-Fernández AJ, Helander BO, Jaspers V, Krone O, Martínez-López E, Mateo R, Movalli P, Sonne C (2014) An overview of existing raptor contaminant monitoring activities in Europe. Environ Int 67:12–21

Goulson D (2014) Ecology: pesticides linked to bird declines. Nature 511:295–296

Green RE, Pain DJ (2012) Potential health risks to adults and children in the UK from exposure to dietary lead in gamebirds shot with lead ammunition. Food Chem Toxicol 50:4180–4190

Gregory MR (2009) Environmental implications of plastic debris in marine settings--entanglement, ingestion, smothering, hangers-on, hitch-hiking and alien invasions. Philos Trans R Soc B 364:2013–2025

Guertin DA, Harestad AS, Ben-David M, Drouillard KG, Elliott JE (2010) Fecal genotyping and contaminant analyses reveal variation in individual river otter exposure to localized persistent contaminants. Environ Toxicol Chem 29:275–284

Guertin DA, Ben-David M, Harestad AS, Elliott JE (2012) Fecal genotyping reveals demographic variation in river otters inhabiting a contaminated environment. J Wildl Manag 76:1540–1550

Gunnarsson L, Jauhiainen A, Kristiansson E, Nerman O, Larsson DGJ (2008) Evolutionary conservation of human drug targets in organisms used for environmental risk assessments. Environ Sci Tech 42:5807–5813

Gunnarsson L, Kristiansson E, Rutgersson C, Sturve J, Fick J, Förlin L, Larsson DGJ (2009) Pharmaceutical industry effluent diluted 1:500 affects global gene expression, cytochrome P450 1A activity, and plasma phosphate in fish. Environ Toxicol Chem 28:2639–2647

Gutleb AC, Cenijn P, Velzen MV, Lie E, Ropstad E, Skaare JU, Malmberg T, Bergman Å, Gabrielsen GW, Legler J (2010) In vitro assay shows that PCB metabolites completely saturate thyroid hormone transport capacity in blood of wild polar bears (*Ursus maritimus*). Environ Sci Tech 44:3149–3154

Hallmann CA, Foppen RP, van Turnhout CA, de Kroon H, Jongejans E (2014) Declines in insectivorous birds are associated with high neonicotinoid concentrations. Nature 511:341–343

Harris CA, Hamilton PB, Runnalls TJ, Vinciotti V, Henshaw A, Hodgson D, Coe TS, Jobling S, Tyler CR, Sumpter JP (2011) The consequences of feminization in breeding groups of wild fish. Environ Health Perspect 119:306–311

Hartwig E, Clemens T, Heckroth M (2007) Plastic debris as nesting material in a Kittiwake (*Rissa tridactyla*) colony at the Jammerbugt, Northwest Denmark. Mar Pollut Bull 54:595–597

Heath JA, Frederick PC (2005) Relationships among mercury concentrations, hormones, and nesting effort of White Ibises (*Eudocimus albus*) in the Florida Everglades. Auk 122:255–267

Hecker M, Hollert H, Cooper R, Vinggaard A-M, Akahori Y, Murphy M, Nellemann C, Higley E, Newsted J, Wu R, Lam P, Laskey J, Buckalew A, Grund S, Nakai M, Timm G, Giesy J (2007) The OECD validation program of the H295R steroidogenesis assay for the identification of in vitro inhibitors and inducers of testosterone and estradiol production. Phase 2: inter-laboratory pre-validation studies. Environ Sci Pollut Res 14(1 Special Issue): 23–30

Hendriks AJ, Maas-Diepeveen JL, Heugens EH, Van Straalen NM (2005) Meta-analysis of intrinsic rates of increase and carrying capacity of populations affected by toxic and other stressors. Environ Toxicol Chem 24:2267–2277

Henny CJ, Kaiser JL, Grove RA, Johnson BL, Letcher RJ (2009) Polybrominated diphenyl ether flame retardants in eggs may reduce reproductive success of ospreys in Oregon and Washington, USA. Ecotoxicology 18:802–813

Henry M, Beguin M, Requier F, Rollin O, Odoux JF, Aupinel P, Aptel J, Tchamitchian S, Decourtye A (2012) A common pesticide decreases foraging success and survival in honey bees. Science 336:348–350

Hoffman DJ, Rattner BA, Burton GA Jr, Cairns J Jr (eds) (2003) Handbook of ecotoxicology. Publisher Lewis Publishers, Boca Raton

Houde M, Muir DCG, Tomy GT, Whittle DT, Teixeira C, Moore S (2008) Bioaccumulation and trophic magnification of short- and medium-chain chlorinated paraffins in food webs from Lake Ontario and Lake Michigan. Environ Sci Tech 42:3893–3899

Houde M, De Silva AO, Muir DCG, Letcher RJ (2011) Monitoring of perfluorinated compounds in aquatic biota: an updated review. Environ Sci Tech 45:7962–7973

Ikemoto T, Kunito T, Tanaka H, Baba N, Miyazaki N, Tanabe S (2004) Detoxification mechanism of heavy metals in marine mammals and seabirds: interaction of selenium with mercury, silver, copper, zinc, and cadmium in liver. Arch Environ Contam Toxicol 47:402–413

Jackson AK, Evers DC, Etterson MA, Condon AM, Folsom SB, Detweiler J, Schmerfeld J, Cristol DA (2011) Mercury exposure affects the reproductive success of a free-living terrestrial songbird, the carolina wren (*Thryothorus ludovicianus*). Auk 128:759–769

Jagannath A, Shore RF, Walker LA, Ferns PN, Gosler AG (2008) Eggshell pigmentation indicates pesticide contamination. J Appl Ecol 45:133–140

Jaspers VL, Covaci A, Deleu P, Neels H, Eens M (2008) Preen oil as the main source of external contamination with organic pollutants onto feathers of the common magpie (*Pica pica*). Environ Int 34:741–748

Jaspers VL, Covaci A, Deleu P, Eens M (2009) Concentrations in bird feathers reflect regional contamination with organic pollutants. Sci Total Environ 407:1447–1451

Jobling S, Williams R, Johnson A, Taylor A, Gross-Sorokin M, Nolan M, Tyler CR, Van Aerle R, Santos E, Brighty G (2006) Predicted exposures to steroid estrogens in UK rivers correlate with widespread sexual disruption in wild fish populations. Environ Health Perspect 114:32–39

Johansen P, Muir D, Asmund G, Riget F (2004) Human exposure to contaminants in the traditional Greenland diet. Sci Total Environ 331:189–206

Karchner SI, Franks DG, Kennedy SW, Hahn ME (2006) The molecular basis for differential dioxin sensitivity in birds: role of the aryl hydrocarbon receptor. Proc Natl Acad Sci U S A 103:6252–6257

Kidd KA, Blanchfield PJ, Mills KH, Palace VP, Evans RE, Lazorchak JM, Flick RW (2007) Collapse of a fish population after exposure to a synthetic estrogen. Proc Natl Acad Sci U S A 104:8897–8901

La Merrill M, Emond C, Kim MJ, Antignac JP, Le Bizec B, Clément K, Birnbaum LS, Barouki R (2013) Toxicological function of adipose tissue: focus on persistent organic pollutants. Environ Health Perspect 121:162–169

Lahr J, Münier B, De Lange HJ, Faber JF, Sørensen PB (2010) Wildlife vulnerability and risk maps for combined pollutants. Sci Total Environ 408:3891–3898.

Landis WG, Chapman PM (2011) Well past time to stop using NOELs and LOELs. Integr Environ Assess Manag 7:vi–viii

Larsson DGJ, de Pedro C, Paxeus N (2007) Effluent from drug manufactures contains extremely high levels of pharmaceuticals. J Hazard Mater 148:751–755

Laskowski R, Bednarska AJ, Kramarz PE, Loureiro S, Scheil V, Kudłek J, Holmstrup M (2010) Interactions between toxic chemicals and natural environmental factors – a meta-analysis and case studies. Sci Total Environ 408:3763–3774

Lavoie RA, Kyser TK, Friesen VL, Campbell LM (2015) Tracking overwintering areas of fish-eating birds to identify mercury exposure. Environ Sci Technol 49:863–872

Laycock I, Lenthall KM, Barratt AT, Cresswell JE (2012) Effects of imidacloprid, a neonicotinoid pesticide, on reproduction in worker bumble bees (*Bombus terrestris*). Ecotoxicology 21:1937–1945

Lazarus RS, Rattner BA, Brooks BW, Du B, McGowan PC, Blazer VS, Ottinger MA (2015) Exposure and food web transfer of pharmaceuticals in ospreys (*Pandion haliaetus*): predictive model and empirical data. Integr Environ Assess Manag 11:118–129

Lemarchand C, Rosoux R, Berny P (2010) Organochlorine pesticides, PCBs, heavy metals and anticoagulant rodenticides in tissues of Eurasian otters (*Lutra lutra*) from upper Loire River catchment (France). Chemosphere 80:1120–1124

Letcher RJ, Bustnes JO, Dietz R, Jenssen BM, Jørgensen EH, Sonne C, Verreault J, Vijayan MM, Gabrielsen GW (2010) Exposure and effects assessment of persistent organohalogen contaminants in arctic wildlife and fish. Sci Total Environ 408:2995–3043

Li Y, Yu S, Wu Q, Tang M, Pu Y, Wang D (2012) Chronic Al 2O 3-nanoparticle exposure causes neurotoxic effects on locomotion behaviors by inducing severe ROS production and disruption of ROS defense mechanisms in nematode *Caenorhabditis elegans*. J Hazard Mater 219–220:221–230

Lilienthal H, Hack A, Roth-Härer A, Wichert Grande S, Talness CE (2006) Effects of developmental exposure to 2,2′,4,4′,5-pentabromodiphenyl ethter (PBDE-99) on sex steroids, sexual development, and sexually dimorphic behaviour in rats. Environ Health Perspect 114:196–201

Liu X, Vinson D, Abt D, Hurt RH, Rand DM (2009) Differential toxicity of carbon nanomaterials in Drosophila: larval dietary uptake is benign, but adult exposure causes locomotor impairment and mortality. Environ Sci Tech 43:6357–6363

López-Antia A, Ortiz-Santaliestra ME, Mateo R (2014) Experimental approaches to test pesticide-treated seed avoidance by birds under a simulated diversification of food sources. Sci Total Environ 496:179–187

López-Antia A, Ortiz-Santaliestra ME, Mougeot F, Mateo R (2015) Imidacloprid-treated seed ingestion has lethal effect on adult partridges and reduces both breeding investment and offspring immunity. Environ Res 136:97–107

Lovvorn JR, Raisbeck MF, Cooper LW, Cutter GA, Miller MW, Brooks ML, Grebmeier JM, Matz AC, Schaefer CM (2013) Wintering eiders acquire exceptional Se and Cd burdens in the Bering Sea: physiological and oceanographic factors. Mar Ecol Prog Ser 489:245–261

Luttik R, Mineau P, Roelofs W (2005) A review of interspecies toxicity extrapolation in birds and mammals and a proposal for long-term toxicity data. Ecotoxicology 14:817–832

Mahler GJ, Esch MB, Tako E, Southard TL, Archer SD, Glahn RP, Shuler ML (2012) Oral exposure to polystyrene nanoparticles affects iron absorption. Nat Nanotechnol 7:264–271

Margalida A, Bogliani G, Bowden CGR, Donázar JA, Genero F, Gilbert M, Karesh W, Kock R, Lubroth J, Manteca X, Naidoo V, Neimanis A, Sánchez-Zapata JA, Taggart MA, Vaarten J, Yon L, Kuiken T, Green RE (2014) One health approach to the use of veterinary pharmaceuticals. Science 346:1296–1298

Martínez-Haro M, Mateo R, Guitart R, Soler-Rodríguez F, Pérez-López M, María-Mojica P, García-Fernández AJ (2008) Relationship of the toxicity of pesticide formulations and their commercial restrictions with the frequency of animal poisonings. Ecotoxicol Environ Saf 69:396–402

Martinez-Haro M, Taggart MA, Rodríguez Martín-Doimeadios R, Green AJ, Mateo R (2011) Identifying sources of Pb exposure in waterbirds and effects on porphyrin metabolism using non-invasive fecal sampling. Environ Sci Tech 45:6153–6159

Mateo R (2009) Lead poisoning in wild birds in Europe and the regulations adopted by different countries. In: Watson RT, Fuller M, Pokras M, Hunt WG (eds) Ingestion of lead from spent ammunition: implications for wildlife and humans. The Peregrine Fund, Boise, pp 71–98

Mateo R, Beyer WN, Spann JW, Hoffman DJ, Ramis A (2003) Relationship between oxidative stress, pathology, and behavioral signs of lead poisoning in mallards. J Toxicol Environ Health A 66:1371–1389

Mateo R, Taggart M, Green AJ, Cristòfol C, Ramis A, Lefranc H, Figuerola J, Meharg AM (2006) Altered porphyrin excretion and histopathology of greylag geese (*Anser anser*) exposed to soil contaminated with lead and arsenic in the Guadalquivir Marshes, SW Spain. Environ Toxicol Chem 25:203–212

Mateo R, Vallverdú-Coll N, López-Antia A, Taggart MA, Martínez-Haro M, Guitart R, Ortiz-Santaliestra ME (2014) Reducing Pb poisoning in birds and Pb exposure in game meat consumers: the dual benefit of effective Pb shot regulation. Environ Int 63:163–168

Matranga V, Corsi I (2012) Toxic effects of engineered nanoparticles in the marine environment: model organisms and molecular approaches. Mar Environ Res 76:32–40

McKinlay R, Plant JA, Bell JNB, Voulvoulis N (2008) Endocrine disrupting pesticides: implications for risk assessment. Environ Int 34:168–183

Meyer JN, Lord CA, Yang XY, Turner EA, Badireddy AR, Marinakos SM, Chilkoti A, Wiesner MR, Auffan M (2010) Intracellular uptake and associated toxicity of silver nanoparticles in *Caenorhabditis elegans*. Aquat Toxicol 100:140–150

Miller DH, Jensen KM, Villeneuve DL, Kahl MD, Makynen EA, Durhan EJ, Ankley GT (2007) Linkage of biochemical responses to population-level effects: a case study with vitellogenin in the fathead minnow (*Pimephales promelas*). Environ Toxicol Chem 26:521–527

Mineau P (2005) A review and analysis of study endpoints relevant to the assessment of "long term" pesticide toxicity in avian and mammalian wildlife. Ecotoxicology 14:775–799

Mommaerts V, Reynders S, Boulet J, Besard L, Sterk G, Smagghe G (2010) Risk assessment for side-effects of neonicotinoids against bumblebees with and without impairing foraging behaviour. Ecotoxicology 19:207–215

Moore CJ (2008) Synthetic polymers in the marine environment: a rapidly increasing, long-term threat. Environ Res 108:131–139

Movalli P, Duke G (eds) (2008) Monitoring for and with raptors. Ambio (Special issue) 37

Murray M (2011) Anticoagulant rodenticide exposure and toxicosis in four species of birds of prey presented to a wildlife clinic in Massachusetts, 2006–2010. J Zoo Wildl Med 42:88–97

Murvoll KM, Skaare JU, Anderssen E, Jenssen BM (2006) Exposure and effects of persistent organic pollutants in European shag (*Phalacrocorax aristotelis*) hatchlings from the coast of Norway. Environ Toxicol Chem 25:190–198

Naidoo V, Wolter K, Cromarty D, Diekmann M, Duncan N, Meharg AA, Taggart MA, Venter L, Cuthbert R (2010) Toxicity of non-steroidal anti-inflammatory drugs to *Gyps* vultures: a new threat from ketoprofen. Biol Lett 6:339–341

Nash JP, Kime DE, Van der Ven LTM, Wester PW, Brion F, Maack G, Stahlschmidt-Allner P, Tyler CR (2004) Long-term exposure to environmental concentrations of the pharmaceutical ethynylestradiol causes reproductive failure in fish. Environ Health Perspect 112:1725–1733

Noël M, Spence J, Harris KA, Robbins CT, Fortin JK, Ross PS, Christensen JR (2014) Grizzly bear hair reveals toxic exposure to mercury through salmon consumption. Environ Sci Tech 48:7560–7567

Oaks JL, Gilbert M, Virani MZ, Watson RT, Meteyer CU, Rideout BA, Shivaprasad HL, Ahmed S, Chaudhry MJI, Arshad M, Mahmood S, Ali A, Khan AA (2004) Diclofenac residues as the cause of vulture population decline in Pakistan. Nature 427:630–633

Oehlmann J, Schulte-Oehlmann U, Kloas W, Jagnytsch O, Lutz I, Kusk KO, Wollenberger L, Santos EM, Paull GC, Van Look KJW, Tyler CR (2009) A critical analysis of the biological impacts of plasticizers on wildlife. Philos Trans R Soc B 364:2047–2062

Pilsner JR, Lazarus AL, Nam D-H, Letcher RJ, Sonne C, Dietz R, Basu N (2010) Mercury-associated DNA hypomethylation in polar bear brains via the LUminometric methylation assay: a sensitive method to study epigenetics in wildlife. Mol Ecol 19:307–314

Poletta GL, Kleinsorge E, Paonessa A, Mudry MD, Larriera A, Siroski PA (2011) Genetic, enzymatic and developmental alterations observed in *Caiman latirostris* exposed in ovo to pesticide formulations and mixtures in an experiment simulating environmental exposure. Ecotoxicol Environ Saf 74:852–859

Prosser P, Hart AD (2005) Assessing potential exposure of birds to pesticide-treated seeds. Ecotoxicology 14:679–691

Raimondo S, Vivian DN, Delos C, Barron MG (2008) Protectiveness of species sensitivity distribution hazard concentrations for acute toxicity used in endangered species risk assessment. Environ Toxicol Chem 27:2599–2607

Rattner BA, Horak KE, Warner SE, Day DD, Meteyer CU, Volker SF, Eisemann JD, Johnston JJ (2011) Acute toxicity, histopathology, and coagulopathy in American kestrels (*Falco sparverius*) following administration of the rodenticide diphacinone. Environ Toxicol Chem 30:1213–1222

Rattner BA, Lazarus RS, Elliott JE, Shore RF, van den Brink N (2014) Adverse outcome pathway and risks of anticoagulant rodenticides to predatory wildlife. Environ Sci Tech 48:8433–8445

Reid N, Thompson D, Hayden B, Marnell F, Montgomery WI (2013) Review and quantitative meta-analysis of diet suggests the Eurasian otter (*Lutra lutra*) is likely to be a poor bioindicator. Ecol Indic 26:5–13

Richards N (ed) (2012) Carbofuran and wildlife poisoning: global perspective and forensic approaches. Wiley-Blackwell, Chichester

Richards NL, Cook G, Simpson V, Hall S, Harrison N, Scot t KS (2011) Qualitative detection of the NSAIDs diclofenac and ibuprofen in the hair of Eurasian otters (Lutra lutra) occupying UK waterways with GC-MS. Eur J Wildl Res 57:1107–1114

Riley SPD, Bromley C, Poppenga RH, Uzal FA, Whited L, Sauvajot RM (2007) Anticoagulant exposure and notoedric mange in bobcats and mountain lions in urban Southern California. J Wildl Manag 71:1874–1884

Ross P, Birbaum L (2003) Integrated human and ecological risk assessment: a case study of Persistent Organic Pollutants (POPs) in humans and wildlife. Human Toxicol Risk Assess 9:303–324

Sánchez-Barbudo IS, Camarero PR, Mateo R (2012) Primary and secondary poisoning by anticoagulant rodenticides of non-target animals in Spain. Sci Total Environ 420:280–288

Sarabia J, Sánchez-Barbudo IS, Siqueira W, Mateo R, Rolla E, Pizarro M (2008) Lesions associated with the plexus venosus subcutaneus collaris of pigeons with chlorophacinone toxicosis. Avian Dis 52:540–543

Scheifler R, Cœurdassier M, Morilhat C, Bernard N, Faivre B, Flicoteaux P, Giraudoux P, Noël M, Piotte P, Rieffel D, de Vaufleury A, Badot P-M (2006) Lead concentrations in feathers and blood of common blackbirds (*Turdus merula*) and in earthworms inhabiting unpolluted and moderately polluted urban areas. Sci Total Environ 371:197–205

Scholz S, Sela E, Blaha L, Braunbeck T, Galay-Burgos M, García-Franco M, Guinea J, Klüver N, Schirmer K, Tanneberger K, Tobor-Kapłon M, Witters H, Belanger S, Benfenati E, Creton S, Cronin MTD, Eggen RIL, Embry M, Ekman D, Gourmelon A, Halder M, Hardy B, Hartung T, Hubesch B, Jungmann D, Lampi MA, Lee L, Léonard M, Küster E, Lillicrap A, Luckenbach T, Murk AJ, Navas JM, Peijnenburg W, Repetto G, Salinas E, Schüürmann G, Spielmann H, Tollefsen KE, Walter-Rohde S, Whale G, Wheeler JR, Winter MJ (2013) A European perspective on alternatives to animal testing for environmental hazard identification and risk assessment. Regul Toxicol Pharmacol 67:506–530

Seitz F, Bundschuh M, Rosenfeldt RR, Schulz R (2013) Nanoparticle toxicity in Daphnia magna reproduction studies: the importance of test design. Aquat Toxicol 126:163–168

Shaw SD, Brenner D, Bourakovsky A, Mahaffey CA, Perkins CR (2005) Polychlorinated biphenyls and chlorinated pesticides in harbor seals (*Phoca vitulina concolor*) from the northwestern Atlantic coast. Mar Pollut Bull 50:1069–1084

Shore RF, Malcolm HM, Mclennan D, Turk A, Walker LA, Wienburg CL, Burn AJ (2006) Did foot-and-mouth disease-control operations affect rodenticide exposure in raptors? J Wildl Manag 70:588–593

Shore RF, Taggart MA, Smits J, Mateo R, Richards NL, Fryday S (2014) Detection and drivers of exposure and effects of pharmaceuticals in higher vertebrates. Philos Trans R Soc (London) B 369:20130570

Simpson VR, Tomlinson AJ, Molenaar FM, Lawson B, Rogers KD (2011) Renal calculi in wild Eurasian otters (*Lutra lutra*) in England. Vet Rec 169:49

Snape JR, Maund SJ, Pickford DB, Hutchinson TH (2004) Ecotoxicogenomics: the challenge of integrating genomics into aquatic and terrestrial ecotoxicology. Aquat Toxicol 67:143–154

Southard GM, Fries LT, Barkoh A (2010) Prymnesium parvum: the Texas experience. J Am Water Resour Assoc 46:14–23

Souza DM, Teixeira RF, Ostermann OP (2015) Assessing biodiversity loss due to land use with life cycle assessment: are we there yet? Glob Chang Biol 21:32–47

Stahl T, Mattern D, Brunn H (2011) Toxicology of perfluorinated compounds. Environ Sci Eur 23:38

Taggart MA, Reglero MM, Camarero PR, Mateo R (2011) Should legislation regarding maximum heavy metal levels in human food also cover large game meat? Environ Int 37:18–25

Tartu S, Goutte A, Bustamante P, Angelier F, Moe B, Clément-Chastel C, Bech C, Gabrielsen GW, Bustnes JO, Chastel O (2013) To breed or not to breed: endocrine response to mercury contamination by an Arctic seabird. Biol Lett 9:20130317

Teuten EL, Saquing JM, Knappe DR, Barlaz MA, Jonsson S, Björn A, Rowland SJ, Thompson RC, Galloway TS, Yamashita R, Ochi D, Watanuki Y, Moore C, Viet PH, Tana TS, Prudente

M, Boonyatumanond R, Zakaria MP, Akkhavong K, Ogata Y, Hirai H, Iwasa S, Mizukawa K, Hagino Y, Imamura A, Saha M, Takada H (2009) Transport and release of chemicals from plastics to the environment and to wildlife. Philos Trans R Soc B 364:2027–2045

Thomas PJ, Mineau P, Shore RF, Champoux L, Martin PA, Wilson LK, Fitzgerald G, Elliott JE (2011) Second generation anticoagulant rodenticides in predatory birds: probabilistic characterisation of toxic liver concentrations and implications for predatory bird populations in Canada. Environ Int 37:914–920

Topping CJ, Odderskær P (2004) Modeling the influence of temporal and spatial factors on the assessment of impacts of pesticides on skylarks. Environ Toxicol Chem 23:509–520

Ucán-Marín F, Arukwe A, Mortensen A, Gabrielsen GW, Fox GA, Letcher RJ (2009) Recombinant transthyretin purification and competitive binding with organohalogen compounds in two gull species (*Larus argentatus* and *Larus hyperboreus*). Toxicol Sci 107:440–450

van Zelm R, Larrey-Lassalle P, Roux P (2014) Bridging the gap between life cycle inventory and impact assessment for toxicological assessments of pesticides used in crop production. Chemosphere 100:175–181

Vandegehuchte MB, Janssen CR (2014) Epigenetics in an ecotoxicological context. Mutat Res Genet Toxicol Environ Mutagen 764–765:36–45

Vasseur P, Cossu-Leguille C (2006) Linking molecular interactions to consequent effects of persistent organic pollutants (POPs) upon populations. Chemosphere 62:1033–1042

Veldhoen N, Skirrow RC, Brown LLY, Van Aggelen G, Helbing CC (2014) Effects of acute exposure to the non-steroidal anti-inflammatory drug ibuprofen on the developing North American bullfrog (*Rana catesbeiana*) tadpole. Environ Sci Tech 48:10439–10447

Wang CY, Ai N, Arora S, Erenrich E, Nagarajan K, Zauhar R, Young D, Welsh WJ (2006) Identification of previously unrecognized antiestrogenic chemicals using a novel virtual screening approach. Chem Res Toxicol 19:1595–1601

Watanabe KH, Andersen ME, Basu N, Carvan MJ, Crofton KM, King KA, Suñol C, Tiffany-Castiglioni E, Schultz IR (2011) Defining and modeling known adverse outcome pathways: domoic acid and neuronal signaling as a case study. Environ Toxicol Chem 30:9–21

Watson RT, Fuller M, Pokras M, Hunt WG (2009) Ingestion of lead from spent ammunition: implications for wildlife and humans. The Peregrine Fund, Boise

Weiss B (2011) Endocrine disruptors as a threat to neurological function. J Neurol Sci 305:11–21

Wellington EMH, Boxall ABA, Cross P, Feil EJ, Gaze WH, Hawkey PM, Johnson-Rollings AS, Jones DL, Lee NM, Otten W, Thomas CM, Williams AP (2013) The role of the natural environment in the emergence of antibiotic resistance in Gram-negative bacteria. Lancet Infect Dis 13:155–165

Whitehead A, Pilcher W, Champlin D, Nacci D (2012) Common mechanism underlies repeated evolution of extreme pollution tolerance. Proc R Soc B Biol Sci 279:427–433

Zoeller RT, Tan SW, Tyl RW (2007) General background on the hypothalamic-pituitary-thyroid (HPT) axis. Crit Rev Toxicol 37:11–53

Zorrilla I, Martinez R, Taggart MA, Richards N (2015) Suspected flunixin poisoning of a wild Eurasian griffon vulture from Spain. Conserv Biol 29:587–592

Zuccato E, Castiglioni S, Bagnati R, Chiabrando C, Grassi P, Fanelli R (2008) Illicit drugs, a novel group of environmental contaminants. Water Res 42:961–968

New Developments in the Study of Species Distribution

Pelayo Acevedo, Alberto Jiménez-Valverde, Pedro Aragón, and Aidin Niamir

Why Should Species Distribution Be Studied?

The distributional range of a species is determined by factors that operate with different intensities and at diverse scales (Gaston 2003). A species is currently found where abiotic conditions are favourable, where an appropriate suite of species enables co-occurrence, and in those places that can be reached in ecological time (Soberón and Peterson 2005); but the evolutionary history of a species is also highly explicative of its current range (Avise 2000; Barve et al. 2011). These factors interact dynamically to produce the complex entity that represents the geographic distribution of the species. Species distribution is a complex expression of its ecology and evolutionary biology (Brown 1995), and the study of distribution patterns therefore have a high potential to provide relevant information that can be used to understand the evolutionary history of a species (e.g. Richards et al. 2007) and guide species management and conservation policies (e.g. Guisan and Thuiller 2005).

P. Acevedo (✉)
SaBio, Instituto de Investigación en Recursos Cinegéticos (IREC), CSIC-UCLM-JCCM, Ciudad Real, Spain

CIBIO (Centro de Investigação em Biodiversidade e Recursos Genéticos), InBio Laboratório Associado, Universidade do Porto, Porto, Portugal
e-mail: pelayo.acevedo@gmail.com

A. Jiménez-Valverde
Departamento de Ciencias de la Vida, Univerdad de Alcalá (UAH), Alcalá de Henares, Madrid, Spain

P. Aragón
Departamento de Biogeografía y Cambio Global, Museo Nacional de Ciencias Naturales (CSIC), Madrid, Spain

A. Niamir
Range Shifts, Vegetation Dynamics and Ecosystem Processes, Biodiversity and Climate Research Centre (BIK-F) Senckenberg, Frankfurt am Main, Germany

R. Mateo et al. (eds.), *Current Trends in Wildlife Research*,
Wildlife Research Monographs 1, DOI 10.1007/978-3-319-27912-1_7

In a broad sense, biogeography is the discipline that studies the distribution of species and ecosystems in geographic space and throughout geological time. In recent decades the availability of extensive biodiversity databases and the advent of powerful tools with which to manage and spatially analyse explicit information have enabled rapid progress in many areas of both pure and applied biogeography (e.g. Williams et al. 2000; Lomolino and Heaney 2004). Conservation biogeography is *the application of biogeographical principles, theories, and analyses, being those concerned with the distributional dynamics of taxa individually and collectively, to problems concerning the conservation of biodiversity* (Whittaker et al. 2005); conservation biogeography thus encompasses some of the most prominent planning frameworks used in wildlife management and conservation.

In this chapter we have reviewed the study of species distribution as a current trend in wildlife research. The chapter is structured in four major sections: biodiversity databases, species distribution modelling, the influence of global change on species distribution, and species distribution models in international conservation initiatives (see Fig. 1).

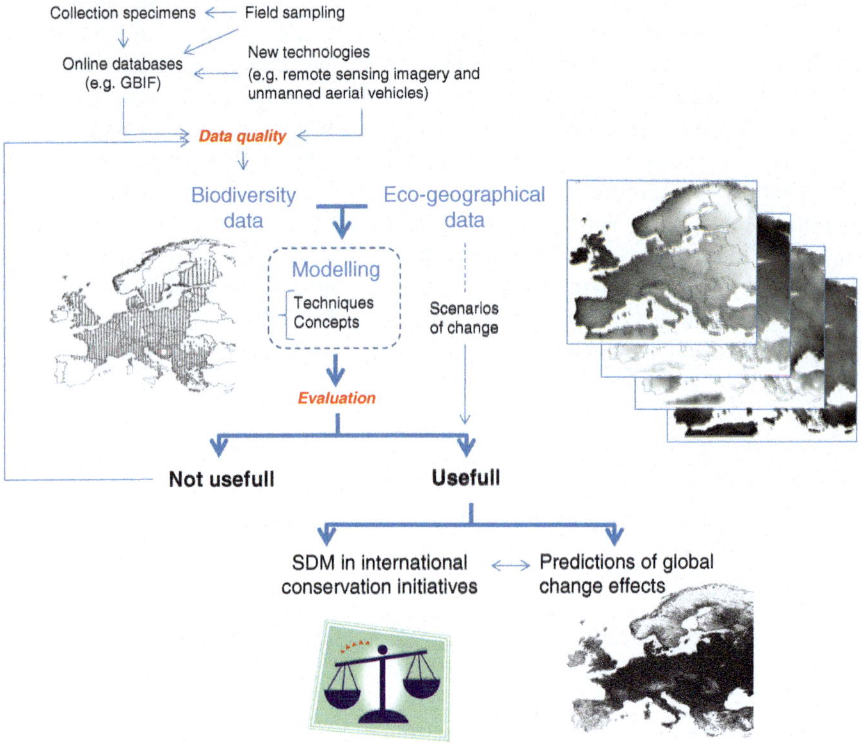

Fig. 1 Flowchart showing the different steps needed for studying species distribution in a context of conservation (SDM, species distribution modelling)

Biodiversity Databases

Primary Data

Wildlife research requires data. The raw data for much of biodiversity research are occurrence records – localities in which species are found in a spatial-temporal context (Gaston 2003). Many of these data were traditionally compiled in natural history collections that are recognized as keepers of the primary information for the biota. More than three billion specimens are estimated to be held in collections throughout the world (e.g. Butler et al. 1998); but compiling these decentralised data has often been an unattainable task for researchers.

In recent decades, primary biodiversity information has started to become accessible by means of the digitalisation of data associated with collection specimens. A good example of this is the Natural History Museum of London which has databased 28 million specimens of its zoological collections (www.nhm.ac.uk/research-curation/scientific-resources/collections/zoological-collections). The needs of biodiversity research to share primary data and the opportunities provided by Internet have converged to place the accumulated information into an electronically available format. In this context, web-portals were created in the last decade and are now a major scientific tool in biodiversity studies (e.g. Shanmughavel 2007).

For biogeographers in general and wildlife researchers in particular, the Global Biodiversity Information Facility (GBIF; www.gbif.org) can be considered the key data infrastructure; but there are similar initiatives such as VertNet (www.vertnet.org/index.php), SpeciesLink (www.splink.cria.org.br), IUCN (www.iucnredlist.org), Ocean Biogeography Information System (www.iobis.org), and others. GBIF is an intergovernmental open data source that allows anyone, anywhere to access data on most plants, animals, fungi and microbes on Earth (Edwards et al. 2000). Since 2004 GBIF's information architecture has made these data accessible and searchable via a single portal. More concretely, it has data for ~440 million occurrences, of more than 1,450,000 species, summarising the information in more than 14,500 databases. Accessing GBIF data is very fast in comparison with compiling data from original sources, thus making large-scale and multitaxon analyses feasible in relatively short timeframes. The power of the data shared by the GBIF infrastructure for research is that much of the information can be mapped spatially, which in turn makes it directly amenable to a vast array of analyses, and principally species distribution modelling in which ranges of species are studied from ecogeographical-based correlations (see section "Species distribution modelling"). Various sound examples of the use of GBIF data for wildlife research can currently be found in literature (e.g. Rissler and Apodaca 2007).

Data Quality

The scarcity of data for some taxa and concerns about data quality and representativeness for others currently limit the use of freely accessible biodiversity databases – including GBIF – and therefore restrict the subsequent benefits for wildlife research (e.g. Yesson et al. 2007). In general, all biodiversity databases contain errors (e.g. Hortal et al. 2007) and data cleaning and quality control procedures must be performed before proceeding (Jiménez-Valverde et al. 2010).

The dynamic nature of taxonomy forces a constant and active updating of databases. Changes in systematic and taxonomy promote outdated records in the databases that should be filtered. Synonyms comprise a high number of names in many databases (e.g. Soberón and Peterson 2004), and the consultation of taxonomic authorities' files is therefore mandatory as regards filtering the available information but, in general, obtaining updated taxonomic files continues to be a major problem for most taxa and they – e.g. Species 2000 (Bisby 2000) – are still far from complete (e.g. Lughadha 2004). What is more important, specimens may have been wrongly identified; this error can be quite frequent and it is difficult to detect and correct without expert participation in the inspection of the original specimens (e.g. Soberón and Peterson 2004), which is a time-expensive task and may be unachievable for large datasets.

Much of the primary data compiled in biodiversity databases include significant bias in the spatial and temporal distribution of collecting efforts, which affects the overall quality of the database. Yesson et al. (2007) evidenced relevant geographical and taxonomical bias in GBIF and concluded that this database is not yet a global biodiversity resource for all species and countries; but this feature is not exclusive to GBIF (see Hortal et al. 2007). Sampling efforts to obtain primary data are usually limited, scattered, and not standardised, and the databases are therefore biased towards easily accessible sampling sites (see Jiménez-Valverde et al. 2010). Spatial bias usually leads to environmental bias because of the over-representation of certain environmental features of the more accessible and extensively surveyed areas, this being quite frequent in scientific databases (Hortal et al. 2008). The study of the distribution patterns represented in spatially and/or environmentally biased databases may therefore lead to inaccurate analyses that may in turn lead to inappropriate management decisions (Moudrý and Šímová 2012). Sampling bias can be addressed by reducing the number of occurrence records in oversampled regions using spatial filtering (Phillips et al. 2009), and other means of incorporating sampling bias into biogeographical analyses have also been proposed (Kramer-Schadt et al. 2013; Byrne et al. 2014).

In addition to the aforementioned errors, errors in georeferencing are also common in biodiversity databases. In this respect, some experiences have been carried out to explore the effects of positional error in studies on species distribution. Graham et al. (2008); Osborne and Leitão (2009) found that some modelling techniques are robust to a moderate level of positional error in sampling localities, and useful analyses of species distributions can consequently be obtained even when some positional error is included. But the impact of positional uncertainty on species occurrences also depends on the spatial autocorrelation in predictors; large ranges of spatial autocorrelation in predictors increase the impact of positional uncertainty when modelling distribution data (Naimi et al. 2011). These authors therefore suggest that the potential impact of positional uncertainty on species occurrences can be understood by comparing it with the spatial autocorrelation range in predictor variables. As a forward step, some authors have explored potential ways in which to manage the positional error in the analyses and have proposed analytical alternatives such as regression calibration procedures (Hefley et al. 2014).

A major peculiarity of most biodiversity databases is that only species occurrences, but not absences, are registered. As described in detail in section "The modelling techniques", some species distribution modelling techniques require presence and absence data. When species that have not been detected during sampling are recorded as absent, although they are in fact present in a given locality (imperfect detection leading to false absences), an additional source of uncertainty emerges in the databases (e.g. Kéry 2011). False absences reduce the predictive accuracy of models that do not account for detection bias and introduce bias into relationships between a species' distribution/abundance and its environment (e.g. Lobo et al. 2010). In general, site-occupancy models – i.e. models designed to account for imperfect detection (e.g. Tyre et al. 2003) – outperform models that do not consider species detection, but only when species detectability is reduced (e.g. Comte and Grenouillet 2013). Site-occupancy models, such as MARK (White and Burnham 1999) and PRESENCE (MacKenzie et al. 2006), and R packages such as 'unmarked' (Fiske and Chandler 2011), are thus promising tools with which to manage imperfect detection bias. However, these models must incorporate further advances as regards modelling procedure (e.g. variable selection) and visualization (e.g. mapping outputs), among others, before they can be a great boost to species distribution modelling (Kéry 2011).

Hence, at a time when gigabytes of primary biodiversity information are becoming available to all, issues related to quality control are more crucial than ever (Moudrý and Šímová 2012). In this context, the introduction of a peer-review system for data publications has been suggested and has begun to be applied as a possible solution to the improvement of data accuracy and a reduction in the need for users to filter errors from the database (Costello et al. 2013). Rigorous sampling protocols have also been described as a means to fill gaps resulting from a lack of knowledge regarding some species and sampling spatial bias (Aranda et al. 2011), which might also increase data use (Costello et al. 2013).

New Technologies and Biodiversity Databases

Obtaining primary data for non-sampled species and/or localities is often labour-intensive and its collection by means of field-based surveys over large spatial extents may even be prohibitive (Turner et al. 2003). Fortunately, recent advances in remotely sensed imagery and related technologies, along with the development of geographic information systems (GIS), have reduced the costs and limitations associated with the collection and processing of species data. Data of interest for wildlife research can currently be remotely measured by using new technologies. In this context, two general approaches can be used to obtain biodiversity information: (i) indirectly, by recording habitat features closely related to the presence of target species, and (ii) directly, by means of species and/or individual identification.

Cutting-edge remote sensing imagery and related technologies can be used to extract independent landscape variables at fine scale which can then be used for wildlife research (Recio et al. 2013). For instance, Hudges et al. (2011) evidenced that the presence of some species can be estimated using Google Earth, always supposing that a suitable signature of species presence can be found in the satellite image. Olea and Mateo-Tomás (2013) similarly used the pictures from Google Street View to survey the vulture's habitat as a cost-effective complementary tool for field surveys. Finally, fine scale independent landscape variables have been obtained for wild mammals by using high resolution remote sensing technology (Recio et al. 2013).

Some studies have also been carried out to monitor biodiversity from space by the direct identification of individuals. The first experiences were carried out with trees (Dennison et al. 2010), but these technologies are now also being applied to monitor wildlife, such as penguins and whales (Fretwell et al. 2012, 2014). At a lower altitude than satellites, unmanned aerial vehicles (UAVs) are transforming remote monitoring by providing perspectives as regards the distribution of wildlife in a cost-effective manner, and at spatial and temporal resolutions that are appropriate to research interests. UAVs are lightweight, low-cost aircraft platforms operated from the ground that can carry different sensors to record georeferenced-environmental information (see Jones et al. 2006; Watts et al. 2010; Anderson and Gaston 2013). Using high resolution onboard sensors, UAVs were able to identify hidden objects under experimental conditions (Martin et al. 2012). These pioneer studies have led to further applications of UAVs in wildlife management, including population monitoring (Vermeulen et al. 2013) and the control of wildlife (Mulero-Pázmány et al. 2014), among others. Although in its infancy, the direct monitoring of species from the air – by both satellite and UAVs – clearly has a relevant potential for wildlife research, is probably the most promising trend for wildlife monitoring, and represents an exciting development for both remote sensing and applied ecology.

Species Distribution Modelling

Modelling the present-day spatial distribution of species' occurrences as a function of ecogeographical variables is termed as Species Distribution Modelling (SDM). Although the study of the response of the species to environmental variables and of the determinants of geographic ranges has a long tradition, the relatively recent developments in computational sciences and GIS technology, and the increased availability of biodiversity data (see section "Biodiversity databases"), have caused SDM to undergo an explosive development (see Fig. 1 in Lobo et al. 2010) and to acquire the category of being a research field in its own right. Yet much debate exists as regards methodological and central conceptual questions, thus highlighting the lack of a unified theoretical basis (Halvorsen 2012). Since extensive reviews can be found in Franklin (2009); Peterson et al. (2011), the remainder of this section will focus on some of the most recent and challenging issues in SDM.

The Modelling Techniques

Numerous techniques exist for the prediction of species distributions, an extensive review of which can be found in Franklin (2009). Using the type of occurrence data on which they are based as a starting point, there are three basic types of modelling methods:

Type 1. The use of presence data only. These models either identify the locations with environmental characteristics that are within the range of values represented by the instances of presence (envelope methods such as BIOCLIM; Busby 1991) or calculate an environmental distance from every territorial unit to those that are occupied (e.g. DOMAIN; Carpenter et al. 1993).

Type 2. These models use both instances of presence and absence and some regressive techniques (e.g. Generalized Linear Models) to estimate a probability of presence (*P*). An interesting modification of *P* has emerged with the favourability function, which was conceived as a means to eliminate the effect of prevalence (the proportion of instances of presences in the sample) from *P* values (Real et al. 2006). Since the prevalence biases the mean probabilities of the presences and absences, *P* values do not represent a measure of environmental quality or favourability, and must therefore be rescaled (Jiménez-Valverde and Lobo 2006). This signifies that the favourability function yields a measure of the degree to which local environmental conditions lead to a local *P* which is higher or lower than that expected by the prevalence of the species in the region under consideration (Acevedo and Real 2012). The favourability function is of the form:

$$F = 1 - \frac{1}{1 + e^{\left(\ln\frac{P}{1-P} - \ln\frac{n_1}{n_o}\right)}}$$

where *P* is the logistic probability, and n_1 and n_0 are the number of instances of presence and absence, respectively, in the sample. Unlike other suitability values obtained with other techniques (see above), *F* is interpretable in absolute terms (Acevedo and Real 2012). Working with *F* rather than *P* thus makes it possible to directly compare the local favourability values between species that differ in their degree of rareness (i.e., prevalence) in the region (Real et al. 2006), and it is therefore possible to combine different F values for multi-species analyses. The characteristics of the favourability function make it a powerful weapon for conservation biogeographers (Acevedo and Real 2012).

Type 3. These models use instances of presence and pseudo-absences (locations in which the focal species has not been found) or background (random sample of locations throughout the territory which may include confirmed presences) data. Some specific techniques and modelling platforms have been developed to work with these data, such as GARP (Stockwell and Noble 1992), ENFA (Hirzel et al. 2002) and MaxEnt (Phillips et al. 2006). What is more, by assuming that pseudo-absences or background data are true absences, this kind of occurrence data is usually used with the techniques that are typically applied in Type 2 models. Type 3 models, which are models of the "used *versus* availability" type (resource selection functions), estimate some sort of suitability values, but not a probability of presence (P) *sensu stricto*.

Key Methodological Issues: Extent of Analysis and Predictors

The delimitation of the extent of analysis (i.e. the geographical area from which the presence-absence data are drawn) is a crucial step in the modelling process that has been overlooked for a long time. Basically, it is a typical sample-selection problem, but in the biogeographic arena it is not a simple question. Several authors have recently shown the dramatic effects that the extent of analysis has on models results. For instance, and not surprisingly, modifying the extent will cause the relative relevance of the predictors to vary while simultaneously altering the geographic projections of the models (i.e. the maps) (VanDerWal et al. 2009a; Anderson and Raza 2010). Working at greater extents will also lead some evaluation measures (see below) to attain higher values since it will be easier to discriminate between occupied and unoccupied locations (Lobo et al. 2008). This last effect is particularly worrisome because it means that by simply choosing one extent or another one can manage to obtain "good and reliable" models. That is why, as stressed by Barve et al. (2011), the selection of the extent of analysis is a crucial step in the modelling process and it has to be justified. Geopolitical boundaries have normally been used; Barve et al. (2011) were the first authors to propose a biological-sound criterion in order to delimit the extent of work. According to these authors, the extent should comprise the parts of the world that have been accessible to the species via dispersal over relevant periods of time. Yet this criterion is fairly inoperable in most situations, thus leading Acevedo et al. (2012a) to devise a functional means to delimit the extent by applying a trend surface analysis that would bound the area in which the focus species is actually interacting with the environment. The delimitation of the extent is undoubtedly an emerging and exciting line of research that promises to give rise to new methods and ideas in forthcoming years.

Another non-trivial question in species distribution modelling is the selection of predictors, although this has in practice and for many years, not been seen as such. The standard procedure is to obtain a set of ecogeographical variables that are available in digital format and to perform some kind of automatic selection procedure. It is assumed that these (distal) variables will be correlated with other proximal variables that exert a more direct influence on species' distributions (Austin 2002). However, different sets of predictors may produce different geographical models even though model performance based on evaluation measures (see below) does not differ very much. This is particularly apparent when the models are transferred in space or time i.e. when they are used to predict regions under invasion risks or future ranges under global change; although models built with different sets of predictors may perform equally well in the training regions, they may yield quite different geographic projections in the new spatial-temporal scenarios (Rödder and Lötters 2009; Synes and Osborne 2011). Braunisch et al. (2013) has even shown that the set of predictors could condition not only the amount of range change in the transferation, but also the direction of change (i.e. contraction versus expansion). Synes and Osborne (2011) showed that, when transferring the models to new climate scenarios, the uncertainty in the predictions associated with the set of variables considered could be higher than the uncertainty associated with the emission scenarios or the general climate models used. Ideally, coherence in the response curves, ecological knowledge based on the natural history of the focus species, and a solid evaluation procedure (see below) should guide the selection of the predictors (Rödder et al. 2009; Halvorsen 2012).

Evaluation of the Models

The evaluation of SDM is a key step since it will provide information about their usefulness and potential applications. Ideally, an evaluation data set that is independent from the training data should be available. "Independence" is a tricky term in a biogeographical context, since occurrence data can hardly be entirely independent. From this point of view, the degree of (in)dependence of the evaluation data set should vary in accordance with the intended application of the model. If new data cannot be gathered (prospective sampling), some kind of partitioning usually takes place in order to keep part of the data for model evaluation (*k*-fold partitioning; Fielding and Bell 1997). This partitioning can be random or spatially structured, depending on whether the model is going to be applied in the same region in which it was parameterized or in a different one (Jiménez-Valverde et al. 2011). Expert opinion can also provide valuable information about models' performance; however, since this information does not represent a "gold standard", special procedures are required for its management (see, for instance, van Zonneveld et al. 2014).

A measure of the capacity of the model to differentiate between instances of presence and instances of absence will typically be of interest. This is called discrimination and can be measured using statistics pertaining to two families: the so-called threshold-dependent and threshold-independent measures (Fielding and Bell 1997). In the first case, the suitability values are divided into two categories: those values that predict the species to be present (1) and those which predict the species to be absent (0). A threshold suitability value is chosen beforehand so that all the values above it are categorized as 1 and the rest as 0. Once this has been done, several statistics can be calculated, such as sensitivity (the proportion of presences correctly predicted), specificity (the proportion of absences correctly predicted), their complementary fractions (false negative and false positive rates, respectively), and the negative and positive predictive values (the rate of predicted absences or presences, respectively, that are real presences or absences), among others (Fielding and Bell 1997). The second family of indexes consists of those that use continuous suitability values. The most widely used statistic pertaining to this category is the AUC (area under the Receiver Operating Characteristic –ROC– curve) (Lobo et al. 2008). In a ROC curve, the sensitivity of the model is plotted against the false positive rate (the proportion of absences incorrectly predicted as presences) across all possible suitability values. The area under this curve i.e. the AUC, is the probability that a presence chosen at random will be assigned a higher suitability value than an absence chosen at random. Thus, although this is usually referred to as a discrimination measure, the AUC is strictly speaking a ranking measure. Jimenez-Valverde (2014) has shown that the AUC is not linearly related to true measures of discrimination capacity, although they positively co-vary. One consequence of this is that the AUC cannot differentiate among high performance models. The AUC has recently received a barrage of other criticisms (Lobo et al. 2008; Hand 2010; Jiménez-Valverde 2012; Smith 2013; but see Pontius and Parmentier 2014) highlighting the risks if it is used as the only measure of models' performance. In spite of the supposed uncertainty associated with the selection of the threshold value, Jimenez-Valverde (2014) has argued that threshold-dependent measures provide more information about models' performance and are of greater value than threshold-independent indexes.

If the suitability values are expressed as the probability of presence *sensu stricto* (Type 2 models), then the calibration of the models is another facet of models' performance that should be evaluated (Pearce and Ferrier 2000). Calibration measures the degree to which the observed proportion of positive cases (empirically estimated probabilities) equates to the models' estimated probabilities. Well-calibrated models provide information about concrete cases and about the uncertainty of the predictions, and from this point of view calibration may therefore be more meaningful than discrimination (Jiménez-Valverde et al. 2013). The calibration plot should be provided together with several indexes (Jiménez-Valverde et al. 2013). In cases in which only information about the presence, not the absence, of the species is available, evaluation becomes challenging since there are usually no *a priori* criteria with which to penalize one model for predicting a greater extent of occurrence than another (Jiménez-Valverde 2012). In a presence-only scenario, the use of typical presence-absence evaluation statistics is usually advised against (Jiménez-Valverde 2012) and other approaches such as predicted/expected curves and the Boyce indices are recommended (Boyce et al. 2002; Hirzel et al. 2006).

However, the evaluation of the models should not be restricted to measuring their calibration or discrimination capacity. Questions such as the quality of the data (Rocchini et al. 2011), the capacity to predict local abundance (VanDerWal et al. 2009b) or functional traits (Thuiller et al. 2009), the degree of transferability (Jiménez-Valverde et al. 2011), or the ecological plausibility of the relationships fitted between the distributional data and the predictors (Halvorsen 2012) are some of the facets of the models that can be examined. It is eventually up to the researcher to choose the way in which to evaluate the models, and what has to be explored is the utility of the model for its intended application; whether it is to locate new populations of an endangered species, to identify suitable places for translocations, in which cases predictive capacity (either of presence/absence data or of functional traits) is important, or simply to increase our general knowledge about a certain species, in which case predictive capacity may not be the most relevant issue (Elith and Leathwick 2009).

The Influence of Global Change on Species Distribution

The International Geosphere-Biosphere Programme defines global change (GC) as the planetary-scale changes in the Earth's system (Earth's interacting physical, chemical, and biological processes). This is an intentionally broad definition that aims to include any processes interlinked at the planetary scale with the potential to have impacts at different scales. During the last two centuries, these changes have been occurring so rapidly and are influenced by human activities to such a great extent that a new geological epoch, the Anthropocene, is being considered as part of the Quaternary period (Zalasiewicz et al. 2011). Various components of GC, such as climate change, land-use change, increasing pollution and increasing transportation networks, are interactively impacting on life on Earth. It has been shown that climate and land-use change have already caused range shifts in many species (Parmesan and Yohe 2003; Gil-Tena et al. 2009). The globalisation of transportation networks has additionally created a human-mediated dispersal vector for invasive species (Tatem and Hay 2007; Wilson et al. 2009). In the last decade, an increasing number of studies have proposed the use of SDM as a means to predict the impact of GC on species distributions; considerable effort has been devoted to their development and to project future species distributions in different GC scenarios (Thomas et al. 2004) or to predicting areas that are susceptible to invasion (Aragón et al. 2010a). As with many other modelling techniques, the use of SDM to predict the geographical responses of species to GC is not free from criticism (e.g. Dormann 2007). While all previously identified limitations of and challenges to SDM are applicable here (Guisan and Thuiller 2005; Araujo and Guisan 2006; Jiménez-Valverde et al. 2008), several are even more problematic or specific to the context of GC. Nevertheless, it is important to note that SDM can be very useful in this context if the results are interpreted appropriately (Jiménez-Valverde et al. 2011; Araújo and Peterson 2012).

Forecasting Range Shifts Caused by Climate and Land-Use Change

In SDM, once the relationships between current species distributions and environmental predictors have been estimated in a georeferenced grid-cell system, the current values for predictor variables are substituted with proposed future values in the model (but see Kearney and Porter 2009, for mechanistic models). These future values are generated in different future change scenarios that are simulated under different assumptions related to future expected socio-economic dynamics (e.g. Carpenter et al. 2005; Solomon et al. 2007). These studies can often be grouped within those that create forecasts for individual species (e.g. Acevedo et al. 2012b) and those that incorporate a huge number of individual forecasts to obtain geographic predictions of species loss and turnover (e.g. Thuiller et al. 2005). Each procedure has its strengths and weaknesses. While individual predictions should incorporate expert knowledge and/or multidisciplinary studies to gain credibility, the inherent complexity of species turnover predictions means that they can only provide clues about general (but informative) trends.

In addition to the uncertainty associated with the use of surrogate predictors, extrapolation exercises, and different modelling techniques (Araújo et al. 2005; Dormann 2007; see section "Species distribution modelling"), a further source of error is the uncertainty associated with climate models and future GC scenarios. This issue is important because a key difference between the latest report from the Intergovernmental Panel on Climate Change (IPCC) and its previous report is that the uncertainties associated with its proposed climate models and regional impacts are higher than expected. For instance, the increasing linear trend in the observed global mean surface temperature has been much smaller over the past 15 years than in previous decades (Flato et al. 2013).

Another important issue that hampers the use of SDM in forecasting range changes is the difficulty involved in integrating species dispersal capacities, which are needed to assess to what extent species will shift their ranges in response to spatial changes in environmental suitability. One common practice is the use of two extreme dispersal scenarios (unlimited dispersal *vs.* no dispersal at all) to cope with geographic changes in environmental suitability and arguing that species will fall somewhere between the two (e.g. Thuiller et al. 2005). However, cellular automata based models have shown that when dispersal parameters are considered, predictions can differ substantially from either unlimited or null dispersal scenarios (Engler and Guisan 2009). Despite the fact that information on the dispersal capabilities of species is often lacking because it is difficult and costly to obtain, individual forecasts can benefit from additional expert knowledge. When there is information about certain parameters, such as dispersal distances and barriers, range shifts resulting from environmental change can be more accurately predicted with the help of specialised software, such as MIGCLIM (Engler et al. 2012).

The challenge for independent model evaluation (see section "Evaluation of the models") in the context of GC is that models cannot be evaluated in the long term, and most of the forecasts are for the years between 2050 and 2100. Interesting exercises have evaluated species' range dynamics for a 20-year time interval in the recent past, and have achieved contrasting results (Araújo et al. 2005; Araújo and Rahbek 2006). However, evaluation through the use of relatively recent past-future events might be only partially useful because model building and data evaluation become less independent (greater temporal autocorrelation) as the time interval between them decreases. Alternatively, although it is less feasible, hindcasting (projecting SDM to previous time periods) provides the possibility of evaluating current SDM with fossil records, which might provide additional insights into the context of model extrapolation and evaluation. However, fossil records are often strongly biased in time and space for various reasons (reviewed in Varela et al. 2011). In spite of the additional limitations inherent to the use of these data in SDM, fossil records provide crucial information about species' associations with the environment that cannot be perceived using only present-day data (Varela et al. 2011). On the other hand, model evaluation using data that are extremely independent in time also has shortcomings because the more distant the temporal projection, the more sources of error there are in the system (e.g. Nogués-Bravo 2009).

Many studies have, until recently, forecasted changes in species distributions by considering the different components of GC separately, which is simplistic because environmental factors may be correlated and/or act jointly. Most of these studies have used only climatic variables to predict species range dynamics, partly because the dynamic land-use change variables were considered to be unreliable or unavailable. Improvements in land use change scenarios are therefore required if more realistic predictions are to be yielded (Martin et al. 2013). For instance, Warren et al. (2001) examined the combined impacts of habitat degradation and climate change on the distribution of British butterflies and found that for three-quarters of the species, their negative responses to habitat loss have outweighed positive responses to climate warming. It has thus been argued that the interactions among predictors should be included in SDM in order to better understand the processes underlying the species' ranges patterns (Guisan and Thuiller 2005; Aragón and Sánchez-Fernández 2013). However, the interactions among different components of GC at various temporal and spatial scales pose a major challenge simply because the strengths and/or directions of these interactions may change throughout time and space. Other studies have used alternative approaches in which the geographic projections of climate change, land-use change (Jetz et al. 2007) and emerging diseases (Hof et al. 2011) are integrated in order to propose potential future combined threats even though the interactions among predictors were not considered during the building of models. Furthermore, just as examining the potential interacting effects of GC is of overwhelming importance, understanding how these components may act independently of each other is as important if we are to understand the underlying processes. Real et al. (2013) developed an approach based on the use of the favourability function (see section "The modelling techniques") to extract the climate component that potentially influences species distributions independently of non-climate factors, and projected species ranges based on either independent or non-independent climate components. This procedure showed that the discrepancies between future species ranges projected with independent and non-independent climate components were species-specific. Aragón et al. (2010b) also found that the independent effect of climate on species distributions differed among functional groups. Taken together, approaches considering combined effects do not consider independent effects and *vice versa*. A future challenge will therefore be to develop a framework that allows for comparisons of range projections with independent effects and those with interacting effects (not only additive effects), which will improve the confidence intervals in predictions and permit a better understanding of the underlying processes.

Prediction of Areas Under Invasion Risk

One of the typical procedures for assessing the risk associated to invasive species is to project the SDM, once it is built, onto areas of interest (but see Guisan et al. 2014 for ordination approaches). Projections onto areas that have already been invaded help to outline the direction of spread from invasion points and to detect potential secondary contact zones in cases of different invasion events. The invasion events are often mediated by global human activity, and this initial phase can therefore be equated to an unlimited dispersal scenario. In such cases, the areas of interest should not be restricted to the zone of invasion because projections elsewhere can provide information about the invasion risk in new areas.

One of the major problems when using SDM to predict areas under invasion risk is that species in the invaded area are often not at equilibrium with the environment because the colonization process is not yet finalised, and the equilibrium postulate is an important assumption for SDM (Guisan and Thuiller 2005). SDM for invasive species should thus aim to characterise the potential distribution rather than the realised distribution. For this and other reasons, the modelling and evaluation of SDM for invasive species should rely more on presences than on absences. Risk maps of potential invasions should originate from predictive models built without absences or their surrogates, as it has been argued that this yields predictions that are closer to the potential distribution (Jiménez-Valverde et al. 2008, 2011; but see Muñoz and Real 2006). For model evaluation, false positives (predictions of suitability in areas in which the species is currently absent) should not be considered to be model failures *a priori* but as potentially colonisable areas in a non-equilibrium scenario. However, there may be false positives in the predictions that could correspond to real errors for other reasons. In an attempt to partially overcome this problem, Aragón et al. (2010a) discriminated between likely real commission errors and potentially colonisable areas through the use of experimentally obtained physiological thresholds.

One of the major criticisms of species distribution projections is that the components of GC might be the drivers of new adaptations that would, in turn, compromise the credibility of such projections (Dormann 2007; Hoffmann and Sgrò 2011). In fact, distinguishing between new evolutionary adaptations and other potentialities not reflected in the observed distributions is difficult to achieve with SDM alone. Because models are built using the realised distributions of species, perfect characterisations of potential distributions are not possible (Jiménez-Valverde et al. 2008). Biological invasions are paradigmatic examples of the inherent complexity; rapid evolutionary change during the process of invasion might occur owing to founder effects, secondary contact zones or new selective pressures (Whitney and Gabler 2008). However, predictions of unfavourable areas that have already been invaded (false negatives) do not necessarily imply a new evolutionary adaptation; it is still possible that the realised native distribution is constrained by factors not included in the model or that the native distribution does not reflect the entire species' potential in terms of phenotypic plasticity. It has been argued that the inclusion of both native and non-native distributions in models may add information about the species' ecological potential (Jiménez-Valverde et al. 2011). Although non-native distributions might not be in equilibrium, their addition to models would increase the possibility of including new adaptations (genetic or plastic) and more accurate predictions of invasion risk could thus be obtained. Guisan et al. (2014) have proposed a niche theory framework with which to detect environmental novelty associated with invaded distributions as an initial step. Despite the fact that applying niche theory is challenging (the source data are realised distributions), this framework would aid in the design of multidisciplinary approaches that integrate SDM, experimentation and genetic characterisation as further steps toward disentangling environmental novelty resulting from evolutionary and ecological changes. In this vein, Hoffmann and Sgrò (2011) have shown that evolutionary changes in traits can be incorporated into mechanistic models, particularly in the case of species with short generation times and large population sizes, traits probably associated with invasive potential.

Final Statement

A final general consideration is that applying niche theory to observed distributions entails a concatenation of assumptions and limitations which, when not addressed, may result in a cascade of sources of error. Ideally, integrating multidisciplinary approaches into specific designs should permit the testing of assumptions and conclusions from SDM by examining whether different areas of research converge on the interpretation of results. However, multidisciplinary in ecology, evolution and conservation is not always feasible owing to limited economic resources. When SDM cannot be integrated into a multidisciplinary design, researchers should bear the correlative nature of SDM in mind when interpreting results.

Species Distribution Models in International Conservation Initiatives

The emergence of the modern quantitative modelling of species distribution has provided new opportunities to support more systematic and evidence-based conservation approaches (Margules and Pressey 2000; Soberón and Peterson 2004). Species distribution models (SDMs) have the potential to play an effective role in multi-scalar policies and to support decisions in different conservation domains

(Guisan and Thuiller 2005; Franklin 2009). Despite an increasing number of peer-reviewed papers related to SDMs and wide-spread claims of its applicability to conservation problems (Guisan and Thuiller 2005; Rodriguez et al. 2007; Elith and Leathwick 2009), examples in which these models are explicitly used to guide decisions or support policies related to natural resources and specifically to conservation biodiversity are often found only in grey literature and are relatively rare in scientific outputs (Guisan et al. 2013).

Below we have briefly investigated the contribution of scientific literature on SDMs in the IPCC subjects. We do not intend to review the application of SDMs nor do we undertake an assessment of the international initiative reports but rather emphasise the importance of expressing the policy context coherently in order to determine where and how SDMs may be even more useful in the new Intergovernmental Science–Policy Platform on Biodiversity and Ecosystem Services (IPBES; www.ipbes.net), and consequently reach the Aichi Biodiversity Targets (www.cbd.int/sp/targets) in 2020.

SDMs in IPCC

The IPCC, one of the most successful international science-policy efforts in history, has recently published its fifth assessment. The IPCC is an example in terms of bringing the scientific community together. IPCC Working Group II (WPII; www.ipcc-wg2.gov) on climate change impacts and adaptation considers the vulnerability and exposure of human and natural systems, the observed impacts and future risks of climate change, and the potential for and limits to adaptation. SDMs have been used to project the potential effects of global changes onto ecosystem properties and species ranges (see section "The influence of global change on species distribution"). However, there is an underlying assumption that effects the contribution of SDMs to integrated climate change scenario assessment (Thuiller 2004, 2007). It has also been indicated that correlative SDMs in general tend to be pessimistic towards species range and thus towards the inferred extinction risk owing to climate change (Settele et al. 2014). This might result from the fact that the exclusion of species interactions has led to the disruption of the biotic community (Visser and Both 2005).

The IPCC WPII fifth assessment report could potentially benefit from the scientific literature on various applications of SDMs in real-world conservation issues. In chapter "Research Priorities and Trends in Infections Shared with Wildlife" of the WPII report, the projected impacts, vulnerabilities and risks of hydrological changes caused by climate change in fresh water ecosystems have been discussed (Jiménez Cisneros et al. 2014). Reference to the SDM related scientific literature, however, was inconspicuous in this chapter. This might be owing to the fact that most species distribution models have used precipitation as proxy to river flow, while this assumption simply ignores the effect of changing flow regimes in the modelling practices (Heino et al. 2009).

Chapter "Wildlife Habitat Requirements: Concepts and Research Approaches" of the WPII report has addressed the vulnerability of terrestrial and fresh water ecosystems to climate change and consequently the impacts on habitat and biodiversity. This chapter has benefitted the most from the peer-reviewed publications in SDMs. The most frequently used SDMs were based on inferring from present spatial patterns in relation to the current climate and making projections to the future distribution under equilibrium conditions. Representing the rate of change during non-equilibrium conditions (see section "The influence of global change on species distribution") requires a mechanistic approach (Keith et al. 2008; Kearney and Porter 2009). However, this approach has received inadequate attention in the report. In chapter "An Overview of Recent Trends in Wildlife Ecotoxicology", ocean systems, peer-reviewed papers on SDMs were used extensively to conclude the effect of future changes in temperature and other physical and chemical oceanographic factors on the distribution of marine fishes and invertebrates (Pörtner et al. 2014). However, uncertainty propagation and specific quantitative projections by SDMs remain imprecise.

The contribution of SDMs to the current IPCC reports is currently limited to impact assessment and principally to biological invasions and species range shifts. They can also be used in other areas; namely adaptation and mitigation of climate change. SDMs can be employed for reserve design and conservation planning (Guisan and Thuiller 2005; Guisan et al. 2013) and can guide the identification of critical habitats over time and space. There are also recent examples that have successfully combined a static SDM with dynamic landscape and population models to forecast the impacts of environmental change on species' status (Franklin 2010). Progress in these topics could provide the next – the sixth – IPCC assessment report with substantial benefits.

SDMs in IPBES

Two thousand and fourteen was an important year for the science–policy platforms that address biodiversity issues (Editorial 2010; Brooks et al. 2014), as IPBES is developing a work programme that includes the preparation of the next global assessment on biodiversity. This calls for an urgent and efficient contribution from the SDM community in biodiversity conservation through the use of practice-oriented case studies. The IPBES work program consists of four objectives. The third objective focuses on strengthening the science-policy interface with regard to thematic and methodological issues. The majority of deliverables of this objective would greatly benefit from the SDMs, e.g. to assess the threat that invasive species pose to biodiversity, and to assess the global status and the trends of this invasion. It is now time to utilise SDMs, to use wide-spread claims of their applicability to conservation policies (Peterson 2011), and to help managers to make better decisions (Sutherland et al. 2004; Sutherland and Freckleton 2012).

Table 1 Summary of the potential application of species distribution modelling for specific conservation goals

Strategic goal	Aichi biodiversity target	Current SDMs tools
A: Address the underlying causes of biodiversity loss by mainstreaming biodiversity throughout government and society	1. Awareness of biodiversity values	Inadequate
	2. Integration of biodiversity values	Inadequate
	3. Incentives	Inadequate
	4. Sustainable production and consumption	Partially adequate
B: Reduce the direct pressures on biodiversity and promote sustainable use	5. Habitat loss, fragmentation and degradation	Adequate
	6. Sustainable exploitation of marine resources	Partially adequate
	7. Biodiversity-friendly agriculture, forestry and aquaculture	Partially adequate
	8. Pollution reduction	Inadequate
	9. Control of invasive alien species	Adequate
	10. Coral reefs and other vulnerable ecosystems	Adequate
C: To improve the status of biodiversity by safeguarding ecosystems, species and genetic diversity	11. Protected areas	Partially adequate
	12. Prevented extinction of threatened species	Adequate
	13. Genetic diversity of socio-economically and culturally valuable species	Partially adequate
D: Enhance the benefits of biodiversity and ecosystem services for all	14. Ecosystem services	Partially adequate
	15. Ecosystem resilience	Partially adequate
	16. Access and benefit sharing	Inadequate
E: Enhance implementation by means of participatory planning, knowledge management and capacity building	17. Policy instrument and communication	Partially adequate
	18. Traditional Knowledge and customary use	Inadequate
	19. Biodiversity knowledge improvement and transfer	Partially adequate
	20. Resource mobilisation	Inadequate

SDMs in Aichi Targets

Parties at the Convention on Biological Diversity (CBD) adopted a Strategic Plan for Biodiversity 2011–2020 consisting of five strategic goals, including twenty Aichi Biodiversity Targets. These parties also committed to using Aichi Biodiversity Targets as a framework for setting national targets and to report on their progress using biodiversity indicators. The task of measuring and monitoring elements of biodiversity and collecting the required data using traditional surveying techniques remains challenging. The records of observed species occurrence do not provide information on locations that have not yet been surveyed and are thus informative only as regards a subset of species range (Rondinini et al. 2006) and in a specific timeframe. SDMs could be employed with regard to their applications as an additional tool in biodiversity monitoring and their potential to support the CBD. The potential application of SDMs in response to the targets is briefly presented in Table 1. These are suggested applications only, and we encourage scientists and modellers to explore the strengths and weaknesses of the distribution modelling approaches and investigate those which might be best suited in their own context. In general, each Target has a biophysical component, e.g. biodiversity loss, and a societal component, e.g. awareness of biodiversity values. Several of the targets may benefit from the SDMs tools in the former component and experience limitations as regards the latter component. In the case of most of the operational indicators, direct complete sampling or earth observation products cannot be employed to measure the indicators directly and should be coupled with SDMs in order to derive measurements or to design cost-effective samplings. Here we briefly present the adequacy of SDMs as regards supporting progress towards each of the Aichi Biodiversity Targets. The potential application for Strategic Goal A is limited, opportunities to contribute to Strategic Goals B and C have already proven to be extensive, whilst recent developments are promising options for Strategic Goals D and E.

We stress the importance of the linkage between SDM research and policy making processes. Researchers should better assess how and when SDMs could be used to guide conservation and biodiversity related policies. We encourage researchers working with SDM to become involved in international conservation initiatives in order to make their technical inputs beneficial to the initiatives' missions.

Acknowledgments All authors contributed equally to this chapter. Pelayo Acevedo and Pedro Aragón were supported by the Ramón y Cajal Programme (RYC-2012-11970 and RYC-2011-07670, respectively). Alberto Jiménez-Valverde was supported by the CSIC JAE-Doc programs, which is partially financed by the European Social Fund. Aidin Niamir was supported by the German Research Foundation (DFG) through the German Federation for the Curation of Biological Data (GFBio) project.

References

Acevedo P, Real R (2012) Favourability: concept, distinctive characteristics and potential usefulness. Naturwissenschaften 99:515–522

Acevedo P, Jiménez-Valverde A, Lobo JM, Real R (2012a) Delimiting the geographical background in species distribution modelling. J Biogeogr 39:1383–1390

Acevedo P, Melo-Ferreira J, Real R et al (2012b) Past, present and future distributions of an Iberian endemic, *Lepus granatensis*: Ecological and Evolutionary Clues from Species Distribution Models. PLoS One 7:e51529

Anderson K, Gaston KJ (2013) Lightweight unmanned aerial vehicles will revolutionize spatial ecology. Front Ecol Environ 11:138–146

Anderson RP, Raza A (2010) The effect of the extent of the study region on GIS models of species geographic distributions and estimates of niche evolution: preliminary tests with montane rodents (genus *Nephelomys*) in Venezuela. J Biogeogr 37:1378–1393

Aragón P, Baselga A, Lobo JM (2010a) Global estimation of invasion risk zones for the western corn rootworm: *Diabrotica virgifera virgifera*: integrating distribution models and physiological thresholds to assess climatic favourability. J Appl Ecol 47:1026–1035

Aragón P, Lobo JM, Olalla-Tárraga MA et al (2010b) The contribution of contemporary climate to ectothermic and endothermic vertebrate distributions in a glacial refuge. Glob Ecol Biogeogr 19:40–49

Aragón P, Sánchez-Fernández D (2013) Can we disentangle predator-prey interactions from species distributions at a macro-scale? A case study with a raptor species. Oikos 122:64–72

Aranda SC, Gabriel R, Borges PAV, de Azevedo EB, Lobo JM (2011) Designing a survey protocol to overcome the Wallacean shortfall: a working guide using bryophyte distribution data on Terceira Island (Azores). Bryologist 114:611–624

Araujo MB, Guisan A (2006) Five (or so) challenges for species distribution modelling. J Biogeogr 33:1677–1688

Araújo MB, Pearson RG, Thuiller W, Erhard M (2005) Validation of species–climate impact models under climate change. Global Change Biol 11:1504–1513

Araújo MB, Rahbek C (2006) How does climate change affect biodiversity? Science 313:1396–1397

Araújo MB, Peterson AT (2012) Uses and misuses of bioclimatic envelope modelling. Ecology 93:1527–1539

Austin MP (2002) Spatial prediction of species distribution: an interface between ecological theory and statistical modelling. Ecol Model 157:101–118

Avise JC (2000) Phylogeography: the history and formation of species. Harvard University Press, Cambridge

Barve N, Barve V, Jiménez-Valverde A et al (2011) The crucial role of the accessibility area in ecological niche modeling and species distribution modeling. Ecol Model 222:1810–1819

Bisby FA (2000) The quiet revolution: biodiversity informatics and the internet. Science 289(5488):2309–2312

Boyce MS, Vernier PR, Nielsen SE, Schmiegelow FKA (2002) Evaluating resource selection functions. Ecol Model 157:281–300

Braunisch V, Coppes J, Arlettaz R, Suchant R, Schmid H, Bollmann K (2013) Selecting from correlated climate variables: a major source of uncertainty for predicting species distributions under climate change. Ecography 36:971–983

Brooks TM, Lamoreux JF, Soberón J (2014) IPBES ≠ IPCC. Trends Ecol Evol 29:543–545

Brown JH (1995) Macroecology. The Chicago University Press, Chicago

Busby JR (1991) BIOCLIM – a bioclimate analysis and prediction system. In: Margules CR, Austin MP (eds) Nature conservation: cost effective biological surveys and data analysis. CSIRO, Melbourne, pp 64–68

Butler D, Gee H, Macilwain C (1998) Museum research comes off list of endangered species. Nature 394:115–117

Byrne AW, Acevedo P, Green S, O'Keeffe J (2014) Estimating badger social-group abundance in the Republic of Ireland using cross-validated species distribution modelling. Ecol Ind 43:94–102
Carpenter G, Gillison AN, Winter J (1993) DOMAIN: a flexible modelling procedure for mapping potential distributions of plants and animals. Biodivers Conserv 2:667–680
Carpenter SR, Pingali PL, Bennett EM et al (2005) Ecosystems and human well-being: scenarios. Island Press, Washington, DC
Comte L, Grenouillet G (2013) Species distribution modelling and imperfect detection: comparing occupancy versus consensus methods. Divers Distrib 19:996–1007
Costello MJ, Michener WK, Gahegan M, Zhang Z-Q, Bourne PE (2013) Biodiversity data should be published, cited, and peer reviewed. Trends Ecol Evol 28:454–461
Dennison PE, Brunelle AR, Carter VA (2010) Assessing canopy mortality during a mountain pine beetle outbreak using GeoEye-1 high spatial resolution satellite data. Remote Sens Environ 114:2431–2435
Dormann CF (2007) Promising the future? Global change projections of species distributions. Basic Appl Ecol 8:387–397
Editorial (2010) Wanted: an IPCC for biodiversity. Nature 465: 525–525.
Edwards JL, Lane MA, Nielsen ES (2000) Interoperability of biodiversity databases: biodiversity information on every desktop. Science 289:2312–2314
Elith J, Leathwick J (2009) Species distribution models: ecological explanation and prediction across space and time. Ann Rev Eco Evol Syst 40:677–697
Engler R, Guisan A (2009) MigClim: predicting plant distribution and dispersal in a changing climate. Divers Distrib 15:590–601
Engler R, Hordijk W, Guisan A (2012) The MIGCLIM R package–seamless integration of dispersal constraints into projections of species distribution models. Ecography 35:872–878
Fielding AH, Bell JF (1997) A review of methods for the assessment of prediction errors in conservation presence/absence models. Environ Conserv 24:38–49
Fiske IJ, Chandler RB (2011) Unmarked: an R package for fitting hierarchical models of wildlife occurrence and abundance. J Stat Softw 43:1–23
Flato G, Marotzke J, Abiodun B et al (2013) Evaluation of climate models. In: Stocker TF, Qin D, Plattner G-K et al (eds) Climate change 2013: the physical science basis. Contribution of Working Group I to the Fifth Assessment Report of the Intergovernmental Panel on Climate Change. Cambridge University Press, Cambridge, pp 741–866
Franklin J (2009) Mapping species distributions. Spatial inference and prediction Cambridge University Press, Cambridge
Franklin J (2010) Moving beyond static species distribution models in support of conservation biogeography. Divers Distrib 16:321–330
Fretwell PT, Larue MA, Morin P et al (2012) An emperor penguin population estimate: the first global, synoptic survey of a species from space. PLoS One 7(4):e33751
Fretwell PT, Staniland IJ, Forcada J (2014) Whales from space: counting southern right whales by satellite. PLoS One 9(2):e88655
Gaston KJ (2003) The structure and dynamics of geographic ranges. Oxford University Press, Oxford
Gil-Tena A, Brotons L, Saura S (2009) Mediterranean forest dynamics and forest bird distribution changes in the late 20th century. Global Change Biol 15:474–485
Graham CH, Elith J, Hijmans RJ, Guisan A, Peterson AT, Loiselle BA (2008) The influence of spatial errors in species occurrence data used in distribution models. J Appl Ecol 45:239–247
Guisan A, Thuiller W (2005) Predicting species distribution: offering more than simple habitat models. Ecol Lett 8:993–1009
Guisan A, Tingley R, Baumgartner JB et al (2013) Predicting species distributions for conservation decisions. Ecol Lett 16:1424–1435
Guisan A, Petitpierre B, Broennimann O et al (2014) Unifying niche shift studies: insights from biological invasions. Trends Ecol Evol 29:260–269
Halvorsen R (2012) A gradient analytic perspective on distribution modelling. Sommerfeltia 35:1–165

Hand DJ (2010) Evaluating diagnostic tests: the area under the ROC curve and the balance of errors. Stat Med 29:1502–1510

Hefley TJ, Baasch DM, Tyre AJ, Blankenship EE (2014) Correction of location errors for presence-only species distribution models. Methods Ecol Evol 5:207–214

Heino J, Virkkala R, Toivonen H (2009) Climate change and freshwater biodiversity: detected patterns, future trends and adaptations in northern regions. Biol Rev 84:39–54

Hirzel AH, Hausser J, Chessel D, Perrin N (2002) Ecological-niche factor analysis: how to compute habitat suitability maps without absence data? Ecology 83:2027–2036

Hirzel A, Le Lay G, Helfer V, Randin C, Guisan A (2006) Evaluating the ability of habitat suitability models to predict species presences. Ecol Model 199:142–152

Hof C, Araújo MB, Jetz W et al (2011) Additive threats from pathogens, climate and land-use change for global amphibian diversity. Nature 480:516–520

Hoffmann AA, Sgrò CM (2011) Climate change and evolutionary adaptation. Nature 470:479–485

Hortal J, Lobo JM, Jiménez-Valverde A (2007) Limitations of biodiversity databases: case study on seed-plant diversity in Tenerife, Canary Islands. Conserv Biol 21:853–863

Hortal J, Jiménez-Valverde A, Gómez JF, Lobo JM, Baselga A (2008) Historical bias in biodiversity inventories affects the observed environmental niche of the species. Oikos 117:847–858

Hughes BJ, Martin GR, Reynolds S (2011) The use of Google Earth (TM) satellite imagery to detect the nests of masked boobies *Sula dactylatra*. Wildl Biol 17:210–216

Jetz W, Wilcove DS, Dobson AP (2007) Projected impacts of climate and land-use change on the global diversity of birds. PLoS Biol 5:e157

Jiménez Cisneros BE, Oki T, Arnell NW, Benito G, Cogley JG, Döll P, Jiang T, Mwakalila SS (2014) Freshwater resources. In: Field CB, Barros VR, Dokken DJ, Mach KJ, Mastrandrea MD, Bilir TE, Chatterjee M, Ebi KL, Estrada YO, Genova RC, Girma B, Kissel ES, Levy AN, Maccracken S, Mastrandrea PR, White LL (eds) Climate change 2014: impacts, adaptation, and vulnerability. Part A: global and sectoral aspects. Contribution of Working Group II to the Fifth Assessment Report of the Intergovernmental Panel of Climate Change. Cambridge University Press, Cambridge/NewYork, pp 229–269

Jiménez-Valverde A, Lobo JM (2006) The ghost of unbalanced species distribution data in geographical model predictions. Divers Distrib 12:521–524

Jiménez-Valverde A, Lobo JM, Hortal J (2008) Not as good as they seem: the importance of concepts in species distribution modelling. Divers Distrib 14:885–890

Jiménez-Valverde A, Lira-Noriega A, Peterson AT, Soberón J (2010) Marshalling existing biodiversity data to evaluate biodiversity status and trends in planning exercises. Ecol Res 25:947–957

Jiménez-Valverde A, Peterson AT, Soberón J et al (2011) Use of niche models in invasive risk assessments. Biol Inv 13:2785–2797

Jiménez-Valverde A (2012) Insights into the area under the receiver operating characteristic curve (AUC) as a discrimination measure in species distribution modelling. Global Ecol Biogeogr 21:498–507

Jiménez-Valverde A, Acevedo P, Barbosa AM, Lobo JM, Real R (2013) Discrimination capacity is species distribution modelling depends on the representativeness of the environmental domain. Global Ecol Biogeogr 22:508–516

Jiménez-Valverde A (2014) Threshold-dependence as a desirable attribute for discrimination assessment: implication for species distribution modelling. Biodivers Conserv 23:369–385

Jones GP, Pearlstine LG, Percival HF (2006) An assessment of small unmanned aerial vehicles for wildlife research. Wildl Soc B 34:750–758

Kearney M, Porter W (2009) Mechanistic niche modelling: combining physiological and spatial data to predict species' ranges. Ecol Lett 12:334–350

Keith DA, Akçakaya HR, Thuiller W et al (2008) Predicting extinction risks under climate change: coupling stochastic population models with dynamic bioclimatic habitat models. Biol Lett 4:560–563

Kéry M (2011) Towards the modelling of true species distributions. J Biogeogr 38:617–618

Kramer-Schadt S, Niedballa J, Pilgrim JD et al (2013) The importance of correcting for sampling bias in MaxEnt species distribution models. Diver Distrib 19:1366–1379
Lobo JM, Jiménez-Valverde A, Real R (2008) AUC: a misleading measure of the performance of predictive distribution models. Global Ecol Biogeogr 17:145–151
Lobo JM, Jiménez-Valverde A, Hortal J (2010) The uncertain nature of absences and their importance in species distribution modelling. Ecography 33:103–114
Lomolino MV, Heaney LR (2004) Frontiers of biogeography: new directions in the geography of nature. Sinauer Associates, Sunderland
Lughadha EN (2004) Towards a working list of all known plant species. Philos Trans R Soc London B 359(1444):681–687
MacKenzie DI, Nichols JD, Royle JA, Pollock KP, Bailey LL, Hines JE (2006) Occupancy estimation and modeling: inferring patterns and dynamics of species occurrence. Academic, San Diego
Margules CR, Pressey RL (2000) Systematic conservation planning. Nature 405:243–253
Martin Y, Van Dyck H, Dendoncker N et al (2013) Testing instead of assuming the importance of land use change scenarios to model species distributions under climate change. Global Ecol Biogeogr 22:1204–1216
Martin J, Edwards HH, Burgess MA et al (2012) Estimating distribution of hidden objects with drones: from tennis balls to manatees. PLoS One 7(6):e38882
Moudrý V, Šímová P (2012) Influence of positional accuracy, sample size and scale on modelling species distributions: a review. Int J Geogr Inf Sci 26:2083–2095
Mulero-Pázmány M, Stolper R, van Essen LD, Negro JJ, Sassen T (2014) Remotely piloted aircraft systems as a rhinoceros anti-poaching tool in Africa. PLoS One 9(1):e83873
Muñoz AR, Real R (2006) Assessing the potential range expansion of the exotic monk parakeet in Spain. Divers Distrib 12:656–665
Naimi B, Skidmore AK, Groen TA, Hamm NAS (2011) Spatial autocorrelation in predictors reduces the impact of positional uncertainty in occurrence data on species distribution modelling. J Biogeogr 38:1497–1509
Nogués-Bravo D (2009) Predicting the past distribution of species climatic niches. Global Ecol Biogeogr 18:521–531
Olea PP, Mateo-Tomás P (2013) Assessing species habitat using google street view: a case study of cliff-nesting vultures. PLoS One 8(1):e54582
Osborne PE, Leitão PJ (2009) Effects of species and habitat positional errors on the performance and interpretation of species distribution models. Divers Distrib 15:671–681
Parmesan C, Yohe G (2003) A globally coherent fingerprint of climate change impacts across natural systems. Nature 421:37–42
Pearce J, Ferrier S (2000) Evaluating the predictive performance of habitat models developed using logistic regression. Ecol Model 133:225–245
Peterson AT (2011) Ecological niche conservatism: a time-structured review of evidence. J Biogeogr 5:816–827
Peterson AT, Soberón J, Pearson RG, Anderson RP, Martínez-Meyer E, Nakamura M, Araújo MB (2011) Ecological niches and geographic distributions. Princeton University Press, Princeton
Phillips SJ, Anderson RP, Schapire RE (2006) Maximum entropy modelling of species geographic distributions. Ecol Model 190:231–259
Phillips SJ, Dudík M, Elith J, Graham CH, Lehmann A, Leathwick J, Ferrier S (2009) Sample selection bias and presence-only distribution models: implications for background and pseudo-absence data. Ecol Appl 19:181–197
Pontius RG, Parmentier B (2014) Recommendations for using the relative operating characteristic (ROC). Landscape Ecol 29:367–382
Pörtner HO, Karl D, Boyd PW et al (2014) Ocean systems. In: Field CB, Barros VR, Dokken DJ, Mach KJ, Mastrandrea MD, Bilir TE, Chatterjee M, Ebi KL, Estrada YO, Genova RC, Girma B, Kissel ES, Levy AN, Maccracken S, Mastrandrea PR, White LL (eds) Climate change 2014: impacts, adaptation, and vulnerability. Part A: global and sectoral aspects. Contribution of Working Group II to the Fifth Assessment Report of the Intergovernmental Panel of Climate Change. Cambridge University Press, Cambridge/NewYork, pp 411–484

Real R, Romero D, Olivero J et al (2013) Estimating how inflated or obscured effects of climate affect forecasted species distribution. PLoS One 8:e53646

Real R, Barbosa AM, Vargas JM (2006) Obtaining environmental favourability functions from logistic regression. Environ Ecol Stat 13:237–245

Recio MR, Mathieu R, Hall GB, Moore AB, Seddon PJ (2013) Landscape resource mapping for wildlife research using very high resolution satellite imagery. Meth Ecol Evol 4:982–992

Richards CL, Carstens BC, Lacey Knowles L (2007) Distribution modelling and statistical phylogeography: an integrative framework for generating and testing alternative biogeographical hypotheses. J Biogeogr 34:1833–1845

Rissler LJ, Apodaca JJ (2007) Adding more ecology into species delimitation: ecological niche models and phylogeography help define cryptic species in the black salamander (*Aneides flavipunctatus*). Syst Biol 56(6):924–942

Rocchini D, Hortal J, Lenygel S, Lobo JM, Jiménez-Valverde A, Ricotta C, Bacaro G, Chiarucci A (2011) Uncertainty in species distribution mapping and the need for maps of ignorance. Prog Phys Geog 35:211–226

Rödder D, Lötters S (2009) Niche shift versus niche conservatism? Climatic characteristics of the native and invasive ranges of the Mediterranean house gecko (*Hemidactylus turcicus*). Global Ecol Biogeogr 18:674–687

Rödder D, Schmidtlein S, Veith M, Lötters S (2009) Alien invasive slider turtle in unpredicted habitat: a matter of niche shift or of predictors studied? PLoS One 4:e7843

Rodriguez JP, Brotons L, Bustamante J, Seoane J (2007) The application of predictive modelling of species distribution to biodiversity conservation. Divers Distrib 13:243–251

Rondinini C, Wilson KA, Boitani L, Grantham H, Possingham HP (2006) Tradeoffs of different types of species occurrence data for use in systematic conservation planning. Ecol Lett 9:1136–1145

Settele J, Scholes R, Betts R, Bunn SE, Leadley P, Nepstad D, Overpeck JT, Taboada MA (2014) Terrestrial and inland water systems. In: Field CB, Barros VR, Dokken DJ, Mach KJ, Mastrandrea MD, Bilir TE, Chatterjee M, Ebi KL, Estrada YO, Genova RC, Girma B, Kissel ES, Levy AN, Maccracken S, Mastrandrea PR, White LL (eds) Climate change 2014: impacts, adaptation, and vulnerability. Part A: global and sectoral aspects. Contribution of Working Group II to the Fifth Assessment Report of the Intergovernmental Panel of Climate Change. Cambridge University Press, Cambridge/NewYork, pp 271–359

Shanmughavel P (2007) An overview on biodiversity information in databases. Bioinformation 1:367–369

Smith AB (2013) On evaluating species distribution models with random background sites in place of absences when test presences disproportionately sample suitable habitat. Divers Distrib 19:867–872

Soberón J, Peterson T (2004) Biodiversity informatics: managing and applying primary biodiversity data. Philos. Trans. R. Soc. London B 359:689–698

Soberón J, Peterson AT (2005) Interpretation of models of fundamental ecological niches and species' distributional areas. Biodiv Informatics 2:1–10

Solomon S, Qin D, Manning M et al (eds) (2007) The physical science basis, Contribution of Working Group I to the Fourth Assessment Report of the Intergovernmental Panel on Climate Change. Cambridge University Press, Cambridge/NewYork

Stockwell DRB, Noble IR (1992) Induction of sets of rules from animal distribution data – a robust and informative method of data analysis. Math Comput Simulat 33:385–390

Sutherland WJ, Freckleton RP (2012) Making predictive ecology more relevant to policy makers and practitioners. Philos Trans R Soc Lond B Biol Sci 367:322–330

Sutherland WJ, Pullin AS, Dolman PM, Knight TM (2004) The need for evidence-based conservation. Trends Ecol Evol 19:305–308

Synes NW, Osborne PE (2011) Choice of predictor variables as a source of uncertainty in continental-scale species distribution modelling under climate change. Global Ecol Biogeogr 20:904–914

Tatem AJ, Hay SI (2007) Climatic similarity and biological exchange in the worldwide airline transportation network. Proc Roy Soc Lond B 274:1489–1496
Thomas CD, Cameron A, Green RE et al (2004) Extinction risk from climate change. Nature 427:145–148
Thuiller W (2004) Patterns and uncertainties of species' range shifts under climate change. Global Change Biol 10:2020–2027
Thuiller W, Lavorel S, Araújo MB et al (2005) Climate change threats to plant diversity in Europe. Proc Natl Acad Sci U S A 102:8245–8250
Thuiller W (2007) Biodiversity: climate change and the ecologist. Nature 448:550–552
Thuiller W, Albert CH, Dubuis A, Randin C, Guisan A (2009) Habitat suitability models do not always relate to variation in species' plant functional traits. Biol Lett 6(1):120–123
Turner W, Spector S, Gardiner N, Fladeland M, Sterling E, Steininger M (2003) Remote sensing for biodiversity science and conservation. Trends Ecol Evol 18(6):306–314
Tyre AJ, Tenhumberg B, Field SA, Niejalke D, Parris K, Possingham HP (2003) Improving precision and reducing bias in biological surveys: estimating false-negative error rates. Ecol Appl 13:1790–1801
van Zonneveld M, Castañeda N, Scheldeman X, van Etten J, Van Damme P (2014) Application of consensus theory to formalize expert evaluations of plant species distribution models. Appl Veg Sci 17:528–542
VanDerWal J, Shoo LP, Graham C, Williams SE (2009a) Selecting pseudo-absence data for presence-only distribution modeling: how far should you stray from what you know? Ecol Model 220:589–594
VanDerWal J, Shoo LP, Johnson CN, Williams SE (2009b) Abundance and the environmental niche: environmental suitability estimated from niche models predicts the upper limit of local abundance. Am Nat 174:282–291
Varela S, Lobo JM, Hortal J (2011) Using species distribution models in paleobiogeography: a matter of data, predictors and concepts. Palaeogeogr Palaeoclimatol Palaeoecol 310:451–463
Vermeulen C, Lejeune P, Lisein J, Sawadogo P, Bouche P (2013) Unmanned aerial survey of elephants. PLoS One 8(2):e54700
Visser ME, Both C (2005) Shifts in phenology due to global climate change: the need for a yardstick. Proc Biol Sci 272:2561–2569
Warren MS, Hill JK, Thomas JA et al (2001) Rapid responses of British butterflies to opposing forces of climate and habitat change. Nature 414:65–69
Watts AC, Perry JH, Smith SE et al (2010) Small unmanned aircraft systems for low-altitude aerial surveys. J Wildl Manag 74:1614–1619
White GC, Burnham KP (1999) ProgramMARK: survival estimation from populations of marked animals. Bird Study 46:120–139
Whitney KD, Gabler CA (2008) Rapid evolution in introduced species: 'invasive traits' and recipient communities: challenges for predicting invasive potential. Divers Distrib 14:569–580
Whittaker RJ, Araújo MB, Jepson P, Ladle RJ, Watson JEM, Willis KJ (2005) Conservation biogeography: assessment and prospect. Divers Distrib 11:3–23
Williams PH, Humphries C, Araújo MB, Lampinen R, Hagemeijer W, Gasc J-P, Mitchell-Jones T (2000) Endemism and important areas for representing European biodiversity: a preliminary exploration of atlas data for plants and terrestrial vertebrates. Belg J Entomol 2:221–246
Wilson JRU, Dormontt EE, Prentis PJ et al (2009) Something in the way you move: dispersal pathways affect invasion success. Trends Ecol Evol 24:136–144
Yesson C, Brewer PW, Sutton T et al (2007) How global is the global biodiversity information facility? PLoS One 2(11):e1124
Zalasiewicz J, Williams M, Haywood A et al (2011) The Anthropocene: a new epoch of geological time? Phil Trans R Soc A 369:835–841

Progresses and Controversies in Invasion Biology

Daniel Sol

Invasion biology is concerned with understanding the causes and consequences of the human-assisted introduction of organisms outside their native ranges. Ever since Elton published the foundational book "The Ecology of Invasions by Animals and Plants" (Elton 1958), the field has gained enormous importance in ecology. A major motivation in developing the discipline has been the growing concern over the environmental, economic and social impact caused by some invaders (Martin and Maron 2012). Non-indigenous species (NIS, hereafter) are an important cause of species extinction, as exemplified by the introduction of the Brown snake in Guam (Fritts and Rodda 1998) or the Nile perch in the African great lakes (Miller 1989); alter key ecosystem functions, like the nutrients cycle and fire regimes (Vitousek et al. 1987; Levine et al. 2004; Callaway and Maron 2006; Vilà et al. 2011); and generate every year millions of euros of economic loses (Pimentel et al. 2001).

Paralleling the concern over the impact of invaders, there has been an increasing appreciation that NIS represent unique opportunities for studying a variety of ecological and evolutionary processes as they occur and at unprecedented spatial and temporal scales. Thus, biological invasions have contributed to reshape important ideas in genetics, behavioral ecology, population dynamics, community ecology and evolutionary ecology (Lodge 1993; Callaway and Maron 2006; Sax et al. 2007).

In developing the discipline, two fundamental questions have almost monopolized the research agenda: Why are some species invasive and others are not? And why are some environments more invaded than others? Despite the enormous progress in answering these questions, the field remains controversial and has been criticized by its dissociation from the rest of ecology and its lack of rigor in adopting and rejecting new hypotheses and theories (Davis 2009). Some have even questioned that the field is useful at all (Valéry et al. 2013; but see Blondel et al. 2013).

D. Sol
CREAF (Centre for Ecological Research and Forestry Applications), CSIC (Spanish National Research Council), Bellaterra, Catalonia E-08193, Spain
e-mail: dsolrueda@gmail.com

R. Mateo et al. (eds.), *Current Trends in Wildlife Research*,
Wildlife Research Monographs 1, DOI 10.1007/978-3-319-27912-1_8

The goal of this chapter is to highlight several conceptual areas that are currently dominating the field. Specifically, I focus on advancing 11 major themes in invasion biology (Table 1), which altogether provides the necessary framework to understanding what make species successful invaders (invasiveness) and what make ecosystems more or less susceptible to invasion (invasibility). While these themes clearly illustrate the progress made in the field, theoretical and empirical advances are needed in all these areas (see also Jeschke et al. 2012). Consequently, I also highlight controversies and underexploited areas that, if addressed, have the potential to reshape the field in the near future.

What Is a Successful Invader?

Progress in invasion biology has been hampered by a lack of an unifying framework to describe the invasion process, particularly between ecologists working in plants and animals (Blackburn et al. 2011). However, some consensus has now been reached (Richardson et al. 2000; Kolar and Lodge 2001; Duncan et al. 2003; Blackburn et al. 2011), which sees the invasion process as a sequence of several stages. To become a successful invader the organism must firstly be deliberately or accidentally transported (transport stage) and introduced (introduction stage) to a new location by humans; next, the organism must establish a self-sustaining population in the novel environment (establishment stage); finally, the population must increase in abundance and expand their geographic area (spread stage).

It is important to recognize the different stages because each stage influences subsequent stages (Kolar and Lodge 2001). The transport/introduction stage, for example, determines the size, structure and genetics of the founder population, which largely affect the likelihood of establishment. Moreover, the underlying processes can vary among stages. For example, while the probability of establishment is driven by population dynamics in the context of small populations and novel

Table 1 Major biological invasions themes discussed in the present chapter

What is a successful invader?
Are biological invasions highly idiosyncratic?
Is the establishment success of invaders a paradox?
What makes a successful invader?
Is growing fast a key feature of successful invaders?
Do communities differ in invasibility?
Are close relatives bad neighbors?
Are introduced species creating an "invasional meltdown"?
Does "enemy release" explain the increase and expansion of invaders?
What is the role of evolution in the invasion process?
Are exotics decreasing, maintaining or increasing biodiversity?

environments, the spread stage is more related to dispersal ability, carrying capacity and ecological interactions (Leung et al. 2012; Sol et al. 2012b). Finally, the impact of the invader can be expressed at several stages. The impact is potentially greater for widespread NIS, yet it also depends on their abundance and per capita effects (Parker et al. 1999) and hence can also be high even when the NIS are localized geographically.

It is widely accepted that a small proportion of species are introduced and establish, and a small proportion of established species spread and becomes a pest. An attempt to give numbers to these proportions is the "tens-rule", which holds that over 10 % of species transition between these different stages (Williamson et al. 1986; Williamson and Fitter 1996). The "tens-rule" has been misinterpreted by some as if it was a fundamental constant of nature, like the gravitational constant in physics. However, this was not the intention of Williamson and co-workers when they proposed the concept, as the number has scarce theoretical basis beyond the heuristic value of arguing that the probability of transition between invasion stages is low.

Why Are Biological Invasions Highly Idiosyncratic?

The search for general rules that govern invasions has often been unfruitful, with many studies -including those related to the search for features that explain invasiveness and invasibility- yielding idiosyncratic results (Williamson et al. 1986; Moles et al. 2012). Several factors can contribute to explain such idiosyncrasies, besides differences in the quality of the studies, yet the main factors relate to the routes toward extinction in introduced populations.

For an invader, a negative population growth is perhaps the most obvious route toward extinction (Figs. 1 and 2). NIS are exposed to novel environmental conditions to which they have had little opportunity to adapt (Figs. 1 and 2). If as a result there is an adaptive mismatch that negatively affects key fitness components, then the population can decrease over time and end up extinct. It follows that the same species can succeed in some environments but not in others, depending on the degree of adaptive matching. Likewise, a region can seem to be less resistant to invaders than others simply because the adaptive matching of the introduced species is higher. This can be accentuated because of the non-random selection of the species used in introductions and the places where they were introduced (Blackburn and Duncan 2001). Because the features of the pool of species introduced may differ from location to location, the search for traits that make species good invaders can yield different results depending on the region investigated.

Even if the population has the adaptations needed to survive and reproduce in the new environment, an introduced population may die out as a result of bad luck (Fig. 1). This is because most introduced populations start with a reduced number of individuals, which make them highly vulnerable to extinction by demographic stochasticity, Allee effects and genetic stochasticity (Fig. 1). Indeed,

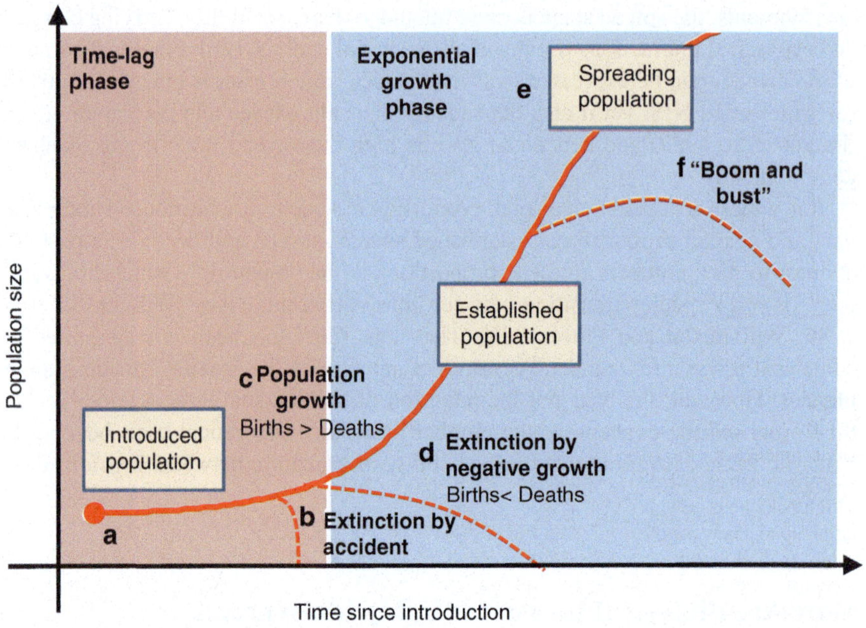

Fig. 1 A framework for the invasion process. The invasion process starts with an introduced population transported from a relatively distant region (*a*), which generally involves a low number of individuals and can have suffered a genetic bottleneck. The population can remain at low numbers for some time, even decades (time-lag phase, in red), during which is highly vulnerable to extinction by accident (i.e. demographic stochasticity, Allee effects and genetic stochasticity) (*b*). The establishment of the species is determined by a positive balance between births and deaths, which largely depends on the species' life history (*c*). However, a number of factors in addition of an appropriate life history can lead to a negative population growth conductive to extinction (*d*), such as an adaptive mismatch or a lack of adaptive plasticity to exploit the available niches. If the population is freed from competitors and enemies, it can increase exponentially in numbers (exponential growth phase, in blue) and start expanding to adjacent areas at a speed limited by dispersal capacity and evolutionary dynamics related to assortative mating (*e*). However, even populations that have initially increased and expanded can sometimes experience a 'boom and bust' in which it experiences a decline and can even go extinct (*f*)

propagule pressure (i.e. variation in the quantity, composition and rate of supply of NIS, sensu Ricciardi et al. 2011) is the most consistent predictor of the probability of establishment in plants and animals (Lockwood et al. 2005). Thus, a same species can have more or less success in a novel environment depending on the number of individuals introduced. Likewise, some systems may look like if they were highly susceptible to be invaded simply because many species have been introduced there and/or the species have been released in larger numbers. Islands, for example, often present a higher number of invaders than continents, which has led to think that they were more vulnerable to invasions. However, quantitative analyses in birds have revealed that the high invasions rates on islands is primarily associated with higher propagule pressure rather than higher invasibility (Sol 2000; Cassey et al. 2004).

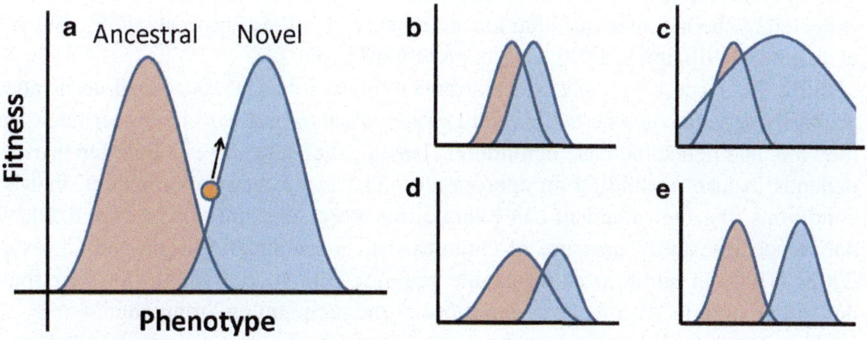

Fig. 2 The problems that a species faces in a novel environment can be metaphorically described in terms of adaptive surfaces, representing mean fitness of a population in the ancestral and novel environments as a function of the organism phenotype (**a**). The first problem is that if there is an adaptive mismatch, the population will see its mean fitness reduced (*red dot*) and hence it will run a high risk of extinction. The second problem is that natural selection is unlikely to move the population up to a new adaptive peak if there is not enough useful heritable variation and the population is too small to resist strong selective pressures. These difficulties are nonetheless reduced when (**b**) there is environmental matching between the region of origin and introduction, and hence the organism already have the necessary adaptations; (**c**) the new adaptive peak is not very demanding in terms of adaptive specializations; and (**d**) the niche of the NIS is so broad that the adaptive peaks of the environments of origin and introductions overlap to a great extent. However, there is also the possibility that niche adaptive peaks of the environments of origin and introductions are substantially different (**e**), and then we need to invoke general adaptations to environmental changes like phenotypic plasticity to understand the persistence of the organism in the novel environment

The idiosyncratic nature of the invasion process implies high uncertainties in predicting the outcome of each invasion stage (Leung et al. 2012). This is unfortunate because accurately assessing invasion success is essential to prevent and mitigate the impact of biological invasions (Kolar and Lodge 2002, Vall-llosera and Sol 2009; Leung et al. 2012). While it is not currently possible to accurately predict the outcome of a particular introduction, ecologists can still make powerful generalizations about the invasion process (Ehrlich 1989; Duncan et al. 2003; Callaway and Maron 2006; Sax et al. 2007). As we will see in the next sections, these generalizations should not be taken as laws, in the sense that physicists use the term, but as patterns that are more or less predominant despite containing some exceptions (Simberloff 2013).

Is the Establishment Success of Invaders a Paradox?

The invasion success of NIS is in a way paradoxical because we do not expect that species that come from distant regions can succeed to establish themselves in environments to which they have had little opportunity to adapt (Sax and Brown 2000); and even more paradoxical is that sometimes the invader attains higher densities than most native species. This is less a paradox however when considering that, as expected by theory, most introduction attempts fail (Williamson et al. 1986; Veltman et al. 1996; Williamson 1996; Haight and Polasky 2010).

Still, the success of some NIS warrants explanation. The success of an invader primarily depends on whether individuals are able to reproduce at a higher rate than they die, and hence increase in numbers. Having such a positive population growth depends in turn on finding an appropriate niche in the new environment, that is, conditions that the organism can tolerate, resources that are not monopolized by native species, and a pressure of enemies that is sustainable (Shea and Chesson 2002). Thus, in addition of propagule pressure, the success of the invader may depend on both its own features and those of the recipient environment.

There are several ways by which exotic organisms can acquire a niche in a novel environment (Table 2). First, if competition for resources is strong and environmental adversity is weak, then the success of the invader relies on being competitively superior to the species with which its niche overlaps to a greater extent (Tilman 2004; Vilà et al. 2005). Second, if competition in the invaded community is weak and environmental adversity is strong, then the invader will only succeed if it has the adaptations needed to survive and reproduce in the novel environment. Third, if both competitive adversity and environmental adversity are weak, then there is no need to invoke adaptations to understand the success; this can be understood by neutral processes in which species are ecologically equivalent (Weiher and Keddy 1995). Finally, if both competitive adversity and environmental adversity are strong, then the invader would need the unlikely combination of high competitive ability and appropriate adaptations to a demanding environment.

Although there is evidence for the case-by-case importance of all these scenarios, except perhaps for the latest scenario, there has been little effort to investigate to what extent they provide a general solution to the invasion paradox (Sol et al. 2012a). Nevertheless, the few existing studies to date provide greater support for the second scenario, in which success depends on being able to fill a niche infra-utilized by native species. In plants, native and alien species often differ in the environments they use, with NIS primarily found in environments that have been modified by vegetation clearance, pasture development and livestock grazing (Pouteau et al. 2015). In birds, many NIS are not only restricted to urban or rural environments, where the diversity and abundance of native species are low (Case 1996; Sol et al.

Table 2 Scenarios accounting for the success of biological invasions

	Competitive adversity	Environmental adversity	Main process	Profile invader
Scenario 1	Strong	Weak	Biotic resistance	Strong competitor
Scenario 2	Weak	Strong	Environmental filtering	Adapted
Scenario 3	Weak	Weak	Neutrality	Ecological equivalence
Scenario 4	Strong	Strong	Biotic-Environmental	None

2012a; Barnagaud et al. 2013), but they overlap little with native species in traits associated with resource use (Sol et al. 2012a; Barnagaud et al. 2013). Although these evidence do not deny the possibility that being superior in contest competition against native species provide advantages in some cases (e.g. Sol et al. 2012a; Hernández-Brito et al. 2014), they do suggest that this is the exception rather than the norm.

What Makes a Successful Invader?

Even when many invaders are restricted to anthropogenic-disturbed environments, the question remains as to why they are able to survive and reproduce in those environments and most native species are not. While understanding failures is relatively easy in light of current theories, what makes some species successful invaders is less obvious. Why should an organism be able to cope with novel environmental pressures to which it has never or rarely been exposed before? One obvious possibility is that there is environmental matching between the place of introduction and that of origin of the NIS, implying that the species already possess adaptations to persist in the novel environment (Table 2, scenario 2; Fig. 2a). Although NIS typically come from distant regions, certain environmental matching is still expected. This is clear for deliberate introductions as humans are more likely to release NIS in environments where they can do better. Some environmental matching is expected even for accidental introductions, as NIS are more likely to be translocated by humans between the environments that people most often frequents. In birds, for example, species that in their places of origin occur in urbanized environments are more likely to succeed when introduced outside their native range (Møller et al. 2015), probably because these species are more readily available for introduction, are more likely to be released close to human settlements, and already possess the adaptations needed to persist in such environments. Climatic matching is also common among successful invaders. In an analysis of 50 terrestrial plant invaders, fewer than 15 % of species had more than 10 % of their invaded distribution outside their native climatic Niche (Petitpierre et al. 2012). Likewise, birds naturalized in Europe occupy a subset of the climatic environments they inhabit in their native ranges (Strubbe et al. 2013).

Another explanation for why some NIS are able to persist in novel environments is that the environment, albeit different from the native one, is little demanding in terms of survival and/or reproduction (Table 2, scenario 3; Fig. 2b), providing for instance abundant resources and few competitors and enemies. In such circumstances, the fitness of the population is not expected to decrease substantially, facilitating establishment. In urban environments, for example, the availability of food deliberately or accidentally provided by humans and the lack of specialized predators might have facilitated the success of some exotic birds escaped from captivity (Shochat et al. 2010; Sol et al. 2012a).

Finally, if the NIS has a broad ecological niche, then it is more likely that it can find the necessary resources and physical conditions in the novel environment (Table 2, scenario 2; Fig. 2c). In birds, species that are either dietary or habitat generalists are more likely to establish themselves successfully in new regions (McLain et al. 1999; Cassey et al. 2004). In Australian acacias and eucalypt trees, invasiveness is so closely associated with environmental tolerance that this feature alone can predict over 90 % of occurrences observed outside of Australia (Higgins and Richardson 2014). Ecological generalism is likely to be a common feature of NIS because generalists are more likely to be abundant close to human settlements than specialists (Evans et al. 2011), and hence more available for introduction. For the same token, communities in anthropogenic environments are often composed primarily by generalists (Sol et al. 2013), which should reduce biotic resistance (see below).

However, when the new adaptive peak is different from the ancestral one (Fig. 2d), then the population is likely to see their fitness reduced. Such 'true' niche shifts differ from the previous scenarios, which are based on 'niche unfilling' (partial filling of the native niche in the invaded range, sensu Petitpierre et al. (2012)). Although substantial niche shifts do not seem to be the most common scenario, at least in terms of climatic tolerance (Petitpierre et al. 2012), two lines of evidence suggest that some degree of shift still occurs. First, the naturalized geographical distributions of some plants and animals are outside those predicted by climatic envelops (Sax et al. 2007; Petitpierre et al. 2012). Second, as discussed later on, many NIS populations have been reported to produce plastic and/or evolutionary responses to the new conditions, suggesting that these differ in some way from the ancestral ones. What allows species to be successful under scenarios of 'true' niche shifts is insufficiently understood, yet evolutionary theory suggests that phenotypic plasticity might play a central role.

Phenotypic plasticity is the capacity of organisms to express different phenotypes in different context; thus, it facilitates that a population can persist in a new environment, thereby bringing it into the realm of attraction of the new adaptive peak (Price et al. 2003). In a meta-analysis, Davidson et al. (2011) showed that invasive plants were more plastic in a variety of traits than non-invasive species, although this plasticity was only evident when resources were abundant. In animals, much attention has been devoted to a form of plasticity, behavioural flexibility. Through innovation and learning, animals can modify their behavior and develop responses to many of the problems that they can encounter in a novel environment, such as finding alternative food sources, developing responses to new predators and accommodating the reproduction to the new environmental conditions (Sol 2003). Indeed, there is evidence for reptiles, birds and mammals that the likelihood of establishment in novel regions increases with the size of the brain (Sol et al. 2005, 2008; Amiel et al. 2011), which mediates innovative propensity and learning (Lefebvre et al. 2004). Evidence is nonetheless lacking for fish (Drake 2007).

Is Growing Fast a General Feature of Successful Invaders?

Life history describes the way organisms allocate time and energy over growth, reproduction and survival, thereby determining how the population grows and fluctuates over time (Stearns 1992). Thus, life history has long been related to the way organisms respond to environmental changes (Roff 2002). The most popular theory, proposed by Lewontin and Cohen (1969) over 40 years ago, is the population growth hypothesis. It argues that species with life histories associated with high reproductive rates should be better invaders because their populations may growth faster and hence can more easily avoid the risk of extinction by accident just after the introduction, when the population is small. Although the population growth hypothesis is based on solid demographic theory and has received wide acceptance, the confidence in the hypothesis is undermined by a lack of empirical support. In birds, where the theory has received much attention, some studies suggest a positive relationship between life history correlates of population growth and establishment success whereas others report a negative relationship or no relationship at all (Blackburn et al. 2009). One limitation of the hypothesis is assuming that demographic stochasticity is the main cause of extinction of introduced populations, when this only operates when the population is extremely small. In addition, the hypothesis ignores that a high reproductive effort may entail costs, like increased mortality, that can counter the advantages of fast population growth (Stearns 1992; Roff 2002). Demographic models show for instance that while species that prioritize reproduction over survival can increase faster in numbers when conditions are favourable, they are at the same time highly exposed to extinction because their populations exhibit greater population fluctuations (Lande et al. 2003).

Recent work also suggests that life history can affect the invasion process by additional mechanisms than those generally considered (see Sol et al. 2012b). While prioritizing current reproductive effort provides benefits in terms of rapid population growth, a strategy that prioritize future reproduction can also afford some advantages for a species exposed to novel environment (Williams 1966; Sol et al. 2012b). In addition of reducing population fluctuations (see above), such a strategy is based on distributing the reproductive effort in a number of reproductive events, thereby reducing the fitness costs of a reproductive failure. This bet-hedging strategy facilitates population persistence when environmental uncertainties increase the probability of reproductive failure due to bad decisions (e.g. settling in an inappropriate habitat). Moreover, such a strategy reduces the fitness costs of skipping a reproduction, allowing the invader to engage in reproductive activities only when conditions are favorable (storage effect). This increases the opportunities for acquiring environmental information and for improving performance on exploiting the resources and avoiding the enemies, particularly in long-lived species with larger brains and enhanced capacity to construct learnt responses. A recent global comparative analysis of avian introductions evidenced that although rapid population growth may be advantageous during invasions under certain circumstances (i.e. low propagule pressure and environmental matching), successful invaders are generally characterized by life-history strategies in which they give priority to future rather than current reproduction (Sol et al. 2012b). Such a strategy of expected future returns is generally achieved by investing in survival and hence attaining a long reproductive lifespan. However, it can also be achieved by reproducing more frequently, which combines the benefits of a higher reproductive effort with lower costs of losing a breeding attempt. This later strategy may explain the invasion success of species like rats and pigeons. Other strategies and mechanisms are likely to emerge in coming years.

Do Communities Differ in Invasibility?

Species-rich communities have long been thought to be more resistant to the establishment of NIS, a theory known as the 'biotic resistance' hypothesis. As more species are present in a community, the niches will be better filled and competition for resources like food, breeding sites or shelter will be stronger (Elton 1958). This would reduce the likelihood of establishment of additional species.

The existence of 'biotic resistance' has been demonstrated in microcosm and mesocosm experiments. Levine (2000), for example, designed a field experiment in which exotic plants were introduced into tussocks where the number of resident plant species had been manipulated in situ. As species richness increased, the likelihood of germinating and surviving the breeding season declined in two out of the three studied invaders.

However, ecological interactions rarely enable communities to resist establishment of NIS but only limit their abundance, at least in plants (Levine et al. 2004). Indeed, stochastic niche theory argues that local diversity is rarely limited by competition (Tilman 2004). Rather, with the addition of large numbers of propagules of novel species, many more species are predicted to coexist locally than ever would occur with natural assembly (Tilman 2004).

Moreover, observations at large spatial scales often show a positive (instead of the predicted negative) correlation between exotic and native species richness. In the tussocks studied by Levine (2000), the natural incidence of all three exotic plants was greater on more diverse tussocks. Such patterns may suggest that other features of the habitat can also be important, if not more important, than biotic resistance. At higher spatial scales the relative effect of biotic interactions diminishes and environmental factors gain importance in shaping regional biodiversity. If the environmental factors that favours higher number of native species also increase niche opportunities for the establishment of NIS, then the existence of a positive correlation between exotic and native species richness do not necessarily deny the importance of biotic resistance (Shea and Chesson 2002).

The nature of such environmental factors is not well-known yet, although some possibilities have been advanced. First, a structurally heterogeneous region may provide a greater array of microenvironments (Davies et al. 2005), increasing the likelihood that the invader encounters a favourable niche not monopolized by native species. In plants, there are more alien plant species and they are more abundant at fragment edges than in the interior of fragments (Vilà and Ibáñez 2011). Second, climatic conditions are also expected to influence both native and exotic species richness by limiting the number of species that can persist in the region when these conditions are most extreme. This can in part explain why invaders are scarcer on the top of mountains (Bartomeus et al. 2011). Finally, disturbance, whether temporal or permanent, is thought to facilitate invasion by simultaneously opening resource opportunities and decreasing competition from resident native species (Tilman 2004). Habitats altered or entirely created by humans may be particularly susceptible to invasion, as the resulting communities have had less time to assemble, and hence to adapt to the local conditions and to each other; moreover, these communities are also more likely to have fewer species with broader niches and lower competitive abilities (Shea and Chesson 2002).

There is indeed evidence that environmental heterogeneity, climate conditions and anthropogenic disturbances can be common factors favouring both native and exotic species richness. Bartomeus et al. (2011), for instance, identified habitat heterogeneity, high precipitation, low altitude, elevated human density and anthropogenic disturbances as common factors favouring both native and exotic plants in Catalonia (see also Pino et al. 2005). However, these common factors did not seem to be the whole explanation for the positive relationship between exotic and native species richness. When these common causes were controlled for with structural equation modelling, the positive relationship did not turn out negative (Bartomeus et al. 2011).

As alternatively, it is possible that the positive correlation between exotic and native species richness also reflects the difficulties of separating invasibility from invasion rate. Human activities may not only create new niche opportunities for both exotic and native species adapted to disturbances, but also facilitate the transport (intentionally or unintentionally), introduction and spread of exotic and native species adapted to such environments. Rejmánek (2003), for example, showed that a positive native–exotic plant richness association previously reported for North American plants turned out negative when human population density was included along with latitude in a model predicting exotic species richness. The analyses of Bartomeus et al. (2011) of plants from Catalonia are also consistent with the importance of human-driven effects. When the comparison was restricted to native plants associated with anthropogenic-disturbed ecosystems (i.e. ruderal plants), the positive relationship between native-exotic species richness became stronger (from an $R^2=0.11$ to 0.52) and the fraction explained by common factors also increased substantially (58.3 %). The important role of human activities in creating a positive native-exotic richness association is not surprising given that biological invasions are induced by human activities, and highlights the need of distinguishing invasiveness from invasion rates when testing the biotic resistance hypothesis.

Darwin Naturalization Hypothesis: Are Close Relatives Bad Neighbors?

In the Origin of species, Darwin (1859) proposed the hypothesis that NIS should be more successful in communities in which their close relatives are absent. The rationale of the hypothesis is that close relatives are more likely to occupy niches that would otherwise facilitate the establishment of NIS, an idea known as Darwin's naturalization hypothesis (DNH). The DNH has been demonstrated experimentally (Jiang et al. 2010). Analysing laboratory bacterial communities, Jiang et al. (2010) showed that the frequency of successful invader establishment was best explained by average phylogenetic distance between the invader and all resident species, consistent with the absence of empty niches. Invader abundance was also related to phylogenetic distance between the invader and its nearest resident relative, possibly indicating reduced availability of the optimal resources.

However, when we move from the controlled conditions of the laboratory to natural conditions the results become less clear, with different studies either supporting or refuting it (Jiang et al. 2010; Sol et al. 2014). This lack of firm support comes in part from using a scale of analyses too large for competition to be relevant or from not considering the possibility that the adaptations to become invasive are little phylogenetically conserved (Sol et al. 2014). In addition, the two basic assumptions of the hypothesis, that competition is a major process involved in biological invasions and that competition is more intense between close-related species, have been called into question by some authors (Duncan et al. 2003; Levine et al. 2004; Gilbert and Lechowicz 2005). As already mentioned, biotic resistance may reduce the establishment of species but rarely enables communities to resist invasion (reviewed in Levine et al. 2004). In addition, exotic species are unlikely to encounter close relatives in the recipient community when they come from distant regions (Valiente-Banuet and Verdú 2007). Finally, the strength of competition will not only depend on the presence of close-relatives, but also on their abundance (Tilman 1997) and the form in which they compete (Sol et al. 2012a; Jones et al. 2013). While competition with close-relatives is expected to increase with exploitative competition, as a result of increase in niche overlap in species with similar phenotypes, interference competition can be stronger among species with different phenotypes (e.g. differences in body size)(Jones et al. 2013).

Moreover, although competition is often assumed to be the primary source of biotic resistance, it is increasingly acknowledged that other mechanisms can also underlie the phenomenon and even be more important (Levine et al. 2004). The pressure from enemies (i.e. pathogens, parasites and predators) appears to be highly influential in this regard. Exotic birds, for example, appear to be more likely to fail on oceanic islands with species-rich mammalian predator assemblages (Cassey et al. 2005).

Not only evidence for the DNH is scarce, but some contrary results have also been accumulating showing that introduced organisms more closely related to native species are more likely to become invasive. Indeed, this possibility was already advanced by Darwin (1859), as NIS can share with their native relatives traits that pre-adapt them to their new environment (Table 2, scenario 2). As example, Duncan et al. (2002) analysed a complete list of seed-plant species introduced to New Zealand and found that those with congeneric relatives were significantly more, not less, likely to naturalize. Again the assumption here is that phylogenetic distance reflects ecological distance, which has been rarely tested. If phylogenetic distance accurately measures similarity in traits related to invasiveness, closely related species should generally exhibit similar invasion potential. However, they often do not (Sol 2007). Although the differences can simply reflect differences in propagule pressure or in place of introduction rather than fundamental intrinsic differences (Fig. 2), evidence that close-relatives exhibit similar invasion potential is lacking.

Are Introduced Species Creating an "Invasional Meltdown"?

The emphasis on biotic resistance has led to under-appreciate the importance of positive interactions in the invasion process. Nevertheless, such perception is changing. In a review of invasions in the Great Lakes, Ricciardi (2001) showed that direct positive (mutualistic and commensal) interactions among introduced species are more common than purely negative (competitive and amensal) interactions. In terrestrial plants, positive interactions between NIS are also common, albeit in this case negative interactions are far more common (Kuebbing and Nuñez 2015).

The importance of mutualistic interactions is exemplified in ectomycorrhizal plants, whose invasion success has been limited in some regions by the absence of appropriate fungal symbionts (Traveset and Richardson 2011). Positive interactions are useful to understand the rapid invasion of some environments, an issue of great importance from a conservation perspective. Simberloff and Holle (1999) coined the term "invasional meltdown" to describe situations in which NIS facilitate one another's invasion instead of limiting invasions as the species accumulation increases biotic resistance (Ricciardi 2001; Simberloff 2006).

However, a full "invasional meltdown", in which interspecific facilitation leads to an accelerating increase in the number of introduced species and their impact, has yet to be conclusively demonstrated (Simberloff 2006). Better supported is nonetheless a weaker version of meltdown that argues that one invader facilitates population persistence of one or more other invaders without itself receiving an evident benefit (Simberloff 2006). For example, Grosholz (2005) used field and laboratory experiments to demonstrate that a recently introduced crab favoured the rapid proliferation of an introduced bivalve that had been rare for nearly 50 years. The effect was not direct but occurred through the positive indirect effects of predation by the introduced crab on native bivalves.

Does "Enemy Release" Explain the Increase and Expansion of Invaders?

One of the most puzzling observations in invasions biology is that some NIS perform better in their new ranges than in their native ones. A recent analysis indeed reported that invasive plants and animals tended on average to be more abundant in their introduced ranges (Parker et al. 2013). A possible explanation for why NIS can proliferate in the new environment is that they are released from the pressure of co-evolved enemies (i.e. pathogens, parasites and predators), an idea known as the "enemy release" hypothesis (ERH, Elton 1958; Maron and Vila 2001). In the case of pathogens and parasites, an introduced host can be released from these enemies if for instance mortality during the transport mostly affects infected or parasitized individuals (Mitchell and Power 2003).

Although there is little doubt that enemies can be important agents of population control, whether NIS are generally released from their enemies remains less clear. At a biogeographic scale all NIS will lose some enemies (Colautti et al. 2004). However, community studies often show that NIS are no generally less affected by enemies than native species in the invaded community (Colautti et al. 2004). For example, Clay (1995) found that grasses native to the United States have, on average, fewer pathogen species than co-occurring NIS. In other cases, evidence either supporting or contradicting the enemy release hypothesis is inconclusive because the study do not quantify the impact of enemies on both NIS and native species, and still more rarely include controls that experimentally exclude enemies (Keane and Crawley 2002). The assumption of the enemy release hypothesis that few specialist enemies shift to attack NIS is also frequently falsified (Keane and Crawley 2002). Many invasive species are generalists unlikely to have co-evolved with enemies.

Some release from the pressure of enemies is expected in anthropogenic disturbed environments, where NIS attain highest success, as these environments typically contain simplified communities in which enemies are scarcer or even controlled for humans. However, this should not only benefit NIS but also native species.

What Is the Role of Evolution in the Invasion Process?

Evolution has often been dismissed as an important factor in the success of invaders because of the idea that genetic variation in small introduced populations should be reduced (see Moles et al. 2012). Indeed, bottlenecks have been demonstrated in many introduced populations. In a literature review, Puillandre et al. (2007) found lower genetic diversity in introduced populations than in native populations in 80 % of the 72 studies they examined. Bottlenecks can limit the success of invaders by inbreeding depression and by reducing genetic variation available for natural selection to adapt the population locally (Allendorf et al. 2013).

However, the predicted genetic reduction is not always observed (Roman and Darling 2007). In fact, genetic variation can be substantially enhanced when propagule pressure is high (Moles et al. 2012), as this increases the probability that individuals come from different source populations. For example, in a review of aquatic invasions, only 16 of 43 invasive species had reduced genetic diversity (Roman and Darling 2007). Genetic variation can also increase by hybridization. (Ellstrand and Schierenbeck 2006). Hybridization has for example been suggested to explain the rapid spread of mosquitos responsible of the transmission of the West Nile virus (Allendorf et al. 2013). In occasions, bottlenecks can themselves contribute to rapid adaptation by providing raw material for natural selection through genetic drift and epistatic interactions (Sax et al. 2007).

Another reason why evolution has often been dismissed as an important factor in biological invasions is the common believe that adaptive change proceeds slowly. However, rapid adaptive evolution has been repeatedly demonstrated in introduced populations (Reznick and Ghalambor 2001; Moles et al. 2012). Reznick and Ghalambor (2001) found that many of the examples of contemporary evolution involves biological invasions. Colautti and Barrett (2013) indeed reported experimental evidence for earlier flowering in the North American invasive plant *Lythrum salicaria*. In this species, northern populations had been found to flower earlier than southern populations. Reciprocal transplant experiments demonstrated the characteristic "home site advantage" in which the organism attains higher fitness in their home region, thereby showing the adaptive nature of earlier flowering.

During the spreading stage, evolution can proceed particularly faster when there is heritable variation in traits affecting dispersal. Individuals at the forefront of the expansion encounter a low density of individuals and hence will tend to mate assortatively with respect to the dispersal trait (Phillips and Suarez 2012), favoring rapid evolutionary divergence. In cane toads (*Bufo marinus*) introduced to Australia, the annual rate of progress of the toad invasion front has increased about fivefold since the toads first arrived (Phillips et al. 2006). This seems to have resulted from selection for longer limbs: Toads with longer legs move faster and are the first to arrive to new areas.

The study of biological invasions has largely contributed to the debate of whether evolution is contingent or follows consistent routes. Several studies on introduced insects and plants have for instance reported the re-establishment of latitudinal clines in life history traits similar to those found in their native ranges. A classic example is the restitution of the Bergman's rule (i.e. increase in body size with latitude) in *Drosophila suboscura* introduced to North American (Huey et al. 2000). In other cases, however, consistent evolutionary routes have been harder to demonstrate. The EICA hypothesis, for instance, argues that because of the enemy release, invaders do not need to invest in defense and can relocate resources to be more efficient and competitive in the novel environment (Blossey and Nötzold 1995). Despite receiving considerable interest, the EICA hypothesis has surprisingly received little unambiguous support. Some studies do provide evidence that introduced species has lost enemy resistance, yet they fail to show that this loss increases fitness (Maron et al. 2004). Even in cases that demonstrate fitness benefits, the explanatory power of the hypothesis appears to be low. Colautti and Barrett (2013), for example, showed that in introduced *L. salicaria* the fitness benefits of earlier flowering in response to shorter growing seasons is significantly higher than those of reducing defense investment.

Despite the progress, our understanding of the exact role of evolution in the invasion process is deficient. Hendry et al. (2008) conducted a meta-analysis and found that plastic responses were more frequent than genetic change when organisms confront human-induced changes (mostly involving NIS). There is thus an urgent need to understand the mechanisms of rapid evolutionary adaptation rather than simply document new cases. We do not even known at what stage of the invasion process is evolution more important, although several lines of evidence suggest that evolution should be more relevant at the last stages of the invasion process. At earlier stages, selection should lead to weaker evolutionary responses, as the population is smaller, and strong selection can lead the population to extinction. Indeed, the often observed time-lags in which the population remains at low numbers before explodes (Sakai et al. 2001) could in part be attributed to insufficient opportunities for local adaption.

Are Exotics Decreasing, Maintaining or Increasing Biodiversity?

NIS can impact on native species through a variety of mechanisms including predation, competition, hybridization and habitat alteration (Vilà et al. 2011). However, their role in reducing biodiversity remains controversial. Biological invasions involve both species additions and extirpations, and hence the resulting regional and local diversity results from the balance between both processes (Case 1996). If some areas hold lower species richness than it could really hold because many have been unable to colonize the area, then human-assisted invasions may increase diversity with little biodiversity loss.

Indeed, empirical work by Sax and collaborators indicate that at the regional level, exotic additions have often increased biodiversity, suggesting that at least at this spatial scale there is no species saturation (Brown and Sax 2007; Sax and Gaines 2008). Except for birds in oceanic islands, in which the number of native extinctions has been largely matched the number of established NIS, an increase in species richness has been reported for plants, freshwater fishes and mammals. Alien plants, for example, have doubled native biodiversity in oceanic islands.

A relevant question is therefore whether alien species richness will continue to increase as new species are added or has instead reached an equilibrium point. According to classical island biogeography theory, a saturation point can be maintained through the balance between extinctions and colonizations, where the number of species that colonize lead to the extinction of an equal number of species already present (extinction-based saturation)(Sax and Gaines 2008). The extinction-based saturation does not seem prevalent, however. In both plants and freshwater fish, species richness has increased because few native species have gone extinct whereas many exotic species have become naturalized. Even in birds from oceanic islands, where colonizations have largely matched extinctions, this is not extinction-based saturation because most extinctions were not caused by the introduced birds but by other factors like human hunting or introduced mammals (Duncan and Young 2000; Owens and Bennett 2000). However, the conclusion that extinction-based saturations are rare should be taken with caution as there are currently high uncertainties in how long extinctions take to manifest, the so-called extinction debt.

Alternatively, a saturation point can be maintained through biotic resistance, where species richness restricts additional introductions. Sax and Gaines (2008) called this later process colonization-based saturation. Although such type of saturation is in line with the observed reduced local extinction observed in plants and fresh-water fish, there is little unambiguous indication for species saturation in natural communities (Sax and Gaines 2008). The only firm evidence currently available comes from the studies of *Anolis* lizards from the Caribbean (Helmus et al. 2014). While no anole has gone extinct on Caribbean islands in the last decades, except possibly one, at least 18 species have established, probably arriving as commensals of humans in cargo shipments. As a result, species richness has increased in average from 4.72 to 5.41 species. Interestingly, islands impoverished in native species have gained the most exotic species, strengthening the species–area relationship by which larger islands can harbour more species. This suggests that the communities have reached a saturation point, although it remains to be tested whether this has been achieved through biotic resistance.

As pointed by Sax and Gaines (2008), a local increase of biodiversity with the addition of NIS is not necessarily good from a conservation perspective as many unique endemic species may have been lost and replaced by more cosmopolitan species. Moreover, rather than the number, the identity of the species and their role in the ecosystem appear more relevant to understand how NIS affect the structure and function of ecosystems. Four possibilities exist here: (1) all species from the community play different roles, so the addition and removal of any species has always certain impact on the ecosystem; (2) species belong to different functional groups, and hence the replacement of a native species for a NIS should have little effect if they belong to the same group (as they do redundant functions) and high effect if belongs to a different group; (3) the NIS becomes a key species in the ecosystem, either because replaces a native species playing such a role or because plays a role different from all other species; and (4) all species are ecologically equivalent and hence there is no effect of the entrance of an invader. Understanding how frequently these alternatives occur is important not only from a conservation perspective, but also to help guarantee ecosystem services that are essential for human societies. While evidence exist for each of these possible outcomes, NIS typically lead to an increase, not a decrease, in ecosystem function (Ehrenfeld 2010; Vilà et al. 2011). Particularly troubling is the existence of examples for the third scenario, which can lead to cascade effects through the whole ecosystem. Zebra mussels (*Dreissena polymorpha*), for example, alters aquatic ecosystems by acting as a powerful filter that increases light and nutrients, allowing the proliferation of plants and algae (Ludyanskiy et al. 1993).

Concluding Remarks

The field of biological invasions has experienced an enormous progress in the last decades. Much of the progress comes from adopting ecological and evolutionary theories as a working framework and from improving the rigor in adopting and rejecting new hypotheses and theories. This has led to reject some ideas that were taken by granted in the past, like the view that islands are easier to invade than continents or that NIS are competitively superior to native organisms. Although many ideas in invasion biology still remain controversial, we now start having a reasonable understanding of the processes of invasiveness and invasibility (Fig. 3). An important advance has been acknowledging that human influences are pervasive and varied throughout the invasion process, and hence that an anthropocene pespective is needed if we aim to understand the biogeography of biological invasions (Stuart et al. 2012).

Valéry et al. (2013) have recently called into question the need of the biological invasions discipline on the grounds that native species should also be called invasive whenever they outbreak. However, as pointed out by Blondel et al. (2013), there is no reason for unifying concepts and terminology to include native species. Rather,

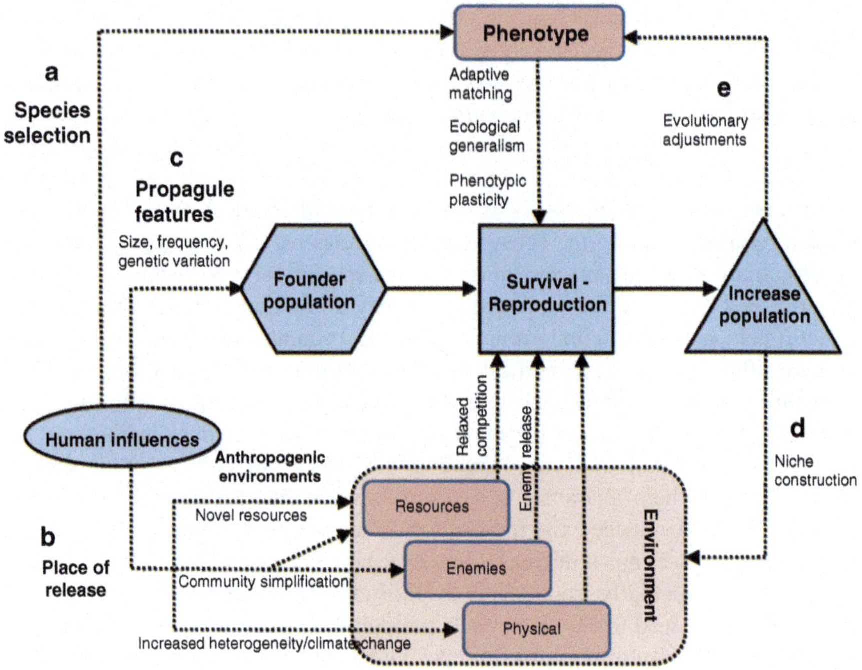

Fig. 3 General framework to understand the success of NIS. The likelihood that a NIS can increase in numbers in the novel environments depends on resource availability, enemies and the physical environment. Human activities can increase this likelihood in three ways: (**a**) selecting the type of organisms that are introduced, which should often favor organisms pre-adapted to the new environment and/or that exhibit high ecological and phenotypic plasticity; (**b**) determining the environments where they are released, which will often exhibit anthropogenic disturbances that increase invasibility through increased environmental heterogeneity, simplification of the native community (releasing the NIS from many enemies and competitors) and enhanced resource opportunities; and (**c**) affecting the size and structure of their founder populations, which determines the importance of Allee effects and demographic and genetic stochasticity. The organism itself can to some extent modify their new environment through niche construction (**d**), as well as improve fitness through an evolutionary rescue (**e**)

as suggests the evidence summarized in this chapter, the discipline of biological invasions departs from other disciplines at two levels. The first is the uniqueness of the processes that investigates, particularly regarding the importance of anthropogenic influences, magnitude of distances at which the organisms are moved, and the extent of which NIS differentiate from native species and exhibit adaptive mismatch respect to the novel environment, all of which requires a specific framework (Fig. 3). The second is the unique focus on preventing and mitigating the impact of organisms outside their native ranges, which makes challenging anticipate consequences. While we are still far from being able to predict the outcome of any introduction event, for some groups like plants and vertebrates we at least can identify situations where the risk is high that the species successfully establishes itself in a new environment (Kolar and Lodge 2002, Vall-llosera and Sol 2009; Leung et al. 2012). Because the entrance of NIS seems inevitable, concern over the impact of invaders will continue being an important reason fueling research on biological invasions.

Yet, ecologist should avoid at the same time committing the "appeal to nature fallacy" of considering that something is good simply because it is natural and bad because it is not (Brown and Sax 2005). Many invaders are innocuous and have come to stay, yet they still have important value for addressing scientific questions. Indeed, as shown throughout the chapter, NIS offer unique opportunities for studying a variety of ecological and evolutionary processes in real time and at an unprecedented scales. Some of the ideas that the discipline has contributed to reshape include community assemblage rules, ecological cascades and the speed of contemporary evolution (Callaway and Maron 2006; Reznick et al. 2008; Helmus et al. 2014). Thus, despite claims for the end of invasion biology, the field still has much to offer and I anticipate that the enormous current interest in the discipline will continue growing in the coming years.

Acknowledgments I'm grateful to Pepe Tella, Oriol Lapiedra, Joan Maspons and Ferran Sayol for reviewing the chapter, and to Louis Lefebvre, Richard Duncan, Tim Blackburn, Phill Cassey, Miquel Vall-llosera, Montse Vilà, Tamas Székely, Joan Pino, Salit Kark, Sven Bacher, Wojciech Solarz, Wolfgang Nentwig, Simon Reader and Diego Vasquez for fruitful discussions over the past years. This work was supported by a Proyecto de Investigación (ref. CGL2013-47448-P) from the Spanish government.

References

Allendorf FW, Luikart G, Aitken SN (2013) Conservation and the genetics of populations. Wiley-Blackwell, Sussex (UK)

Amiel JJ, Tingley R, Shine R (2011) Smart moves: effects of relative brain size on establishment success of invasive amphibians and reptiles. PLoS One 6:e18277

Barnagaud J, Barbaro L, Papaïx J (2013) Habitat filtering by landscape and local forest composition in native and exotic New Zealand birds. Ecology 95:78–87

Bartomeus I, Sol D, Pino J, Vicente P, Font X (2011) Deconstructing the native-exotic richness relationship in plants. Glob Ecol Biogeogr 21:524–533

Blackburn TM, Cassey P, Lockwood JL (2009) The role of species traits in the establishment success of exotic birds. Glob Chang Biol 15:2852–2860
Blackburn TM, Duncan RP (2001) Establishment patterns of exotic birds are constrained by non-random patterns in introduction. J Biogeogr 28:927–939
Blackburn TM, Pyšek P, Bacher S, Carlton JT, Duncan RP, Jarošík V, Wilson JRU, Richardson DM (2011) A proposed unified framework for biological invasions. Trends Ecol Evol 26:333–339
Blondel J, Hoffmann B, Courchamp F (2013) The end of invasion biology: intellectual debate does not equate to nonsensical science. Biol Invasions 16:977–979
Blossey B, Nötzold R (1995) Evolution of increased competitive ability in invasive nonindigenous plants: a hypothesis. J Ecol 83:887–889
Brown BJH, Sax DF (2007) Do biological invasions decrease biodiversity? Conserv Mag 8:16–17
Brown JH, Sax DF (2005) Biological invasions and scientific objectivity: reply to Cassey et al. (2005). Austral Ecol 30:481–483
Callaway RM, Maron JL (2006) What have exotic plant invasions taught us over the past 20 years? Trends Ecol Evol 21:369–374
Case TJ (1996) Global patterns in the establishment and distribution of exotic birds. Biol Conserv 78:69–96
Cassey P, Blackburn TM, Duncan RP, Gaston KJ (2005) Causes of exotic bird establishment across oceanic islands. Proc R Soc Lond B 272:2059–2063
Cassey P, Blackburn TM, Sol D, Duncan RP, Lockwood JL (2004) Global patterns of introduction effort and establishment success in birds. Proc R Soc Lond B 271:S405–S408
Colautti RI, Barrett SCH (2013) Rapid adaptation to climate facilitates range expansion of an invasive plant. Science (New York, NY) 342:364–366
Colautti RI, Ricciardi A, Grigorovich IA, MacIsaac HJ (2004) Is invasion success explained by the enemy release hypothesis? Ecol Lett 7:721–733
Clay K (1995) Correlates of pathogen species richness in the grass family. Canadian Journal of Botany 73:S42–S49
Darwin C (1859) The origin of species. J. Murray, London
Davidson AM, Jennions M, Nicotra AB (2011) Do invasive species show higher phenotypic plasticity than native species and, if so, is it adaptive? A meta-analysis. Ecol Lett 14:419–431
Davies KE, Chesson P, Harrison S, Inouye BD, Melbourne BA, Rice KJ (2005) Spatial heterogeneity explains the scale dependence of the native–exotic diversity relationship. Ecology 86:1602–1610
Davis M (2009) Invasion biology. Oxford Univ Press, Oxford
Drake JM (2007) Parental investment and fecundity, but not brain size, are associated with establishment success in introduced fishes. Funct Ecol 21:963–968
Duncan RP, Blackburn TM, Sol D (2003) The ecology of bird introductions. Annu Rev Ecol Evol Syst 34:71–98
Duncan RP, Williams PA, Wikelski M, Wong V, Chevalier B, Rattenborg N, Snell HL, Janvier P, Arsenault M (2002) Darwin's naturalization hypothesis challenged. Nature 417:608–609
Duncan RP, Young JR (2000) Determinants of plant extinction and rarity 145 years after European settlement of Auckland, New Zealand. Ecology 81:3048–3061
Ehrenfeld JG (2010) Ecosystem consequences of biological invasions. Annu Rev Ecol Evol Syst 41:59–80
Ehrlich PR (1989) Attributes of invaders and invading processes: vertebrates. In: Drake JA, DiCastri F, Groves RH, Kruger FJ, Mooney HA, Rejma´nek M, Williamson MH, (eds) Biological invasions: a global perspective. Wiley, New York, USA, pp 315–328
Ellstrand NC, Schierenbeck KA (2006) Hybridization as a stimulus for the evolution of invasiveness in plants? Proc Natl Acad Sci U S A 148:35–46
Elton CS (1958) The ecology of invasions by animals and plants. Wiley, New York
Evans K, Chamberlain D, Hatchwell BENJ, Gregory RD, Gaston KJ (2011) What makes an urban bird? Glob Chang Biol 17:32–44
Fritts TH, Rodda GH (1998) The role of introduced species in the degradation of island ecosystems: a case history of Guam. Ann Rev Ecol Syst 29:113–140

Gilbert B, Lechowicz MJ (2005) Invasibility and abiotic gradients: the positive correlation between native and exotic plant diversity. Ecology 86:1848–1855
Grosholz ED (2005) Recent biological invasion may hasten invasional meltdown by accelerating historical introductions. Proc Natl Acad Sci U S A 102:1088–1091
Haight RG, Polasky S (2010) Optimal control of an invasive species with imperfect information about the level of infestation. Res Energy Econ 32:519–533
Helmus MR, Mahler DL, Losos JB (2014) Island biogeography of the anthropocene. Nature 513:543–546
Hendry AP, Farrugia TJ, Kinnison MT (2008) Human influences on rates of phenotypic change in wild animal populations. Mol Ecol 17:20–29
Hernández-Brito D, Carrete M, Popa-Lisseanu AG, Ibáñez C, Tella JL (2014) Crowding in the city: losing and winning competitors of an invasive bird. PLoS One 9:e100593
Higgins SI, Richardson DM (2014) Invasive plants have broader physiological niches. Proc Natl Acad Sci U S A 111:10610–10614
Huey RB, Gilchrist GW, Carlson ML, Berrigan D, Serra L (2000) Rapid evolution of a geographic cline in size in an introduced fly. Science (New York, NY) 287:308–309
Jeschke J, Gómez Aparicio L, Haider S, Heger T, Lortie C, Pyšek P, Strayer D (2012) Support for major hypotheses in invasion biology is uneven and declining. NeoBiota 14:1–20
Jiang L, Tan J, Pu Z (2010) An experimental test of Darwin's naturalization hypothesis. Am Nat 175:415–423
Jones EI, Nuismer SL, Gomulkiewicz R (2013) Revisiting Darwin's conundrum reveals a twist on the relationship between phylogenetic distance and invasibility. Proc Natl Acad Sci U S A 110:20627–20632
Keane RM, Crawley MJ (2002) Exotic plant invasions and the enemy release hypothesis. Trends Ecol Evol 17:164–170
Kolar CK, Lodge DM (2001) Progress in invasion biology: predicting invaders. Trends Ecol Evol 16:199–204
Kolar CS, Lodge DM (2002) Ecological predictions and risk assessment for alien fishes in North America. Science (New York, NY) 298:1233–1236
Kuebbing SE, Nuñez MA (2015) Negative, neutral, and positive interactions among nonnative plants: patterns, processes and management implications. Global Change Biol 21:926–934
Lande R, Engen S, Saether B-E (2003) Demographic and environmental stochasticity. Stochastic population dynamics in ecology and conservation. Oxford University Press, Oxford
Lefebvre L, Reader SM, Sol D (2004) Brains, innovations and evolution in birds and primates. Brain Behav Evol 63:233–246
Leung B, Roura-Pascual N, Bacher S, Heikkilä J, Brotons L, Burgman MA, Dehnen-Schmutz K, Essl F, Hulme PE, Richardson DM, Sol D, Vilà M, Rejmanek M (2012) TEASIng apart alien species risk assessments: a framework for best practices. Ecol Lett 15:1475–1493
Levine JM (2000) Species diversity and biological invasions: relating local process to community pattern. Science 288:852–854
Levine JM, Adler PB, Yelenik SG (2004) A meta-analysis of biotic resistance to exotic plant invasions. Ecol Lett 7:975–989
Lewontin RCC, Cohen D (1969) On population growth in a randomly varying environment. Proc Natl Acad Sci U S A 62:1056–1060
Lockwood JL, Cassey P, Blackburn TM (2005) The role of propagule pressure in explaining species invasions. Trends Ecol Evol 20:223–228
Lodge DM (1993) Biological invasions: lessons for ecology. Trends Ecol Evol (Personal Edn) 8:133–137
Ludyanskiy ML, McDonald D, MacNeill D (1993) Impact of the Zebra Mussel, a Bivalve Invader. Bioscience 43:533–544
Maron JL, Vila M (2001) When do herbivores affect plant invasion? Evidence for the natural enemies and biotic resistance hypotheses. Oikos 95:361–373
Maron J, Vilà M, Arnason J (2004) Loss of enemy resistance among introduced populations of St. John's Wort (Hypericum perforatum). Ecology 85:3243–3253

Martin TE, Maron JL (2012) Climate impacts on bird and plant communities from altered animal–plant interactions. Nat Climate Change 2:195–200

McLain DK, Moulton MP, Sanderson JG (1999) Sexual selection and extinction: the fate of plumage-dimorphic and plumage-monomorphic birds introduced onto islands. Evolutionary Ecol Res 1:549–565

Miller DJ (1989) Introductions and extinction of fish in the African great lakes. Trends Ecol Evol (Personal Edn) 4:56–59

Mitchell CE, Power AG (2003) Release of invasive plants from fungal and viral pathogens. Nature 421:625–627

Moles AT, Flores-Moreno H, Bonser SP, Warton DI, Helm A, Warman L, Eldridge DJ, Jurado E, Hemmings FA, Reich PB, Cavender-Bares J, Seabloom EW, Mayfield MM, Sheil D, Djietror JC, Peri PL, Enrico L, Cabido MR, Setterfield SA, Lehmann CER, Thomson FJ (2012) Invasions: the trail behind, the path ahead, and a test of a disturbing idea. J Ecol 100:116–127

Møller AP, Díaz M, Flensted-Jensen E, Grim T, Ibáñez-Álamo JD, Jokimäki J, Mänd R, Markó G, Tryjanowski P (2015) Urbanized birds have superior establishment success in novel environments. Oecologia. doi:10.1007/s00442-015-3268-8

Reznick DN, Ghalambor CK (2001) The population ecology of contemporary adaptations: what empirical studies reveal about the conditions that promote adaptive evolution. Genetica 112–113:183–198

Owens IPF, Bennett PM (2000) Ecological basis of extinction risk in birds: habitat loss versus human persecution and introduced predators. Proc Natl Acad Sci U S A 97:12144–12148

Parker IM, Simberloff D, Lonsdale WM, Goodell K, Whonham M, Kareiva PM, Williamson MH, Von Holle B, Moyle PB, Byers JE, Goldwasser L (1999) Impact: toward a framework for understanding the ecological effects of invaders. Biol Invasions 1:3–19

Parker JD, Torchin ME, Hufbauer RA, Lemoine NP, Alba C, Blumenthal DM, Bossdorf O, Byers JE, Dunn AM, Heckman RW, Hejda M, Jarosík V, Kanarek AR, Martin LB, Perkins SE, Pysek P, Schierenbeck K, Schlöder C, van Klinken R, Vaughn KJ, Williams W, Wolfe LM (2013) Do invasive species perform better in their new ranges? Ecology 94:985–994

Petitpierre B, Kueffer C, Broennimann O, Randin C, Daehler C, Guisan A (2012) Climatic niche shifts are rare among terrestrial plant invaders. Science 335:1344–1348

Phillips BL, Brown GP, Webb JK, Shine R (2006) Invasion and the evolution of speed in toads. Nature 439:803

Phillips B, Suarez A (2012) The role of behavioural variation in the invasion of new areas. In: Candolin U, Wong B (eds) Behavioural responses to a changing world. Oxford Univerisity Press, Oxford, pp 190–200

Pimentel D, McNair S, Janecka J, Wightman J, Simmonds C, O'Connell C, Wong E, Russel L, Zern J, Aquino T, Tsomondo T (2001) Economic and environmental threats of alien plant, animal, and microbe invasions. Agricul Ecosyst Environ 84:1–20

Pino J, Font X, Carbó J, Jové M, Pallarès L (2005) Large-scale correlates of alien plant invasion in Catalonia (NE of Spain). Biol Conserv 122:339–350

Pouteau R, Hulme PE, Duncan RP (2015) Widespread native and alien plant species occupy different habitats. Ecography 38:462–471

Price TD, Qvarnström A, Irwin DE (2003) The role of phenotypic plasticity in driving genetic evolution. Proc R Soc Lond B 270:1433–1440

Puillandre N, Dupas S, Dangles O, Zeddam J-L, Capdevielle-Dulac C, Barbin K, Torres-Leguizamon M, Silvain J-F (2007) Genetic bottleneck in invasive species: the potato tuber moth adds to the list. Biological Invasions 10:319–333

Rejmánek M (2003) The rich get richer. Front Ecol Environ 1:122–123

Reznick DN, Ghalambor CK, Crooks K (2008) Experimental studies of evolution in guppies: a model for understanding the evolutionary consequences of predator removal in natural communities. Mol Ecol 17:97–107

Ricciardi A (2001) Facilitative interactions among aquatic invaders: is an "invasional meltdown" occurring in the Great Lakes? Can J Fisheries Aquat Sci 58:2513–2525

Ricciardi A, Jones LA, Kestrup ÅM, Ward JM (2011) Expanding the propagule pressure concept to understand the impact of biological invasions. In: Richardson DM (ed) Fifty Years of Invasion Ecology: The Legacy of Charles Elton, Blackwell, Oxford, pp 225–235

Richardson DM, Pysek P, Rejmanek M, Barbour MG, Panetta FD, West CJ (2000) Naturalization and invasion of alien plants: concepts and definitions. Divers Distribut 6:93–107

Roff DA (2002) Life history evolution. Sinauer Associates, INC, Sunderland

Roman J, Darling JA (2007) Paradox lost: genetic diversity and the success of aquatic invasions. Trends Ecol Evol 22:454–464

Sakai AKK, Allendorf FWW, Holt JSS, Lodge DMM, Molofsky J, With KAA, Baughman S, Cabin RJJ, Cohen JEE, Ellstrand NCC, Cauley DE, O'Neil P, Parker IM, Thompson JN, Weller SG (2001) The population biology of invasive specie. Ann Rev Ecol Syst 32:305–332

Sax DF, Brown JH (2000) The paradox of invasion. Glob Ecol Biogeogr 9:363–371

Sax DF, Stachowicz JJ, Brown JH, Bruno JF, Dawson MN, Gaines SD, Grosberg RK, Hastings A, Holt RD, Mayfield MM, O'Connor MI, Rice WR (2007) Ecological and evolutionary insights from species invasions. Trends Ecol Evol 22:465–471

Sax D, Gaines S (2008) Species invasions and extinction: the future of native biodiversity on islands. Proc Natl Acad Sci U S A 105:11490–11497

Shea K, Chesson P (2002) Community ecology theory as a framework for biological invasions. Trends Ecol Evol 17:170–176

Shochat E, Susannah B, Warren PS, Faeth S (2010) Invasion, competition, and biodiversity loss in urban ecosystems. Bioscience 60:199–208

Simberloff D (2006) Invasional meltdown 6 years later: important phenomenon, unfortunate metaphor, or both? Ecol Lett 9:912–919

Simberloff D (2013) Invasive species what everyone needs to know. Oxford University Press, New York

Simberloff D, Von Holle B (1999) Positive interactions of nonindigenous species: invasional meltdown? Biol Invasions 1:21–32

Sol D (2000) Are islands more susceptible to be invaded than continents? Birds say no. Ecography 23:687–692

Sol D (2003) Behavioural flexibility: a neglected issue in the ecological and evolutionary literature? In: Reader SM, Laland KN (eds) Animal innovation. Oxford University Press, Oxford, pp 63–82

Sol D (2007) Do successful invaders exist? Pre-adaptations to novel environments in terrestrial vertebrates. In: Nentwig W (ed) Biological invasions. Springer, Heidelberg, pp 127–141

Sol D, Bacher S, Reader SM, Lefebvre L (2008) Brain size predicts the success of mammal species introduced into novel environments. Am Nat 172 Suppl:S63–S71

Sol D, Bartomeus I, Griffin AS (2012a) The paradox of invasion in birds: competitive superiority or ecological opportunism? Oecologia 169:553–564

Sol D, Duncan RP, Blackburn TM, Cassey P, Lefebvre L (2005) Big brains, enhanced cognition, and response of birds to novel environments. Proc Natl Acad Sci U S A 102:5460–5465

Sol D, Lapiedra O, Gonzalez-Lagos C (2013) Behavioural adjustments for a life in the city. Anim Behav 85:1101–1112

Sol D, Lapiedra O, Vila M (2014) Do close relatives make bad neighbors? Proc Natl Acad Sci 111:E534–E535

Sol D, Maspons J, Vall-llosera M, Bartomeus I, Garcia-Pena GE, Pinol J, Freckleton RP (2012b) Unraveling the life history of successful invaders. Science 337:580–583

Stearns SC (1992) The evolution of life histories. Oxford University Press, Oxford

Strubbe D, Broennimann O, Chiron F, Matthysen E (2013) Niche conservatism in non-native birds in Europe: niche unfilling rather than niche expansion. Glob Ecol Biogeogr 22:962–970

Stuart YE, Losos JB, Algar AC (2012) The island-mainland species turnover relationship. Proceedings. Biolog Sci/R Soc 279:4071–4077

Tilman D (1997) Community invasibility, recruitment limitation, and grassland diversity. Ecology 78:81–92

Tilman D (2004) A stochastic theory of resource competition, community assembly and invasions. Proc Natl Acad Sci U S A 101:10854–10861

Traveset A, Richardson DM (2011) Mutualisms: key drivers of invasions … key casualties of invasions. In: Richardson DM (ed) Fifty years of invasion ecology: the legacy of Charles Elton. Blackwell Publishing Ltd, Oxford, pp 143–160

Valéry L, Fritz H, Lefeuvre JC (2013) Another call for the end of invasion biology. Oikos 122:1143–1146

Valiente-Banuet A, Verdú M (2007) Facilitation can increase the phylogenetic diversity of plant communities. Ecol Lett 10:1029–1036

Vall-llosera M, Sol D (2009) A global risk assessment for the success of bird introductions. J Appl Ecol 46:787–795

Veltman CJ, Nee S, Crawley MJ (1996) Correlates of introduction success in exotic New Zealand birds. Am Natural 147:542–557

Vilà M, Espinar JL, Hejda M, Hulme PE, Jarošík V, Maron JL, Pergl J, Schaffner U, Sun Y, Pyšek P (2011) Ecological impacts of invasive alien plants: a meta-analysis of their effects on species, communities and ecosystems. Ecol Lett 14:702–708

Vilà M, Gómez A, Maron JL (2003) Are alien plants more competitive than their native conspecifics? A test using Hypericum perforatum L. Oecologia 137:211, 2005, 145:175

Vilà M, Ibáñez I (2011) Plant invasions in the landscape. Landsc Ecol 26:461–472

Vitousek PM, Walker LR, Whiteaker LD, Mueller-Dombois D, Matson PA (1987) Biological invasion by Myrica faya alters ecosystem development in Hawaii. Science (New York, NY) 238:802–804

Weiher E, Keddy P (1995) Assembly rules, null models, and trait dispersion: new questions from old patterns. Oikos 74:159–164

Williams GC (1966) Adaptation and natural selection. Princeton University Press, Princeton

Williamson MH (1996) Biological invasions. Chapman and Hall, London

Williamson MH, Brown KC, Holdgate MW, Kornberg H, Southwood R, Mollison D (1986) The analysis and modelling of British invasions. Philos Transac R Soc Lond B Biol Sci 314:1167

Williamson MH, Fitter A (1996) The varying success of invaders. Ecology 77:1661–1666

High-Throughput DNA Sequencing and the Next Generation of Molecular Markers in Wildlife Research

Stuart B. Piertney

The ability to exploit the inherent differences in the genetic make-up of individuals, populations and species to design and develop molecular markers is now a well-established practice in wildlife biology. Indeed, the entire sub-discipline of molecular ecology has been built upon isolating and characterising genetic diversity and understanding how this is apportioned both spatially and temporally within and between populations (Avise 2004; Hoglund 2009; Beebee and Rowe 2004).

The use of molecular markers has revolutionised our understanding of a broad range of issues that cover the entire evolutionary spectrum and span the complete taxonomic range. These include defining the phylogenetic relationships among species (e.g., Baldauf 2003), elucidating the phylogeographic history of populations (e.g., Piertney et al. 2005), identifying population structure (e.g., Wenzel et al. 2012), characterising dispersal (e.g., Zalewski et al. 2009), resolving relationships and relatedness amongst individuals (e.g., Shorey et al. 2000; Piertney et al. 2008), and understanding the mechanistic links between genotype and phenotype that affects the proper development and functioning of organisms (Wenzel et al. 2015). It is testament to the power and importance of molecular markers that they have found extensive and pervasive application over the last two decades or so. They have moved from being the niche tool of specialist molecular ecologists operating in dedicated laboratories to an essential part of the toolkit serving a broad community of behavioural, evolutionary and population biologists, all comfortable with using molecular techniques alongside more traditional analytical approaches.

The type, scope and utility of molecular markers that are currently considered the tools of choice in molecular ecology and wildlife biology is changing. This is primarily in response to the advent of high-throughput, next-generation DNA sequencing (NGS), which provides an inroad for rapidly resolving genome-wide

S.B. Piertney
School of Biological Sciences, University of Aberdeen, Aberdeen, UK
e-mail: s.piertney@abdn.ac.uk

R. Mateo et al. (eds.), *Current Trends in Wildlife Research*,
Wildlife Research Monographs 1, DOI 10.1007/978-3-319-27912-1_9

diversity in organisms and is yielding a growing body of genomic information for non-model species from natural populations.

The aim of this chapter to provide an overview of the new molecular tools that are based around NGS, and highlight several emerging or developing research areas where they have already, or soon will, deliver a step-change in our understanding of the functioning of natural populations. These include issues associated with conservation genetics, identifying genes associated with ecologically important traits, the form and function of epigenetic variation, and characterising communities. For each area exemplar case studies are provided that cover a broad taxonomic spectrum of animals, but all of which are taken from natural populations of non-model organisms. This is to emphasise that despite the step-change that has accompanied the advent of NGS and associated technologies, it remains the case that a relatively small suite of common tools can provide powerful insight into wildlife biology for practitioners. That said, some of the potential challenges and problems associated with the use of NGS are highlighted, and a commentary provided on how molecular markers may further develop in the short and medium term futures.

Molecular Markers: A Historical Perspective

All molecular markers require the common starting point of resolving DNA sequence polymorphism in homologous regions of the genome across individuals. Historically this was achieved from simple proxies such as chromosomal variation or phenotypic variation assumed to have an underlying genetic basis (Powell 1994). This assumption however was not always warranted, and the separation of genetic and environmental components of variance was difficult to ascertain. The large scale application of allozyme electrophoresis to natural populations from the mid-1960s onwards provided a hitherto untapped suite of genetically defined characters to use as molecular markers (Powell 1994). Allozymes are enzyme variants encoded by the same genetic locus that differ in structure to affect mobility across a coarse electrophoretic gel (May 1992). This allelic variation was sufficient to provide more robust estimators of inter – and intra-species evolutionary divergence in a phylogenetic or population genetic context. Allozymes remained the mainstay of population genetic analyses for over two decades. Their potential, however, was limited given only a small fraction of the genome could ever be screened, and the variation observed on gels did not necessarily reflect the true underlying variation in genotype given that synonymous DNA variants do not affect the protein that is transcribed then translated by a particular locus. Moreover, many non-synonymous mutations do not result in changes in the protein structure that underpins mobility differences among allozymes. A further issue was that the most easily scored allozyme loci required biopsies from tissues such as liver, which made the process destructive and so limited in the context of screening species of conservation concern.

The development of the Polymerase Chain Reaction (PCR) in the 1980s allowed more direct assessment of DNA sequence variation, and prompted the development of a suite of different molecular marker approaches such as restriction fragment length polymorphism (RFLP; Saiki et al. 1985), random amplification of polymorphic DNA (RAPD; Lynch and Milligan 1994), amplified fragment length polymorphism (AFLP; Vos et al. 1995) and microsatellite length polymorphisms (Goldstein and Schlotterer 1999). The relative merits and trends in the use of these approaches have been described in detail elsewhere (Avise 2004; Hoglund 2009). Microsatellites have been by far the most popular molecular marker of choice through the 1990s and 2000s. The high levels of diversity that could be resolved from microsatellites, their co-dominant Mendelian mode of inheritance, putative neutrality and logistical ease of scoring meant they not only underpinned population genetics analyses, but also ushered in an increasing number of studies examining relatedness and social structure within populations (e.g., Shorey et al. 2000; Piertney et al. 2008). Moreover, given they were PCR derived, a broader range of both non-destructive and non-invasive samples could be utilised in analysis, enhancing the scope into conservation monitoring and wildlife research. The only perceived downside to microsatellite markers was that the development of PCR primers for a given species was technically non-trivial, and the use of markers developed in a closely related species was invariably problematic because of the inherent problems associated with null alleles (Pemberton et al. 1995) and ascertainment bias. The development of microsatellite enrichment cloning protocols obviated some of these issues and made microsatellite isolation and characterisation relatively routine and economical (Nunome et al. 2006).

In parallel, direct Sanger DNA sequencing has become an increasingly important tool in molecular ecology. The ability to resolve intraspecific phylogenies in a phylogeographic context provided a new historical component to population genetic studies (Avise 2000, 1998), and extended the use of DNA sequencing beyond systematics and phylogenetics. A corollary of DNA sequencing was the development of single nucleotide polymorphisms (SNPs) as a molecular marker (Morin et al. 2004). SNPs offered considerable scope for resolving genetic diversity given their frequency across genomes, and also allowed for the development of markers within the neutral component of the genome or directly within genes to allow assessment of how selection shapes patterns of diversity or the relationship between genotype and phenotype at the genic level. High throughput SNP genotyping protocols made characterising marker loci easier (e.g., Hinten et al. 2007), and the technology developed to incorporate microarray technology through different types of SNP arrays (LaFramboise 2009). This allowed the simultaneous screening of large numbers of loci that could deliver the same power as microsatellites and equivalent markers with inherently more allelic variation per locus.

The primary impediment to Sanger sequencing was that it had limited capacity, recovering only about 1 kb of sequence data from a single specimen, with a maximum of 96 individuals being analysed at any one time on an automated sequencing run. The step change that has accompanied the commercial development of different NGS sequencing devices is the potential to generate tens of millions of sequence reads in parallel.

Several different high-throughput DNA sequencing platforms have been developed since NGS first became commercially available in 2005. They can be roughly divided into two types of technologies. The first involves PCR based technology, and includes the Roche 454 pyrosequencing platforms, the Applied Biosystems SOLiD system by Life Technologies, the Illumina HiSeq/MiSeq machines and the Ion Personal Genome Machine. The second group involves so-called single molecule sequencing (SMS) technologies which do not include any amplification steps prior to sequencing. These include the PacBio RS SMRT (Pacific Biosciences) and HeliScope (Helicos Biosciences) sequencing systems. The underlying mechanics of the different approaches and their relatives merits have been described elsewhere (e.g., Metzker 2010; Shokralla et al. 2012), so will not be detailed further here. The most important differences among the different platforms is their average read length, their inherent error rates, the maximum number of reads that can be generated per run and hence the total sequencing output relative to cost and their run times. Read lengths vary among platforms from around 50 base pair reads through to >1500 base pairs on the PacBio system. Total output can range from a few Mega base pairs up to several 100 gigabases, and the time for a run ranging from an hour to a day. With this type of throughput, the technology has now reached a point of being able to deliver a "genome in a day" as well as the "$1000 genome" (Hayden 2014), goals which were considered fanciful during the initial push to sequence the human genome.

For studies in wildlife biology, the capacity of NGS is currently best exploited not by whole genome sequencing across single genomes but by constructing mixtures of tagged DNA templates that allows for multiplexing of a single polymorphic marker from multiple individuals and populations, or multiplexing of different gene targets from a single bulk sample. This facilitates the simultaneous genotyping of hundreds of molecular markers simultaneously which can then be assigned at a population or individual level and used in the same way as more classical molecular markers characterised by PCR and electrophoresis.

On top of the ongoing advances with NGS technology there has been several parallel developments that enhance sequencing workflow protocols, several of which are of utility in a wildlife research context. The first is so-called "Sequence Capture" and involves targeted isolation of regions of the genome prior to sequencing (Tewhey et al. 2009). DNA is hybridised to oligonucleotide probes either immobilised onto microarrays or in solution and contain the genome regions of interest. Non-specific hybrids are removed and the targeted DNA is eluted for sequencing where polymorphism can be examined. Commercial kits are available that provide oligonucleotide arrays for exome sequencing of model organisms, or can be used to develop bespoke sequence capture arrays for non-model species. Sequence capture allows for SNP genotyping across a large number of loci in multiple individuals, which provide considerable power for examining diversity within and divergence between individuals, populations or species. Moreover because of growing genomic resources from genome sequencing programmes across different species, the capacity to target specific genes or types of marker is considerably enhanced.

A second major recent advance has followed the development and widespread use of Restriction-site associated DNA (RAD) markers (Baird et al. 2008). This is a reduced representation sequencing approach that facilitates the large scale genotyping of multiple SNPs simultaneously from homologous regions across multiple individuals. RAD-markers are produced by digesting genomic DNA with restriction endonucleases, ligating specific adaptors to the overhangs, shearing the DNA into smaller fragments and identifying and genotyping SNPs within the DNA sequence that flanks the original restriction site. The length of the sequence associated with each RAD tag can be sufficient to BLAST the sequence to potentially determine sequence identity, or use the sequences in a shotgun sequencing approach to construct larger contigs for mapping. Several protocols that are variations on the RAD theme exist to allow RAD marker sequencing on specific NGS platforms, or fine tune the number of markers that are resolved depending on the question being asked (e.g., double digest RAD; Peterson et al. 2012). Such is the utility of RAD-seq for genotyping, relatively user-friendly softwares have emerged that facilitate rapid analysis of population divergence or diversity (e.g., Stacks; Catchen et al. 2013). The perception is that RAD will become the approach of choice for large scale genotyping studies in non-model systems at least in the short term.

Related genotyping-by-sequencing approaches have been developed that perform equivalent functions to RAD genotyping, in that they generate large numbers of SNP polymorphisms without any recourse to underlying genome structure or function (Mamanova 2010). These can then be used in a mapping, population genetics, or genetical genomics framework depending on the questions being asked. Several commercial companies are now offering genotyping by sequencing as a service, allowing efficient outsourcing for laboratories without specialist infrastructure.

In the following sections the use of NGS technology and data is described in the context of key or emerging research areas in wildlife biology.

DNA Barcoding and Environmental DNA Analysis

Molecular marker systems that are targeted at the level of species were developed primarily for phylogenetic analyses geared towards describing the evolutionary relationships among taxa from differences in DNA sequence. Mitochondrial DNA has long been a popular target for such analysis given it is maternally inherited and is haploid. This reduces the effective population size of the mitochondrial genome relative to that of the nuclear genome, which facilitates phylogenetic evolutionary relationships between even closely related taxa to be resolved. Different regions of the mitochondrial genome mutate at different rates allowing for both deep and shallow evolutionary relationships between putative groups to be identified effectively (Brown et al. 1979). Moreover a lack of recombination in mtDNA greatly simplifies analysis and interpretation of the sequence data.

A demonstration that different species produced diagnostic mitochondrial DNA sequences, or in the least were monophyletic with less evolutionary divergence within species than between, has underpinned the development of DNA barcoding (Hebert et al. 2003) and provided the motivation for the many DNA barcoding initiatives that are currently ongoing (e.g., the Barcoding of Life initiative; Hajibabaei et al. 2007). DNA sequences that are diagnostic for a species are deposited in open access databases alongside species metadata that thus allow species identification of any unknown sample simply by deriving a DNA sequence and comparing against reference sequences. This precludes the need to involve expertise in more classical taxonomy for species identification or delineation, which may be lacking for certain groups or confused by phenotypic plasticity or ontogenetic variation between life history stages. DNA barcoding has been extremely useful in a wildlife forensic context (Ogden et al. 2009), determining the provenance of foodstuffs (Galimberti et al. 2013) and the identification of morphologically cryptic taxa (Nolan et al. 2007).

Traditionally, DNA barcoding has been achieved using a Sanger sequencing approach with DNA being extracted directly from the unknown sample then PCR amplified with primers for a targeted gene such as mitochondrial cytochrome c oxidase I (COI), internal transcribed spacer regions or 16S ribosomal RNA (Hebert et al. 2003). More recently however, NGS technologies have been exploited to allow the genetic characterisation of multiple individuals from a single bulk DNA extraction. This can be achieved with prior isolation of organisms to provide a mix of relatively high-quality, pure DNA ("DNA metabarcoding"), or by extraction of DNA from environmental samples such as soil, water or air without any attempt to isolate individual organisms ("eDNA metabarcoding"). The latter attempts to identify all of the different organisms that contribute DNA to the environmental sample, which will be derived from the living cells or organisms in the sample and also extracellular DNA from natural cell death and subsequent breakdown. eDNA metabarcoding is technically challenging given the DNA obtained is generally highly degraded, there is a high level of variation in copy number between the organisms contributing DNA to the overall sample, and the presence of numerous contaminants that can affect downstream analyses. Such issues can be circumvented by blocking abundant or contaminant DNA at the PCR step to allow identification of rare DNA templates, focusing on very short barcoding targets when DNA extracts are degraded, and the use of bespoke bioinformatics tools that help in the effective translation of a list of sequences into a list of species in situation where reference barcodes may or may not be available (Coissac et al. 2012).

eDNA metabarcoding has proven invaluable for charactering microbial, meiofaunal and macrofaunal communities, assessing biodiversity change in relation to environmental variation, determining extinct community structure from ancient DNA samples, assessing ecosystem health from biodiversity assessment, understanding food web interactions, defining species distributions, environmental assessment using bioindicator taxa and high resolution diet analysis (Yoccoz 2012). Its utility as a broad tool in ecological research is well illustrated by Ando et al. (2013) who used eDNA metabarcoding of faecal samples to characterise the diet of the critically endangered red headed wood pigeon (*Columba janthina nitens)* that is endemic to oceanic islands off Japan. The analysis identified 44 plant taxa from across 48 samples, which was many more plant species in the diet that was estimated from microhistology. The analysis also allowed for broad-brush estimation of relative abundance of different plant species in the diet, and how this changed month by month. Importantly the work highlighted that the birds were consuming introduced as well as native species, which is crucial for nature restoration plans for the island given eradication of invasive species needs to be balanced with the restoration of native plants to ensure sufficient food resources are available to sustain the pigeon population.

In an analogous way, eDNA metabarcoding based on small subunit 16S ribosomal RNA allowed the first characterisation of microbial communities in the oceanic hadal zone below 6000 m (Nunoura et al. 2015). It was shown that the hadal microbial community was enriched with heterotrophic populations whereas chemolithotrophic populations were more abundant in the neighbouring abyssal depths. This highlights that the hadal microbial biosphere must be supported by endogenous recycling of organic matter associated with trench geomorphology.

Environmental metagenomics is a progression of eDNA metabarcoding and moves away from single gene targets for species ID and instead focusses on providing DNA sequences of a large unbiased sample of all genes present from all members from a sampled community (Eisen 2007). Deconvoluting genes into genomes has proven analytically challenging, but has been successfully achieved to understand the most important biochemical functions operating under any given environmental context via the analysis of coding genes, assembling whole genomes of uncultivable microbes and defining host-microbe interactions involved in different disease states.

Conservation Genomics

An ability to accurately assay levels of genetic diversity within species and examine how this is apportioned spatially among populations is fundamental to conservation genetics. This is for two primary reasons. First, maintaining high levels of allelic diversity and heterozygosity in natural populations is important to provide populations and species with the capacity to adapt and evolve to a changing environment, and stave off the effects of inbreeding depression mediated through the expression of deleterious recessive mutations and reduced heterosis (Hoglund 2009; Frankham et al. 2002). As such, the conservation of genetic resources is now frequently built into broader species management plans, and a strategy for prioritising populations for conservation has been to focus on those populations that are genetically depauperate or show a declining trend in genetic variation over time (Moritz 2002). Secondly, the ability to demarcate where a population begins and ends is essential for developing and executing management and conservation strategies geared towards maintaining viable populations and maximising population growth rates. Comparison of allele or mitochondrial haplotype frequencies between putative populations, or deconstructing a mixed group of individuals into constituent populations from genetic data has proven invaluable for identifying population isolation or the levels of connectivity mediated through dispersal for multiple species. Indeed the application of molecular markers have underpinned the development of some of the commonly discussed conservation units, such as the evolutionary significant unit (ESU) and the management unit (MU). An ESU is generally defined as a population or group of populations that warrant separate management or high priority for conservation because of genetic or ecological distinctiveness, whereas at a smaller scale an MU is a population that is demographically independent (Funk et al. 2012).

An enduring issue in conservation genetics however has been how to accurately and effectively assay genetic diversity in natural populations (Fraser and Bernatchez 2001). Traditionally, genetic variation can be partitioned into its neutral and adaptive components. Neutral genetic diversity is influenced by stochastic microevolutionary processes such as random genetic drift and migration whilst functional genetic diversity directly arises from adaptation through natural selection, though levels of extant diversity in natural populations are also greatly influenced by drift. Determining levels of variation for adaptive and neutral components has involved different strategies. Neutral genetic diversity has been assayed from markers such as microsatellites, whereas adaptive polymorphism is assessed through variation at quantitative traits (Frankham et al. 2002). Both approaches are intuitively attractive yet are also far from ideal. Variation in quantitative traits does reflect the adaptive evolutionary response of populations, but can be difficult to assay given a requirement of a large, well-resolved pedigree. Also, in some cases it has proven difficult to separate genetic versus environmental effects on the focal trait of interest. Conversely, whilst neutral molecular markers are relatively easy to assay, it is unclear the extent to which neutral genetic diversity mirrors functional variation and hence the adaptive evolutionary potential of populations. Meta-analyses suggest that neutral markers are a poor predictor of quantitative trait variation, especially for morphological traits, so should not be used as a proxy for total genetic variation (Merila and Crnokrak 2001; Reed and Frankham 2001).

The application of genome-wide information derived from NGS can provide a radically different approach for partitioning the neutral from adaptive components of genetic variation. Recent emphasis has been placed on so-called outlier analyses that take genome-wide SNP genotyping data derived from RAD-genotyping or genome resequencing among individuals to estimate genetic differentiation using the F_{ST} statistic for a large number of loci across the genome. The aim is to identify loci that display an elevated genetic divergence relative to the genomic average or neutral models (Luikart et al. 2003). Such loci show high divergence through the effects of divergent selection, so by definition must be adaptive loci or be in close linkage disequilibrium to a locus under selection. Once identified these markers can be removed from subsequent analysis if focus is on neutral variation, or analysed in isolation if the adaptive component of the genome is the target. Both classes of marker can then be utilised in classical population genetics analyses for visualising population relationships such as ordination, *Structure*-based analysis (Pritchard et al. 2000) or estimation of genetic divergence among a priori defined groups.

These types of analyses have the potential to revolutionise how conservation units are identified and prioritised both because of the greater resolution that will be afforded by using more markers and also by incorporating information on adaptive genetic variation. Given adaptive variation is affected by selection, it can structure and group populations in a different way to neutral variation depending on the relative contributions of stochastic and deterministic microevolutionary processes in shaping the overall patterns of genetic variation. Significant adaptive genetic differences would be expected among populations that experience stark environmental variation across their range or among habitat patches within a metapopulation, have large effective population sizes such that selection overpowers drift, or have inherently low migration rates (Li 1978; Hohenlohe et al. 2010). In reality, the former might be the case for many populations of conservation concern where broad geographical ranges have been fragmented, and the latter for some pelagic or oceanic fisheries stocks (Cano et al. 2008).

In practice, however, the power of genome-wide sequence information will derived from an ability to examine both neutral and adaptive genetic diversity simultaneously. A combined analysis over both classes of marker allows the relative effects of drift and selection operating in natural populations on standing genetic variation to be ascertained, and is more appropriate for identifying ESU's given these will have been shaped by historical isolation and adaptive processes such as divergent selection. MUs reflect demographically independent units defined by restricted dispersal, so neutral markers may prove more effective for identifying their boundaries. Once those are established, adaptive genetic differentiation from adaptive polymorphisms can be used to examine the interrelationships among MUs, and in doing so identify any drivers of adaptive divergence through association with known environmental variation (Funk et al. 2012).

The main concern with the use of outlier loci is that they can be prone to false positives caused by demography, low recombination rates, background selection through proximity to centromeres, or areas of low diversity generated by linked purifying selection (Narum and Hess 2011). Notwithstanding, given the motivation for an outlier analysis is to characterise the response to multiple dimensions of environmental variation, averaging across outlier loci that include some false positives and likely exclude some false negatives will still reveal genome wide signatures of local adaptation useful for defining conservation units. Moreover, putative outlier loci can be confirmed using other approaches for identifying signatures of selection such as nucleotide substitution rates (Oleksyk et al. 2010).

Understanding the patterns of adaptive genetic differentiation among populations is also useful for informing supplementation, translocation, induced hybridisation or assisted recolonisation programmes by matching as closely as possible donor and recipient populations at adaptive loci. Similarly, prior understanding of how environment shapes underlying patterns of adaptive variation is useful for identifying the most appropriate populations to act as sources for the colonisation of uninhabited habitat (Edmands 2007). The effectiveness of such conservation actions could be enhanced further by using NGS-based eDNA profiling to examine pathogen load in habitats prior to release, and select individuals for translocation accordingly based upon disease resistance or maximised adaptive variation. Moreover the effectiveness of translocation programmes can be monitored through genomic approaches. Miller et al. (2012) used genome-wide polymorphisms to examine the success of migrant individuals translocated into an insular population of big horn sheep to increase genetic diversity and reverse the effects of inbreeding depression. The migrant alleles from translocated individuals increased in frequency over time and both longevity and number of offspring was positively affected by migrant ancestry.

The large numbers of polymorphic molecular markers provided by genome scans derived from NGS also allows for much more precise understanding of population demography and demographic history (Luikart et al. 2003). Estimation of migration, occurrence and duration of population bottlenecks, historical population perturbations and fluctuations, effective population sizes and population introgression can be gleaned using an increasingly sophisticated set of analytical approaches involving likelihood based methods (e.g., Beerli and Felsenstein 2001), Approximate Bayesian Computation (ABC; Csilléry et al. 2010) and site frequency spectrum analysis computed on large genomic datasets (Excoffier et al. 2013).

The ability to use high-throughput sequencing approaches on ancient DNA samples either from environmental samples or archived material also allows for direct comparison of extant levels of genetic diversity with historical estimates. From this, baseline levels of genetic parameters can be established and trends through time assessed. Bi et al. (2013) used this approach to test for changes in genomic diversity accompanying a climate related range retraction in alpine chipmunks (*Tamias alpinus*). They compared diversity between extant samples and early twentieth century museum skins and found an increase in population subdivision consistent with the effects of fragmentation during range contraction.

It is noteworthy that the utility of NGS in a conservation context is vigorously debated among the conservation genetics community (Shafer et al. 2015). There is a school of thought that NGS technology should be limited to providing more accurate estimates of population genetic structure and associated parameters, rather than attempt to identify any key genetic regions that should be a focus for genetic restoration or a gauge for overall genetic diversity. This debate echoes that from several years ago when variation at the major histocompatibility complex (MHC) was considered a panacea for conservation genetic diversity (Edwards and Potts 1996; Piertney and Oliver 2006). We now appreciate that such emphasis was misplaced, and can actually be as damaging as it is beneficial because it ignores the complex and multifaceted nature of genomic variation that includes pleiotrophic and epistatic effects as well as environmental context dependence. The challenge, now that we have the tools to assay genome-wide variation through NGS, is to develop pipelines for genetic management of natural populations that mirror advances in personalized medicine models in healthcare.

Genetic Basis of Ecologically Meaningful and Adaptively Important Traits

Whilst genome scans and outlier analyses are useful for identifying genes under selection and the broad signature of adaptation to local environmental conditions, they represent a relatively blunt tool for identifying the allelic variants that underlie phenotypic variation and divergence for traits that are ecologically or adaptively important. Such information is important to understand issues such as the heritability of traits, the potential for micro- or macro-evolutionary change, the significance of pleiotrophy and epistastis across genes, and the relative roles of genotypic versus cis-regulatory effects on phenotypic variation (Rockman 2012). Moreover from a conservation perspective, a focus on identifying the genetic basis of fitness related traits can identify target markers that can be used to predict gene dynamics in relation to demographic or environmental change and the knock-on consequences for individual fitness or population viability.

A number of approaches are well suited to achieving this aim of identifying the genetic basis of complex phenotypic traits, and the application of NGS data has considerably enhanced their power and scope. A traditional approach in non-model species has been to examine the extent to which phenotypic variance is explained by genotype at a set of candidate genes with *a priori* evidence for a functional link to the ecological character of interest (Piertney and Webster 2010). This has been well illustrated for the genes of the Major Histocompatibility Complex (MHC). The MHC plays a fundamental role in the vertebrate immune system, recognising foreign antigens and initiating an immune response (Piertney and Oliver 2006). There is an obvious link between diversity at MHC genes and fitness given that individuals with more sequence variation at MHC loci can identify and process a larger number of pathogenic antigens and hence combat a wider range of immune insults.

An increasing number of studies are showing links between MHC genotype and parasite abundance. In the water vole (*Arvicola terrestris*) different alleles at the MHC DQA locus explain variance in ectoparasite loads of mites, fleas and tick nymphs, with MHC heterozygote voles having the lowest overall parasite load in mixed infections (Oliver et al. 2009). It would be predicted therefore that populations with reduced genetic diversity caused by demographic perturbation should have a reduced fitness and compromised viability. However, examination of MHC gene dynamics through a population bottleneck highlighted that balancing selection operating on the MHC was sufficient to prevent the loss of allelic diversity and cause heterozygote excesses even in the face of potent genetic drift (Oliver and Piertney 2012). Such observations caution against using neutral markers as proxies of adaptive genetic diversity.

Similar links between phenotype and genotype have emerged for several other candidate genes that cover a broad range of physiological and behavioural traits. These include calmodulin (Abzhanov et al. 2006), interferons (Coltman et al. 2001), heat shock proteins (Deane and Woo 2004), melanocortin 1 receptor (Majerus and Mundy 2003; Nachman et al. 2003) and toll-like receptors (Gavan et al. 2015).

What these all have in common is that they were developed using a "bottom up" approach (*sensu* Piertney and Webster 2010) whereby they were identified in a model organism and predicted to have a similar influence on phenotype in another. As such, genotypic variation was resolved in the focal species of interest and related to phenotype in the hope that the gene would explain some of the phenotypic variation observed.

A different strategy altogether is to identify candidate genes derived NGS-based transcriptomic or genomic assays in the target species in a "top down" manner. These candidates thus have known association with the phenotypic trait of interest but it is unclear whether genotypic variation explains trait variation sufficient to infer a causal genotypic link. The utility of these "top down" approaches are well illustrated across several studies attempting to identify the genetic basis of resistance of red grouse (*Lagopus l. scotica*) to its primary gastrointestinal parasite *Trichostrongylus tenuis*. Red grouse are an economically important gamebird species inhabiting heather moorlands of Scotland and northern England. *T. tenuis* is highly prevalent in red grouse where it maintains a direct life cycle and imposes major fitness costs including poor physiological condition and compromised survival and fecundity (Martínez-Padilla et al. 2014). The vast majority of grouse are infected (Wilson 1983) and although some parasite-directed immune responses are mounted (Webster et al. 2011a), grouse generally cannot purge the infection such that they must cope with high parasite burdens for life (Shaw and Moss 1989). Classical candidate immune genes such MHC, TLR and IFN that influence parasite resistance in other avian species do not explain any variation in *T. tenuis* load in red grouse.

Webster et al. (2011a) sequenced the messenger RNA molecules found in immunologically active tissues of grouse with experimentally manipulated parasite loads to maximise the contrast in burden. The occurrence of an mRNA transcript in the high parasite treatment that is absent in the low parasite birds would indicate that that particular gene is being expressed in response to the actual parasite insult. A subsequent microarray analysis of gene expression identified that 52 gene transcripts were significantly upregulated under conditions of chronic parasite load or under acute parasite stress (Webster et al. 2011b). These markers thus represent a suite of "top down" candidate genes that may contribute to the genetic basis of parasite resistance. Some of these genes are associated with immune functioning, including specific cytokines, toll receptors, ß-defensins, nitric oxide synthase, TNF receptors and cytokines (Diez-Tascon et al. 2005) whereas others were associated with other physiological processes outwith immune response. It was subsequently shown that genotypic variants in nine of these genes were significantly associated with parasite load, with effect sizes accounting for differences of 514–666 worms per bird (Wenzel and Piertney 2015). As such, genotypic variation at these loci are responsible for explaining at least some of the variation in parasite resistance in grouse.

An alternative to using a candidate gene approach to identify genes associated with complex phenotypic traits is to explore and derive the genetic architecture of traits *de novo* using statistical association between phenotypic state and mapped marker alleles. This can be achieved through a classical quantitative trait locus (QTL) framework in families or pedigrees are available. RAD-genotyping or genome resequencing can provide the large number of polymorphic markers that facilitate a highly resolved genetic map and thus offer the most scope for resolving the number and location of genes affecting any trait of interest. However, in many cases obtaining a suitable pedigree can be an impediment to QTL based analyses, especially for natural populations. An alternative is to use a genome wide association study (GWAS) that does away with a pedigree and focusses solely on the statistical association between genotypic and phenotypic variation across different individuals.

The utility of GWAS is illustrated in identifying a further set of genes that explain variance in *T. tenuis* parasite load in red grouse. In parallel to the candidate gene studies described previously, Wenzel et al. (2015) used RAD-genotyping to identify a suite of SNP polymorphisms spaced across the grouse genome, and association between SNP variation and parasite load examined across 695 individuals' Five additional SNPs were identified beyond those indicated by candidate gene analysis that accounted for differences of up to 499 worms per bird.

One conceptual concern when attempting to identify the genetic basis of traits of ecological importance is whether the character of interest is actually controlled by a multitude of epistatically interacting genes each with minute and possibly undetectable effect sizes. The links between polymorphism at genes such as MC1R and patterns of feather, fur and hair colouration across several species clearly illustrate that some large effect genic "gold nuggets" (*sensu* Rockman 2012) do exist. However, there remains considerable debate over the generality of large effect causal variants, especially given that in a number of cases the genotypic variants that are identified from GWAS or QTL analyses explain only a small fraction of the phenotypic variance even when the phenotype is highly heritable (Manolio et al. 2009). This may be a particular issue for more complex phenotypes, which are likely to be those that interest to wildlife biologists that aim to understand factors affecting issues such as survival, reproductive success and other components of fitness. For example, Santure et al. (2013) found no significant QTL or GWAS based associations with clutch size and egg mass in wild populations of great tit (*Parus major*) despite a large number of markers being screened across a considerable number of pedigreed individuals.

One way to address this issue is to move focus from individual SNP associations to examine whether large linkage groups of SNPs together explain more phenotypic variance that expected under a polygenic null model where all markers contribute equally. Such "genome partitioning" approaches do not identify causal SNPs but have proven useful for retrieving some of the missing heritability for complex phenotypes such as nematode parasite burden in sheep, and can flag regions of the genome for finer scale mapping to identify causal variants. In the case of the great tit example above (Santure et al. 2013), genome partitioning identified that the amount of variance explained by each chromosome scaled with its size, consistent with the assertion that the clutch size and egg mass traits are underpinned by a large number of genes of small individual effect.

Speciation Genomics

A long-standing goal in evolutionary biology has been to identify the genes that underlie reproductive isolation and speciation. Such information is required to clarify the processes involved in speciation and adaptive radiation, understand the origins of biodiversity, reconcile gradual adaptive change with sudden bursts of population divergence leading to adaptive radiation and help define and identify species as biologically meaningful entities (Nosil and Schluter 2011). However, we are still a long way from properly understanding how many genes are involved in the speciation process, what their relative effect sizes are, whether these genes are consistent across speciation events in different taxa, what their phenotypic effects are, and how they are influenced by specific forms of mutation and stochastic versus deterministic microevolutionary processes.

The nascent field of speciation genomics is making important contributions to our understanding of the mechanics of speciation (Seehausen et al. 2014), and much of that is underpinned by the pervasive application of NGS.

Reproductive barriers between incipient species can be generated by disruptive or divergent natural selection leading to extrinsic reproductive isolation, or by the accumulation of genetic incompatibilities that causes intrinsic reproductive isolation. Most attention has traditionally been given to the genes causing hybrid sterility or inviability which can be readily assessed in the laboratory and stresses the isolation aspects that are central to the biological species concept (Mayr 1988). Several key genes have been identified from QTL mapping studies and confirmed using experimental approaches such knockout or transgenic manipulation.

However, more recent studies that utilise high throughput sequencing are focussed on putative ecological speciation caused by extrinsic isolation. This is partly a consequence of the emergence of a number of studies that have demonstrated incipient speciation in the presence of gene flow from DNA sequence data used in a coalescent framework under an isolation-with-migration model (e.g., Niemiller et al. 2008). These implicate ecological adaptation as a driver of speciation, and provided momentum to examine underlying genetic mechanisms that would allow reproductive isolation to accumulate despite the homogenising effects of gene flow.

Under the genic view of speciation species boundaries are porous, and reproductive isolation is an emergent property of divergence at a few gene loci under strong divergent selection (Wu 2001). The remainder of the neutral or weakly-selected genome is homogenized by gene flow such that there is a heterogeneous landscape of differentiation across the genome, with those adaptive loci under selection displaying minimal introgression and high divergence. Over time, the size of these so-called islands of elevated genomic differentiation increases through divergence and genome hitch-hiking until the whole genome becomes fully incompatible.

NGS provides an approach for directly investigate genome wide divergence along this speciation continuum (Ellegren et al. 2012). Patterns of genetic differentiation across the genome can be determined from SNP variation among individuals and visualised using approaches such as F_{ST} kernel density analysis and Manhattan plots to identify regions of elevated divergence. Several studies have resolved a small number of large genomic islands, whilst others have resolved many physically unlinked genomic regions as small as single genes or point mutations dotted throughout the genome. These possibly reflect a transition process across the speciation continuum, and together highlight that selection can cause strong isolation of small genomic regions between incipient species, and that when reproductive isolation is sufficiently strong that unlinked genomic regions not directly affected by selection can similarly become differentiated.

Genome wide data can also be used in alternative ways to inform aspects of ecological speciation, such as testing for signatures of introgression using approaches such as the ABBA-BABA test (Martin et al. 2014); identify genomic regions carrying the signatures of divergent selection between incipient species using outlier analyses; and mapping the genes involved in reproductive isolation either within a classical QTL framework using a phenotypically variable pedigree or using admixture mapping where incipient species distributions are sympatric or overlap. QTL mapping can be exploited further to gain a better understanding of functional effects of genes involved in speciation if combined with RNA sequencing approaches that characterise gene expression levels across genes from mRNA sequence abundance identified from NGS. By using gene transcript abundance as a quantitative, continuously variable phenotype and assessing whether a particular marker co-segregates with mRNA expression levels in a conventional QTL analysis, the *cis*- or *trans*- acting gene regions that affect steady state levels of each transcript can be identified (Chen et al. 2007). These eQTL can then be viewed with respect to genomic regions of elevated divergence to assess the effects of variation in gene expression relative to protein coding polymorphism for influencing reproductive isolation.

The scope and potential of NGS in the context of speciation genomics is epitomised by the recent analysis of the evolutionary relationships among the Darwin's finches (Lamichhaney et al. 2015). Whole-genome re-sequencing of 120 individuals from across all of the species and two closely related Tanager species was undertaken to analyse patterns of intra- and interspecific genome diversity and phylogenetic relationships among the species. The phylogeny that was generated dated the rapid radiation of ground and tree finches at around 100,000–300,000 years bp. It highlighted several differences to previous phylogenetic analyses and large discrepancies with the phenotype-based taxonomy that could be explained by extensive introgressive hybridisation throughout the radiation. The study resolved considerable nucleotide diversity within species that estimate effective population sizes between 6000 and 60,000 individuals, with extensive sharing of genetic variation such that there are no fixed differences between species. Genome wide-scans of population genetic differentiation between species that are closely related but show differences in beak morphology identified a number of regions of elevated divergence, several of which contained genes associated with craniofacial and/or beak development in mammals or birds. This would suggest a polygenic basis for beak diversity. Most notable was a 240 Kb region that harboured the ALX1 gene and revealed a deep divergence between blunt and pointed beak birds in phylogenetic analysis, driven by 335 fixed nucleotide differences. In concert, these findings have shed considerable light on the evolutionary pattern and process driving diversity in this group, and provided a level of resolution that is beyond the more targeted DNA based analyses that have come previously.

Epigenetics

The central dogma that phenotypic variation and evolutionary change is underpinned solely by DNA sequence variation is under challenge. A growing focus on epigenetics has highlighted several molecular mechanisms that can affect gene function and phenotypic diversity through the modulation of gene expression in the absence of underlying DNA sequence polymorphism (Jablonka and Raz 2009; Massicotte et al. 2011). The most widely studied epigenetic mechanism is enzyme-mediated attachment of a methyl group to cytosine nucleotides (Angers et al. 2010). This can have a number of effects depending on where it occurs within a gene or genome. Increased methylation in gene promoters is often associated with gene silencing of those genes (Jones 2012) and methylation in gene bodies or non-coding regions may silence transposable elements (Suzuki and Bird 2008; Zemach et al. 2010), provide mutational hotspots through increased deamination rate of methylated compared to unmethylated cytosine (Jones 2012), or recruit protein factors that are involved in chromatin remodelling (Bannister and Kouzarides 2011). DNA methylation is taxonomically widespread, but its prevalence and genomic distribution varies widely. In vertebrates methylation occurs throughout the genome whereas in invertebrates there is a mosaic pattern with only repetitive DNA and actively transcribed sequences being affected.

Methylation is a key mechanism influencing ontogenetic development by mediating cell differentiation and allowing environmental factors to influence phenotype (Skinner 2011; Feil and Fraga 2012). There is also evidence that patterns of methylation change throughout life having direct effects on phenotypic plasticity that can be transmitted vertically either through meiotic persistence of methylation or by transmission of extra-genomic molecules in gametes (Jablonka and Raz 2009; Smith and Ritchie 2013).

Recent studies have demonstrated a role for methylation across a broad suite broad eco-evolutionary processes including biological invasion (Richards et al. 2012), sexual selection (Crews et al. 2007), domestication (Xiang et al. 2013), inbreeding depression (Vergeer et al. 2012), seasonal timing of physiology (Stevenson and Prendergast 2013), transition between maturation stages (Morán and Pérez-Figueroa 2011), reproductive labour division in social insects (Amarasinghe et al. 2014), osmotic stress (Morán et al. 2013), nutritional plasticity (Herrera et al. 2012; Snell-Rood et al. 2013), water and temperature tolerance (Paun et al. 2010) and stress response (Verhoeven et al. 2010).

At a population level, differentiation of methylation states is frequently observed among individuals in different environments (Schulz et al. 2013), implying a potential role for local adaptation and speciation (Smith and Ritchie 2013).

Genome-wide screens for epigenetic methylation states in natural populations of non-model organisms is possible via several approaches. The initial route was using methylation sensitive amplified fragment length polymorphism (MS-AFLP) analysis (Reyna-Lopez et al. 1997) that involves comparison of the fingerprint profiles obtained when methylation sensitive and insensitive isochizomers are utilised in parallel digests from the same individual. The presence or absence of a band across paired profiles indicates methylation has occurred at the restriction site the two enzymes target. Variation in epiloci across individuals can then be assessed relative to variation in phenotype, or population level estimates of epigenetic variation can be obtained that can be used in a conservation or population genetics context. This approach was used by Wenzel and Piertney (2014) to identify epiloci that explain variation in *T. tenuis* parasite load in red grouse to complement the candidate gene and GWAS approaches described previously. In this case, 35 epiloci were identified that were significantly associated with parasite load and could be linked to gene function. These included genes associated with regulation of cell cycle, immune system, rRNA methylation and cell signalling.

A more comprehensive screen of genome-wide methylation can be obtained using whole genome bisulphite sequencing (Deng et al. 2009). Treatment of DNA with bisulphite converts cytosine residues to uracil, but will not change those cytosines that are methylated. As such, bisulphite treatment changes the nucleotide composition of DNA based on its methylation state that can then be easily identified by comparison of bisulphite treated and untreated DNA sequences. As with all NGS approaches costs for bisulphite sequencing are continually reducing, whilst the quality of data is increasing in terms of the numbers of false negative and positive sites being detected. As such, there is the expectation of more studies that examine patterns of methylation in wildlife populations and relate polymorphism to phenotypic pattern and process.

Future Perspectives

A major contributing factor to the broad application of the more traditional molecular marker systems such as microsatellite DNA polymorphisms and Sanger-derived mitochondrial DNA sequences in wildlife biology is that they possess inherently high information content, yet are inexpensive and technically relatively simple to screen, analyse and interpret. Indeed, the convenience of DNA extraction kits for isolating DNA from a range of source materials, the accessibility of cost-effective commercial genotyping services for outsourcing, and off-the-shelf software packages mean that molecular analysis has become decoupled from specialist research labs. This allowed the broad uptake of molecular markers across many biological disciplines, and more importantly saw the translation of molecular techniques into applied conservation and wildlife management practice by practitioners and end-users working independently of academic collaborators.

We are a long way from that being the situation with genome-wide markers derived from NGS. The infrastructure to generate data remains expensive, some of the procedures involved are technically demanding, analysis can be challenging from the low quantity or poor quality non-destructive or non-invasive DNA sources that frequently are the focus in studies on natural populations, and the vast quantities of "big data" generated by NGS can be unwieldy to analyse. It is probably the latter that represents the biggest impediment to the pervasive application of genomics data in wildlife biology. Data analysis is both computationally intensive, and in most cases requires the development of bespoke analytical pipelines that can only be developed with skills in computer coding. Moreover, many areas of research are still actively involved in vibrant debate and ongoing studies to identify the best approach to distill biological meaningful insight from genome wide data. This is clearly a concern for more applied wildlife biologists as it returns the onus to having specialist back-up support, not only from specialist molecular laboratories but also from dedicated bioinformaticians and biostatisticians who can wrangle the large amounts of data being generated and provide the answers to questions that are actually required in the field for effective conservation and management.

To bridge his gap and ensure that genomic analysis is translated into applied fields will require recursive interaction between molecular ecologists and end-users. The former need to communicate the promise of next generation sequencing and how the scaling up to genome-wide data can provide greater or better insight for practitioners. This is best achieved by trumpeting the case studies where NGS has been successfully utilised. In turn, practitioners need to embrace the opportunities afforded by high-throughput sequencing which may require a degree of re-engagement with academic collaborators. Communication between molecular practitioners and end-users has not always been easy or proven effective (Hoban et al. 2013; Shafer et al. 2015), but requires continues effort and emphasis on both sides. Clearly, given some of the issues associated with getting some end-user groups to embrace traditional molecular markers as an applied tool in particular contexts, there are challenges moving forward to highlight that NGS is more than just a fashionable technology with esoteric utility.

What is clear is that next generation sequencing technology has not yet reached any form of stable equilibrium, and there will continue to be a push for an increasing amount of data, and further miniaturization of the hardware that generates it. We are already a long way from the banks of large automated DNA sequencing machines that dominated large dedicated laboratories when the first model genomes were produced. Bench-top personalised DNA sequencers are already the norm, and are available in ever decreasing sizes capable of producing ever increasing amounts of data in shorter and shorter run times. This trend will continue, with USB plug-and-play sequencers already available. This will certainly excite molecular ecologists working in natural populations given the portability of the infrastructure required to undertake analysis, and make the next generation of molecular markers more accessible for wildlife practitioners in the same way that been observed for PCR-based markers this past decade.

Clearly, there are challenges associated with seamlessly integrating next-generation sequencing into studies of natural populations on non-model species, and we are a way off seeing genome wide data used as widely and openly as more traditional molecular markers. Notwithstanding, the promise of having orders of magnitude more polymorphisms to utilize to better understand and manage biodiversity and our natural capital means these are exciting times to be involved in wildlife biology.

Acknowledgments I would like to thank Marius Wenzel, Carles Vilà, Heather Ritchie, Alex Douglas and Martha Gavan for comment, content and discussion. Support was provided through a Leverhulme Trust Research Fellowship.

References

Abzhanov A, Kuo WP, Hartmann C et al (2006) The calmodulin pathway and evolution of elongated beak morphology in Darwin's finches. Nature 442:563–567

Amarasinghe HE, Clayton CI, Mallon EB (2014) Methylation and worker reproduction in the bumble-bee (*Bombus terrestris*). Proc R Soc Lond B Biol Sci 281:20132502

Ando H, Setsuko S, Horikoshi K et al (2013) Diet analysis by next-generation sequencing indicates the frequent consumption of introduced plants by the critically endangered red-headed wood pigeon (*Columba janthina nitens*) in oceanic island habitats. Ecol Evol 3:4057–4069

Angers B, Castonguay E, Massicotte R (2010) Environmentally induced phenotypes and DNA methylation: how to deal with unpredictable conditions until the next generation and after. Mol Ecol Res 19:1283–1295

Avise JC (1998) The history and purview of phylogeography: a personal reflection. Mol Ecol 7:371–379

Avise JC (2000) Phylogeography. Harvard University Press, Cambridge

Avise JC (2004) Molecular markers, natural history & evolution. Chapman & Hall, New York

Baird NA, Etter PD, Atwood TS et al (2008) Rapid SNP discovery and genetic mapping using sequenced RAD markers. PLoS One 3:1–7. doi:10.1371/journal.pone.0003376

Baldauf SL (2003) Phylogeny for the faint of heart: a tutorial. Trends Genet 19:345–351. doi:10.1016/S0168-9525(03)00112-4

Bannister AJ, Kouzarides T (2011) Regulation of chromatin by histone modifications. Cell Res 21:381–395

Beebee T, Rowe G (2004) An introduction to molecular ecology. OUP, Oxford

Beerli P, Felsenstein J (2001) Maximum liklihood estimation of a migration matrix and effective population sizes of n subpopulations by using a coalescent approach. Proc Natl Acad Sci U S A 98:4563–4568

Bi K, Linderoth T, Vanderpool D et al (2013) Unlocking the vault: next-generation museum population genomics. Mol Ecol 22:6018–6032. doi:10.1111/mec.12516

Brown W, George MJ, Wilson A (1979) Rapid evolution of animal mitochondrial DNA. Proc Natl Acad Sci U S A 76:1967–1971

Cano J, Shikano T, Kuparinen A, Merila J (2008) Genetic differentiation, effective population size and gene flow in marine fishes: implications for stock management. JIFS 5:1–10

Catchen J, Hohenlohe PA, Bassham S et al (2013) Stacks: an analysis tool set for population genomics. Mol Ecol 22:3124–3240

Chen LS, Emmert-Streib F, Storey JD (2007) Harnessing naturally randomized transcription to infer regulatory relationships among genes. Genome Biol 8:R219. doi:10.1186/gb-2007-8-10-r219

Coissac E, Riaz T, Puillandre N (2012) Bioinformatic challenges for DNA metabarcoding of plants and animals. Mol Ecol 21:1834–1847

Coltman DW, Wilson K, Pilkington JG et al (2001) A microsatellite polymorphism in the gamma interferon gene is associated with resistance to gastrointestinal nematodes in a naturally-parasitized population of Soay sheep. Parasitology 122:571–582

Crews D, Gore AC, Hsu TS et al (2007) Transgenerational epigenetic imprints on mate preference. Proc Natl Acad Sci U S A 104:5942–5946

Csilléry K, Blum MGB, Gaggiotti OE, François O (2010) Approximate Bayesian Computation (ABC) in practice. Trends Ecol Evol 25:410–418. doi:10.1016/j.tree.2010.04.001

Deane EE, Woo NYS (2004) Differential gene expression associated with euryhalinity in sea bream (Sparus sarba). Am J Physiol Regul Integr Comp Physiol 287:55–65

Deng J, Shoemaker R, Xie B (2009) Targetef bisulphite sequencing reveals changes in DNA methylation associated with nuclear programming. Nat Biotechnol 27(4):353–360

Diez-Tascon C, Keane OM, Wilson T et al (2005) Microarray analysis of selection lines from outbred populations to identify genes involved with nematode parasite resistance in sheep. Physiol Genomics 21:59–69

Edmands S (2007) Between a rock and a hard place: evaluating the relative risks of inbreeding and outbreeding for conservation and management. Mol Ecol 16:463–475

Edwards SV, Potts WK (1996) Polymorphism of Mhc genes: implications for conservation genetics of vertebrates. In: Smith TB, Wayne RK (eds) Molecular genetic approaches to conservation. Oxford University Press, Oxford, pp 214–237

Eisen J (2007) Environmental shotgun sequencing: its potential and challenges for studying the hidden world of microbes. PLoS Biol 5:e82

Ellegren H, Smeds L, Burri R et al (2012) The genomic landscape of species divergence in Ficedula flycatchers. Nature 491(7426):756–760. doi:10.1038/nature11584

Excoffier L, Dupanloup I, Huerta-Sánchez E et al (2013) Robust demographic inference from genomic and SNP data. PLoS Genet. doi:10.1371/journal.pgen.1003905

Feil R, Fraga MF (2012) Epigenetics and the environment: emerging patterns and implications. Nat Rev Genet 13:97–109

Frankham R, Ballou JD, Briscoe DA (2002) An introduction to conservation genetics. CUP, Cambridge

Fraser DJ, Bernatchez L (2001) Adaptive evolutionary conservation: towards a unified concept for defining conservation units. Mol Ecol 10:2741–2752

Funk C, McKay J, Hohenlohe PA, Allendorf F (2012) Harnessing genomics for delineating conservation units. Trends Ecol Evol 27:489–496

Galimberti A, DeMattia F, Losa A et al (2013) DNA barcoding as a new tool for food traceability. Food Res Int 50:55–63

Gavan MK, Oliver MK, Douglas A, Piertney SB (2015) Gene dynamics of Toll-like receptor 4 through a population bottleneck in an insular population of water voles (*Arvicola amphibius*). Conserv Genet 16(5):1181–1193

Goldstein DB, Schlotterer C (1999) Microsatellites: evolution and application. OUP, Oxford

Hajibabaei M, Singer G, Clare E, Hebert P (2007) Design and applicability of DNA arrays and DNA barcodes in biodiversity monitoring. BMC Biol 5:24

Hayden E (2014) Technology: the $1000 genome. Nature 507:294–295

Hebert P, Cywinska A, Ball S, DeWaard J (2003) Biological identifications through DNA barcodes. Proc R Soc London B 270:313–321

Herrera CM, Pozo MI, Bazaga P (2012) Jack of all nectars, master of most: DNA methylation and the epigenetic basis of niche width in a flower-living yeast. Mol Ecol Res 21:2602–2616

Hinten GN, Hale MC, Gratten J et al (2007) SNP-SCALE: SNP scoring by colour and length exclusion. Mol Ecol Notes 7:377–388

Hoban SM, Hauffe HC, Pérez-Espona S et al (2013) Bringing genetic diversity to the forefront of conservation policy and management. Conser Genet Resour 5:593–598

Hoglund J (2009) Evolutionary conservation genetics. OUP, Oxford

Hohenlohe P, Bassham S, Etter PD et al (2010) Population genomics of parallel adaptation in threespine stickleback using sequenced RAD tags. PLoS Genet 6:e1000862

Jablonka E, Raz G (2009) Transgenerational epigenetic inheritance: prevalence, mechanisms, and implications for the study of heredity and evolution. Q Rev Biol 84:131–176

Jones PA (2012) Functions of DNA methylation: Islands, start sites, gene bodies and beyond. Nat Rev Genet 13:484–492

LaFramboise T (2009) Single nucleotide polymorphism arrays: a decade of biological computational and technological advances. Nucl Acids Res 37:4181–4193

Lamichhaney S, Berglund J, Almén MS et al (2015) Evolution of Darwin's finches and their beaks revealed by genome sequencing. Nature 518:371–375

Li W (1978) Maintainance of genetic variability under the joint effects of mutation selection and random drift. Genetics 90:349–382

Luikart G, England P, Tallmon D et al (2003) The power and promise of population genomics: from genotyping to genome typing. Nat Rev Genet 4:981–994

Lynch M, Milligan BG (1994) Analysis of population genetic structure with RAPD markers. Mol Ecol 3:91–99

Majerus MEN, Mundy NI (2003) Mammalian melanism: natural selection in black and white. Trends Genet 19:585–590

Mamanova L (2010) Target-enrichment strategies for next generation sequncing. Nat Methods 7:111–118

Manolio T, Collins F, Cox N (2009) Finding the missing heritability of complex diseases. Nature 461:747–753

Martin S, Davey J, Jiggins C (2014) Evaluating the use of ABBA-BABA statistics to locate introgrerssed loci. Mol Biol Evol 32:244–257

Martínez-Padilla J, Redpath SM, Zeineddine M, Mougeot F (2014) Insights into population ecology from long-term studies of red grouse *Lagopus lagopus scoticus*. J Anim Ecol 83:85–98

Massicotte R, Whitelaw E, Angers B (2011) DNA methylation: a source of random variation in natural populations. Epigenetics Off J DNA Methylation Soc 6:421–427

May B (1992) Starch gel electrophoresis of allozymes. In: Hoelzel AR (ed) Molecular genetics analysis of populations. OUP, Oxford, pp 1–27

Mayr E (1988) The how and why of species. Biol Philos 3:431–441

Merila J, Crnokrak P (2001) Comparison of genetic differentiation at marker loci and quantitative traits. J Evol Biol 14:892–903

Metzker M (2010) Sequencing technologies – the next generation. Nat Rev Genet 11:31–46

Miller J, Poissant J, Hogg J, Coltman D (2012) Genomic consequences of genetic rescue in an insular population of bighorn sheep (Ovis canadensis). Mol Ecol 21:1583–1596

Morán P, Pérez-Figueroa A (2011) Methylation changes associated with early maturation stages in the Atlantic salmon. BMC Genet 12:86

Morán P, Marco-Rius F, Megías M et al (2013) Environmental induced methylation changes associated with seawater adaptation in brown trout. Aquaculture 392:77–83

Morin PA, Luikart G, Wayne RK, group SNP workshop (2004) SNPs in ecology, evolution and conservation. Trends Ecol Evol 19:208–216

Moritz C (2002) Strategies to protect biological diversity and the evolutionary processes that sustain it. Syst Biol 51:238–254

Nachman MW, Hoekstra HE, D'Agostino SL (2003) The genetic basis of adaptive melanism in pocket mice. Proc Natl Acad Sci U S A 100:5268–5273

Narum S, Hess J (2011) Comparison of FST outlier tests for SNP loci under selection. Mol Ecol Res 11:184–194

Niemiller ML, Fitzpatrick BM, Miller BT (2008) Recent divergence with gene flow in Tennessee cave salamanders (Plethodontidae: Gyrinophilus) inferred from gene genealogies. Mol Ecol 17:2258–2275. doi:10.1111/j.1365-294X.2008.03750.x

Nolan D, Carpenter S, Barber J et al (2007) Rapid diagnostic PCR assays for members of the *Culicoides obsoletus* and *Culicoides pulicaris* species complexes, implicated vectors of bluetongur virus in Europe. Vet Microbiol 124:82–94

Nosil P, Schluter D (2011) The genes underlying the process of speciation. Trends Ecol Evol 26:160–167. doi:10.1016/j.tree.2011.01.001

Nunome T, Negoro S, Miyatake K et al (2006) A protocol for the construction of microsatellite enriched genomic library. Plant Mol Biol Re 24:305–312

Nunoura T, Takaki Y, Hirai M, et al (2015) Hadal biosphere: insight into the microbial ecosystem in the deepest ocean on Earth. Proc Natl Acad Sci 201421816. doi:10.1073/pnas.1421816112

Ogden R, Dawney N, McEwing R (2009) Wildlife DNA forensics – bridging the gap between conservation genetics and law enforcment. Endanger Species Res. doi:10.3354/esr00144

Oleksyk TK, Smith MW, O'Brien SJ (2010) Genome-wide scans for footprints of natural selection. Philos Trans R Soc Lond B Biol Sci 365:185–205. doi:10.1098/rstb.2009.0219

Oliver MK, Piertney SB (2012) Selection maintains MHC diversity through a natural population bottleneck. Mol Biol Evol. doi:10.1093/molbev/mss063

Oliver MK, Telfer S, Piertney SB (2009) Major histocompatibility complex (MHC) heterozygote superiority to natural multi-parasite infections in the water vole (*Arvicola terrestris*). Proc R Soc Lond B Biol Sci 276:1119–1128

Paun O, Bateman RM, Fay MF et al (2010) Stable epigenetic effects impact adaptation in allopolyploid orchids (Dactylorhiza: Orchidaceae). Mol Biol Evol 27:2465–2473

Pemberton JM, Slate J, Bancroft D, Barrett J (1995) Nonamplifying alleles at microsatellite loci: a caution for parentage and population studies. Mol Ecol 4:249–252

Peterson BK, Weber JN, Kay EH et al (2012) Double digest RADseq: an inexpensive method for de novo SNP discovery and genotyping in model and non-model species. PLoS One. doi:10.1371/journal.pone.0037135

Piertney SB, Oliver MK (2006) The evolutionary ecology of the major histocompatibility complex. Heredity (Edinb) 96:7–21

Piertney S, Webster LMI (2010) Characterising functionally important and ecologically meaningful genetic diversity using a candidate gene approach. Genetica 138:419–432

Piertney SB, Stewart WA, Lambin X et al (2005) Phylogeographic structure and postglacial evolutionary history of water voles (*Arvicola terrestris*) in the United Kingdom. Mol Ecol 14:1435–1444. doi:10.1111/j.1365-294X.2005.02496.x

Piertney SB, Lambin X, Maccoll ADC et al (2008) Temporal changes in kin structure through a population cycle in a territorial bird, the red grouse *Lagopus lagopus scoticus*. Mol Ecol 17:2544–2551. doi:10.1111/j.1365-294X.2008.03778.x

Powell J (1994) Molecular techniques in population genetics: a brief history. In: Streit B, Wagner G, DeSalle R, Schierwater B (eds) Molecular ecology and evolution: approaches and applications. Birkhauser, Basel, pp 131–156

Pritchard JK, Stephens M, Donnelly P (2000) Inference of population structure using multilocus genotype data. Genetics 155:945–959. doi:10.1111/j.1471-8286.2007.01758.x

Reed DH, Frankham R (2001) How closely correlated are molecular and quantitative measures of genetic variation? A meta-analysis. Evol (N Y) 55:1095–1103

Reyna-Lopez GE, Simpson J, Ruiz-Herrera J (1997) Differences in DNA methylation patterns are detectable during the dimorphic transition of fungi by amplification of restriction polymorphisms. Mol Gen Genet MGG 253:703–710

Richards CL, Schrey AW, Pigliucci M (2012) Invasion of diverse habitats by few Japanese knotweed genotypes is correlated with epigenetic differentiation. Ecol Lett 15:1016–1025

Rockman MV (2012) The QTN program and the alleles that matter for evolution: all that's gold does not glitter. Evolution 66:1–17

Saiki R, Scharf S, Faloona F et al (1985) Enzymatic amplification of B-Globin genomic sequences and restriction site analysis for diagnosis of sickle cell anaemia. Science 230:1350–1354

Santure AW, De Cauwer I, Robinson MR et al (2013) Genomic dissection of variation in clutch size and egg mass in a wild great tit (*Parus major*) population. Mol Ecol 22:3949–3962. doi:10.1111/mec.12376

Schulz B, Eckstein RL, Durka W (2013) Scoring and analysis of methylation-sensitive amplification polymorphisms for epigenetic population studies. Mol Ecol Res 13:642–653

Seehausen O, Butlin RK, Keller I et al (2014) Genomics and the origin of species. Nat Rev Genet 15:176–192. doi:10.1038/nrg3644

Shafer AB, Wolf JBW, Alves PC et al (2015) Genomics and the challenging translation into conservation practice. Trends Ecol Evol 30:78–87
Shaw JL, Moss R (1989) Factors affecting the establishment of the cecal threadworm *Trichostrongylus tenuis* in red grouse (*Lagopus lagopus scoticus*). Parasitology 99:259–264
Shokralla S, Spall J, Gibson J, Hajibabaei M (2012) Next generation sequencing technologies for environmental DNA reseach. Mol Ecol 21:1794–1805
Shorey L, Piertney SB, Stone J, Hoglund J (2000) Fine-scale genetic structuring on *Manacus manacus* leks. Nature 408:352–353
Skinner MK (2011) Environmental epigenetic transgenerational inheritance and somatic epigenetic mitotic stability. Epigenetics 6:838–842
Smith G, Ritchie MG (2013) How might epigenetics contribute to ecological speciation? Curr Zool 59:686–696
Snell-Rood EC, Troth A, Moczek AP (2013) DNA methylation as a mechanism of nutritional plasticity: limited support from horned beetles. J Exp Zool Part B Mol Dev Evol 320:22–34
Stevenson TJ, Prendergast BJ (2013) Reversible DNA methylation regulates seasonal photoperiodic time measurement. Proc Natl Acad Sci U S A 110:16651–16656
Suzuki MM, Bird A (2008) DNA methylation landscapes: provocative insights from epigenomics. Nat Rev Genet 9:465–476
Tewhey R, Nakano M, Wang X (2009) Enrichment of sequencing targets from the human genome by solution hybridisation. Genome Biol 10:R116
Vergeer P, Wagemaker NC, Ouborg NJ (2012) Evidence for an epigenetic role in inbreeding depression. Bio Lett 8:798–801
Verhoeven KJF, Jansen JJ, van Dijk PJ, Biere A (2010) Stress-induced DNA methylation changes and their heritability in asexual dandelions. New Phytol 185:1108–1118
Vos P, Hogers R, Bleeker M et al (1995) AFLP: a new technique for DNA fingerprinting. Nucleic Acids Res 23:4407–4414
Webster LMI, Mello LV, Mougeot F et al (2011a) Identification of genes responding to nematode infection in red grouse. Mol Ecol Res 11:305–313
Webster LMI, Paterson S, Mougeot F et al (2011b) Transcriptomic response of red grouse to gastro-intestinal nematode parasites and testosterone: implications for population dynamics. Mol Ecol Res 20:920–931
Wenzel MA, Piertney SB (2014) Fine-scale population epigenetic structure in relation to gastro-intestinal parasite load in red grouse (*Lagopus lagopus scotica*). Mol Ecol Res 23:4256–4273
Wenzel MA, Piertney SB (2015) Digging for gold nuggets: uncovering novel candidate genes for variation in gastrointestinal nematode burden in a wild bird species. J Evol Biol 28(4):807–825
Wenzel MA, Webster LMI, Blanco G et al (2012) Pronounced genetic structure and low genetic diversity in European red-billed chough (*Pyrrhocorax pyrrhocorax*) populations. Conserv Genet 13:1213–1230. doi:10.1007/s10592-012-0366-6
Wenzel MA, James MC, Douglas A, Piertney SB (2015) Genome-wide association and genome partitioning reveal novel genomic regions underlying variation in gastrointestinal nematode burden in a wild bird. Molecular Ecology 24:4175–4192
Wilson GR (1983) The prevalence of cecal threadworms (Trichostrongylus tenuis) in red grouse (Lagopus lagopus scoticus). Oecologia 58:265–268
Wu CI (2001) The genic view of the process of speciation. J Evol Biol 14:851–865. doi:10.1046/j.1420-9101.2001.00335.x
Xiang H, Li X, Dai F et al (2013) Comparative methylomics between domesticated and wild silkworms implies possible epigenetic influences on silkworm domestication. BMC Genomics 14:646
Yoccoz N (2012) The future of enviornmental DNA in ecology. Mol Ecol 21:1848–1863
Zalewski A, Piertney SB, Zalewska H, Lambin X (2009) Landscape barriers reduce gene flow in an invasive carnivore: geographical and local genetic structure of American mink in Scotland. Mol Ecol 18:1601–1615. doi:10.1111/j.1365-294X.2009.04131.x
Zemach A, McDaniel IE, Silva P, Zilberman D (2010) Genome-wide evolutionary analysis of eukaryotic DNA methylation. Science 328:916–919

Meat from the Wild: Extractive Uses of Wildlife and Alternatives for Sustainability

Nathalie van Vliet, Daniel Cornelis, Harald Beck, Peter Lindsey, Robert Nasi, Sébastien LeBel, Jessica Moreno, José Fragoso, and Ferran Jori

Introduction

Hunting and gathering remained the main mode of subsistence of humanity for hundreds of thousands of years, beginning some 1.8 million years ago, and until the Neolithic Revolution (some 10,000 years ago), when agriculture gradually spread through human societies (Marlowe 2005). Hunter-gatherer societies obtained their food directly from "natural" ecosystems, by hunting wild animals and collecting wild plants (Richerson et al. 1996). Early agrarian societies started planting desired crops on suitable lands, competing with wildlife for space and resources. As agrarian societies evolved, techniques for planting and harvesting became technologically more advanced and more efficient (Richerson et al. 1996). Innovations thus allowed the human population to grow and to colonize nearly every terrestrial ecosystem type on Earth.

However, along with the alteration of natural ecosystems, came a huge loss of biodiversity. Since the 17th century, it is estimated that 2.1 % of mammals and

N. van Vliet (✉) • R. Nasi
Bushmeat Research Initiative, Center for International Forestry Research, Bogor, Indonesia
e-mail: vanvlietnathalie@yahoo.com

D. Cornelis • S. LeBel • F. Jori
AGIRs, CIRAD, Montpellier, France

H. Beck
Biological sciences, Towson University, Towson, MD, USA

P. Lindsey
Panthera, Harare, Zimbabwe

J. Moreno
Fundación Science International, Bogotá, Colombia

J. Fragoso
Biology, Stanford University, San Francisco, CA, USA

R. Mateo et al. (eds.), *Current Trends in Wildlife Research*,
Wildlife Research Monographs 1, DOI 10.1007/978-3-319-27912-1_10

1.3 % of birds have gone extinct on the planet (Primack 2002). Furthermore, a 52 % decline was observed in a representative sample of mammals, birds, reptiles, amphibians and fish since 1970 (WWF 2014). Human populations are therefore increasingly confronted with the question of how to balance their space and food needs and the preservation of biodiversity.

Human history has given birth to a wide variety of wildlife extraction models (e.g. hunting-gathering, subsistence and commercial hunting, sport hunting). Presently, the motivations for and perceptions of extractive use are thus extremely varied, and often questioned by contemporary urban societies.

This chapter introduces extractive uses of wildlife and explores the potential for sustainable use. The first section provides a glance of the different types of extractive use and motivations for hunting. The second section discusses the drivers and impacts of unsustainable use on wildlife populations and ecosystems. The last section highlights current methodological caveats for measuring sustainability in a holistic manner and the difficulty of managing for uncertainty in the system. Some of the more promising alternatives for sustainable use are presented. This chapter focuses on terrestrial wildlife, mainly mammals, and although covering different functions of hunting, the focus is on the use of meat from the wild.

From Subsistence Hunting to International Wildlife Trade

The Multi-Functionality of Hunting

In prehistoric times, early humans essentially survived through hunting, fulfilling most of their nutritional needs and a significant part of their other requirements (e.g. rituals, clothing, tools made of bone, etc.) (Grayson 2001). Although still playing a key role for the food security of several contemporary rural societies, hunting is now also practised for a variety of reasons throughout the world. The multiple functions of hunting can be generally summarized using a framework based on three categories: (a) ecological, (b) economic and (c) socio-cultural (Fisher et al. 2013):

Ecological functions Human-wildlife conflicts have increased dramatically worldwide in recent decades due to land-use changes and high human population growth around protected areas (Woodroffe et al. 2005). In many temperate areas, hunting is regarded as a management tool for the achievement of non meat procuring objectives, reducing herbivory by wildlife to allow the regeneration of forests (for conservation or production purpose), controlling the spread of zoonoses, or reducing pests. Open public hunts for carnivores in many countries are touted a population control and property protection measure (Wilkie and Carpenter 1999; Mincher 2002; Bartel and Brunson 2003; Heberlein 2008; Campbell and Mackay 2009). Recreational hunting can play an important role in buffering development and other pressures through the maintenance of restricted use areas around core protection zones. It can also constitute a sustainable development option for developing peripheral areas

(Fig. 1). The ecological functions of hunting can be complementary, synergistic or in competition with the other functions (Rossing et al. 2007): for example culling of certain species to reduce competition with farming acts in synergy with the other uses of the landscape, but in other cases managing to maintain biodiversity and ecosystem balance might reduce the economic profits generated by hunting.

In addition to its direct ecological role, hunting also contributes indirectly to conservation through the sale of hunting licenses, tags, and stamps. For example, in the United States, hunting revenues are the primary source of funding for most state wildlife conservation efforts (U.S. Fish and Wildlife Service 2004). In southern Africa, potential income from trophy hunting was the primary driver behind the conversion of vast areas of livestock farms to wildlife ranches, resulting in major increases in wildlife populations (Bond et al., 2004; Lindsey et al. 2013a).

Economic functions There are two primary economic functions of hunting: (a) a contribution to livelihoods directly through the provision of meat and other products for consumption or the legal/illegal sale, and (b) financial income from the legal recreational industry (Fig. 2). Hunting also strongly contributes to local livelihoods, particularly in developing countries. Hunting can play a role in poverty eradication as well as contributing to a social safety net or serving as a com-

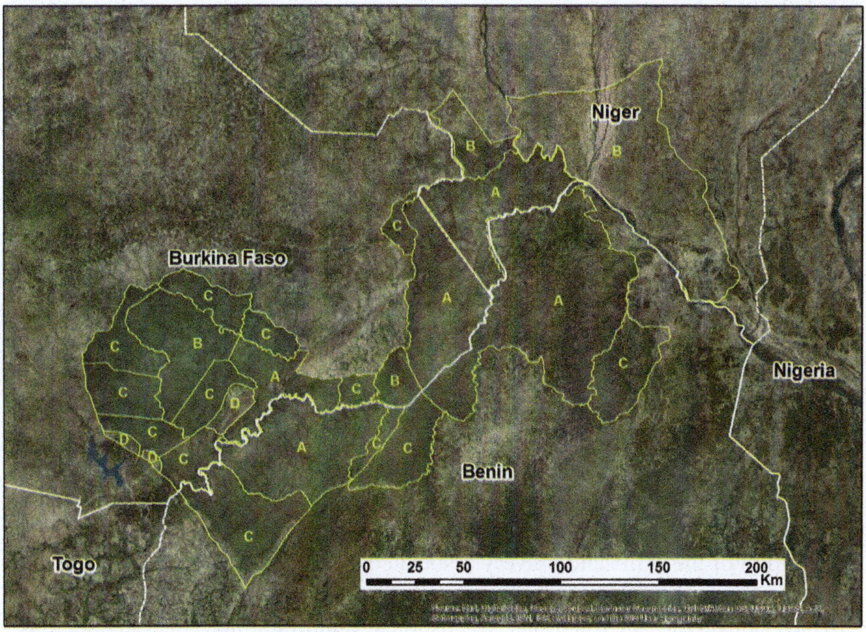

Fig. 1 WAP transfrontier complex of protected areas and their contiguous hunting blocks (Burkina Faso, Benin and Niger) (**a**) National parks, (**b**) Partial or total reserves, (**c**) Hunting blocks, (**d**) Enclave villages.This map emphasizes the role of both protected areas and recreational hunting blocks in the conservation of vegetation cover, in a context of pervasive land conversion (Source: ESRI World imagery (satelitte base map); EU ECOPAS Program (administrative contours))

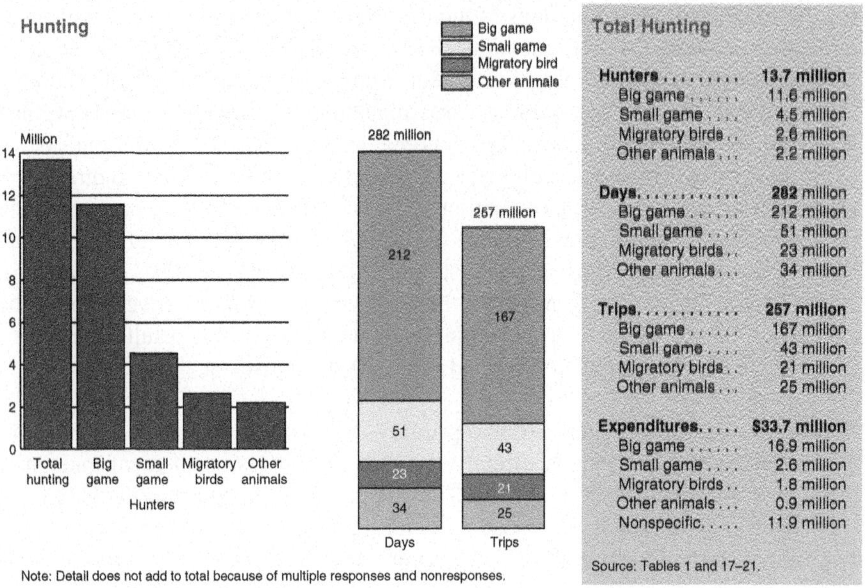

Fig. 2 Total hunting whithin the United States in 2011 (Source: U.S. Fish and Wildlife Service et al. (2011)

plement to farming activities (Brown 2003). Hunting can be legal or illegal but is mostly seen as a legitimate activity by the societies where it is practiced. In western countries, hunting is a business that generates both upstream and downstream industries and creates substantial employment and revenue. Economic benefits from recreational hunting benefit the landowners and their staff (e.g. professional stalkers), and thus allow employment in remote rural areas (MacMillan and Leitch 2008).

Socio-cultural functions Social functions of hunting relate predominantly to the development and maintenance of social capital (Putnam 2000) and respect, prestige and status, i.e. symbolic capital (Bourdieu 1977). Hunting is sometimes a culturally important activity and has important bonding functions by providing opportunities for camaraderie through what is sometimes both a physically demanding and dangerous outdoor pursuit (MacMillan and Leitch 2008). In many communities, bushmeat hunters derive elevated social status from hunting through recognition of the skills and bravery associated with hunting and through the profits derivable from selling animal products (Lowassa et al. 2012; Lindsey et al. 2013a). Conversely, in some places, bushmeat hunting is not generally seen as a high status activity – on the contrary, villagers refer to hunting as a poor man's activity (Fisher et al. 2013).

Hunting for ceremonies or festivities is another category of hunting with special characteristics (McCorquodale 1997; Peres and Nascimento 2006). For example, the Canelos Kichwa indigenous people of the Ecuadorian Amazon hunt for

Picture 1 Bushmeat (*Mazama Americana*) sold in the open market of Caballococha, Amazonas, Peru (Daniel Cruz)

Picture 2 Python meat sold in the openmarket in Makokou, Gabon (Nathalie van Vliet)

ceremonial purposes as part of the hista festival (Siren 2012). Walters et al (2014) have described the many ceremonies that are still practiced in some form by the Teke tribe in Gabon and how those still influence their beliefs about wildlife abundance, scarcity and plantation raiding.

Hunting for Livelihoods: Subsistence and Trade

In tropical countries, several authors have argued that hunting for consumption purposes represents a multibillion-dollar business, which although largely ignored in official trade and national statistics, plays a crucial role in the economies of numerous countries (Fargeot 2009). Even where wild meat is used to satisfy basic subsistence requirements, many families also hunt commercially to meet short-term cash needs. For hunters, the distinction between subsistence and commercial use is often blurred, with meat from the forest supplementing both diets and incomes (Table 1) (Nasi et al. 2008). Hunting households are not the only beneficiaries of the wild meat trade. In some cases, bushmeat hunting has become highly commercialised and is practised primarily to obtain and sell meat, often to urban markets (Lindsey et al. 2013a). From first harvest to final sale, the trade in wild meat for local, national or regional trade represents an important part of a "hidden economy". However, in many instances, bushmeat harvests are not sustainable and the economic and social benefits are likely to wane (Lindsey et al. 2013a). Furthermore, unsustainable bushmeat hunting forecloses opportunities for more sustainable use, deriving people of jobs, meat and income from legal forms of other wildlife based land use (Lindsey et al. 2013a).

Table 1 Composition of the catch in Central Africa

Country	Location	Ungulates	Primates	Rodents	Other	Source
CONGO						
DRC	Ituri forest	60–95	50–40	1	1	Hart (2000)
Gabon	Makokou	58	19	14	9	Lahm (1993)
	Dibouka, Baniati	51,3	10,6	31		Starkey (2004)
	Dibouka, Kouagna	27	8,3	48,7		Coad (2007)
	Ntsiete	65	23,5	9		van Vliet (2008)
Congo	Diba, Congo	70	17	9	4	Delvingt et al. (1997)
	Oleme, Congo	62	38			Gally and Jeanmart (1996)
	Ndoki and Ngatongo	81–87	11–16	2–3		Auzel and Wilkie (2000)
CAR	Dzanga – Sangha	77–86	0	11–12	2–12	Noss (1995)
Equatorial Guinea	Bioko and Rio Muni	36–43	23–25	31–37	2–4	Fa et al. (1995)
	Sendje	30	18	32		Fa and Yuste (2001)
	Sendje	35	16	43		Kümpel (2006)
Cameroon	Dja	88	3	5	4	Dethier (1995)
	Ekim	85	4	6	5	Delvingt et al. (1997)
	Ekom	87	1	6	6	Ngnegueu and Fotso (1996)

Source: Nasi et al. (2011)

In tropical Africa, hunting provides a very important source of income, often more important than the income generated by the trade of agricultural products (Starkey 2004; Wright and Priston 2010; Kumpel et al. 2010). In Africa, communities often prefer to harvest wild animals for food and reserve livestock as a form of money in the bank (Lindsey et al. 2013a). In South America, wild meat reduces the consumption of domestic livestock such as goats and cattle, key economic reserves that can be easily converted into cash for poor country dwellers (Altrichter 2006). In some cases, hunting tends to be relied on more by some community members such as seasonal migrant labourers who have less time to plant family gardens or for livestock husbandry (van Vliet et al. 2014). Animal-based remedies for zootherapy are also important drivers of that trade. In Latin America, at least 584 animal species, distributed in 13 taxonomic categories, are used in traditional medicine (Alves and Alves 2011). In South East Asia, increasing affluence in major consumer markets, particularly in China, coupled with improvements in transport infrastructure has led to increasing demand for many rare wild animal species. For example, pangolins and turtles used for meat and in traditional Chinese medicine are frequently seized from illegal traders in the region (TRAFFIC 2008) with major markets in Hong Kong, China, Singapore and Malaysia.

Recreational Hunting

In Africa, vast game reserves were delineated during the colonial period to limit the pressure of commercial hunting practised by European settlers. In the 20 countries or so where game hunting is permitted, an average of 10 % of the land is dedicated to this purpose (Roulet 2004), and in southern and parts of East Africa, often much more (Lindsey et al. 2007). Protected area networks in Africa comprise both fully protected parks and in many countries, large blocks where the primary land uses is trophy hunting. Recreational hunting and protected areas respectively represent 15 % and 9 % of the total land area in the 11 main big game hunting countries in Africa (UICN 2009). Recreational hunting is managed by private (safari hunting) operators, granted hunting rights for concessions by the governments (or delegate authorities) for periods of 5–25 years (Table 2). Hunts are organized by approximately 1,300 Safari hunting operators that employ around 3,400 guides and 15,000 local staff (IUCN 2009). Around 18 500 tourist-hunters hunt in Africa every year, primarily from North America and Europe (Lindsey et al. 2007). Southern African countries and Tanzania attract the largest number of customers. Big game hunting primarily targets medium to large mammals and is generally practised in natural or restored ecosystems, whereas bird shooting (usually involving waterfowl, terrestrial wildfowl or doves) occurs primarily in agro-ecosystems (inhabited and partially cultivated areas). The average contribution to the countries' GDP is 0.06 % for the 11 main big game hunting countries (maximum 0.3 % in Tanzania) (Lindsey et al. 2007). As game hunting areas are generally established in the periphery of protected areas, they play a key role in buffering human pressure on core conservation areas.

Table 2 Gross Domestic Product (GDP) in absolute terms, per unit of surface area and per capita. for the main big game hunting countries

Country	Contribution of big game hunting to GDP as a %	% of national territory covered by hunting areas	GDP per hectare in $US	GDP from hunting in per hectare in $US
South Africa	0.04	13.1	2092	2.1
Namibia	0.45	11.4	76	13.9
Tanzania	0.22	26.4	135	0.7
Botswana	0.19	23.0	186	12.7
Zimbabwe	0.29	16.6	142	1.4
Zambia	0.05	21.3	145	0.4
Cameroon	0.01	8.4	386	0.1
Republic of Central Africa	0.10	31.5	24	0.3
Ethiopia	0.01	0.8	118	0.02
Burkina Faso	0.02	3.4	221	0.07
Benin	0.01	3.6	423	0.05

Source: UICN/PACO (2009). Note: It can be noted that the GDP values per hectare in Benin and Burkina Faso are close to those obtained by agricultural production (around $US300/ha)

They also benefit from collecting animals dispersing from the protected areas. However, trophy hunting can confer negative impacts on the populations of some species (notably lions and leopards) if quotas are set too high (Fig. 3) (Jorge et al. 2013; Lindsey et al. 2013a).

In North America, hunting is practiced at the same time for recreational and regulation purposes (Dale et al. 2000). In 2010, 14.4 million hunting licenses were sold and 4.7 % of the population hunted to some extent (Winkler and Warnke 2013). In a context where most large predators have been eradicated, hunting by humans is a low-cost method for maintaining wildlife populations (e.g. white-tailed deer) at levels within habitat carrying capacity or for eliminating exotic species such as feral pigs (Hayes et al. 2009). Wildlife conservation and management costs are mainly funded by hunters, though licence fees and special taxes on hunting equipment (this amounts to about 65 % of state wildlife agency budgets (Mahoney 2009)). However, the long-term viability of this strategy is currently challenged the number of hunters is declining across the United States (the number of hunting licences issued dropped by 9 % between 1982 and 2010) (Winkler and Warnke 2013).

In the European Union (EU), hunting is generally considered a recreational activity and status symbol in high-income states, but also plays a role of food supply in lower income countries. Approximately 13 million EU citizens (2.7 %) hunt, with participation ranging from as little as 0.2 % in the Netherlands to 12.4 % in Italy (Schulp et al. 2014). Hunting occurs across about 65 % of the European land surface, though such land is also used for a variety of other activities and uses. A total of 97 species are hunted in the EU and 38 of these provide meat (26 birds and 12 mammals). Hunting in the EU also is a business that generates substantial revenue and creates both upstream and downstream industries. Hunting supports the equivalent of 70,000 full time jobs in United Kingdom and hunters spend £2 billion

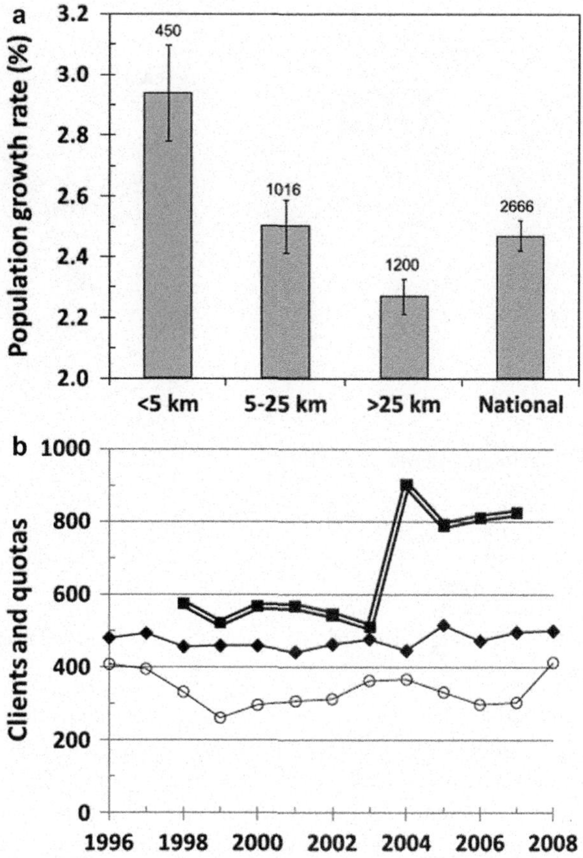

Fig. 3 Human population growth and demand for lion and leopard trophies in Tanzania. (**a**) Annual population growth from 1988 to 2002 in wards located each distance from national parks and game reserves (numbers above bars, number of wards; lines, SE). Wards <5 km from protected areas grew faster than those 5–25 or >25 km away ($p<0.001$). (**b**) Total number of 21-day safaris (*double line, solid squares*) and total quotas for lions (*solid diamonds*) and leopards (*open circles*) across all of Tanzania's hunting blocks (Source: Packer et al. 2010)

each year on goods and services (PACEC 2006). Hunting in the EU also plays a key role in maintaining habitats favourable to some wildlife species and regulating populations in a landscape matrix mainly composed of agricultural lands and production forests, where natural predators are absent (Gordon et al. 2004; Scherrer 2002).

International Wildlife Trade

The illegal killing and poaching of wild animals threatens the viability of many species worldwide (Gavin et al. 2010; Agnew et al. 2009; Fulton et al. 2011; Hilborn et al. 2006; Redpath and Thirgood 2009). A universal problem in the assessment of poaching impacts is the absence of rigorous estimates of its effects relative to other sources of mortality (Fig. 4) (Gavin et al. 2010). The poaching of wildlife for body parts and skins receives significant publicity and poses a major threat to the species affected. For example, ivory poaching is having exceptionally deleterious impacts in Central Africa, where forest elephant populations declined by 62 % between 2002

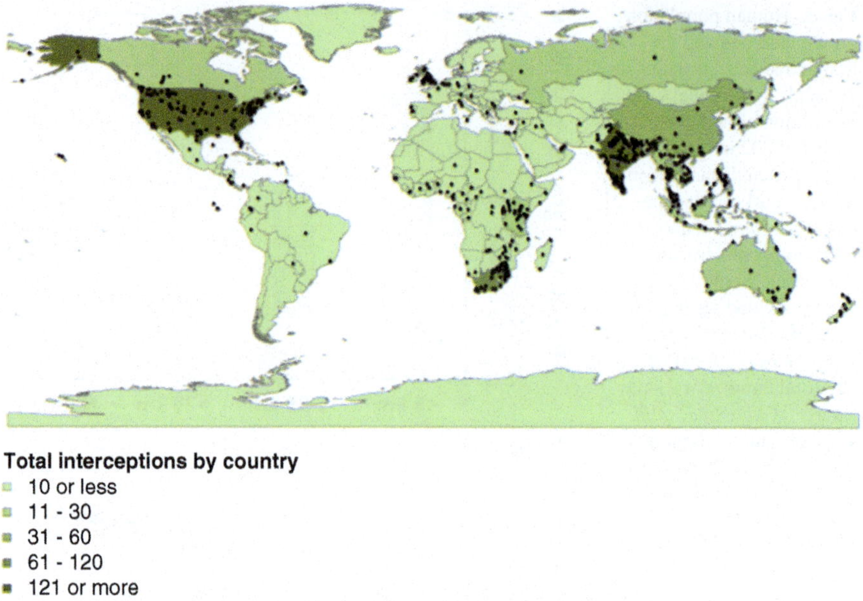

Fig. 4 Wildlife interceptions per country (Source: Sonricker et al. 2012)

and 2011 (Maisels et al. 2013) and ape populations declined by 50 % between 1984 and 2000. Although most species of large carnivores are now legally protected, poaching for international trade or trophy hunting in some cases, remains a widespread problem for their conservation. Some species are commercially poached for pelts or body parts used in traditional medicine (Gratwicke et al. 2008) but many are killed because of conflicts with human interests, such as competition for game, depredation of livestock and threats to human safety (Treves and Karanth 2003). Predators are also affected by hunting for bushmeat, either directly or by being caught unintentionally by-catch in snares set for other species, or by experiencing reduced prey populations. Skins of spotted carnivores such as leopards (*Panthera pardus*) and genets (*Genetta* spp.) fetch high prices. In addition expanded trade of wildlife parts such as the recent practice of selling lion (*Panthera leo*) as tiger (*Panthera tigris*) bones in Asian markets is an indication that the international trade may increase in future (Lindsey et al. 2012). The poaching of more common wildlife species for bushmeat also represents a severe problem that, in some instances, has a component of international trade (e.g. Europe, Chaber et al. 2010; US, Bair-Blake et al. 2014).

In many regions, poaching is intimately linked with national conflicts and international security interests. For example, wildlife poaching plays a role in financing the activities of belligerent groups and catalysing social conflict (Douglas and Alie 2014). Wildlife poaching is often managed by criminal, mafia-type organizations and the actual structure of the value chains are largely unknown (Warchol 2004). One can infer that the poaching of wildlife for products destined for international trade is controlled by wealthy urban people and generally executed by generally

Picture 3 Hunting bag and the hunter's family in Ovan, Gabon (Nathalie van Vliet)

Picture 4 Hunter resting during a night hunting trip in Ovan, Gabon (Daniel Cornelis)

poor people who take the most risks while getting only a small share of the profits. In Africa, contemporary illegal wildlife trade uses village hunters to secure tusks, meat and skins. Such individuals are often armed with military or heavy calibre sporting weapons by individuals or syndicates operating from outside the area who pay villagers for supplying wildlife products (Abernethy et al. 2013). Meat and

Picture 5 Juvenile duiker (*C. Dorsalis*) in captivity along the Kisangani-Ituri road, Democratic Republic of Congo (Daniel Cornelis)

Picture 6 Small diurnal monkey (*Saimiri sciureus*) being sold as pet in Caballococha, Amazonas, Peru (Nathalie van Vliet)

ivory pass via highly organized trade chains to their destinations in the cities of the region and overseas.

Poaching also fuels the medicinal and pet trade. In Brazil, in spite of being illegal, 295 bird species and 47 species of reptiles are captured and sold in the local and international market (Nobrega Alves et al. 2012, 2013). In many parts of South East Asia, wild meat from species such as deer, pangolin and snakes is consumed as delicacies or 'tonic' food items, rather than for subsistence needs (Drury 2009, 2011).

Drivers and Impacts of Unsustainability

Impacts on Hunted Populations

'Defaunation' is often cited as the most evident impact of hunting, resulting in the so-called "Empty forest" syndrome (Redford 1992) and increasingly, the 'empty savannah' syndrome as well (Lindsey et al. 2013a). Defaunation can be defined as the local or regional population decline or species extirpation including arthropods, fish, reptile, bird, and mammal species (Dirzo 2001). Because defaunation is solely driven by human activities, it is also referred to as "*anthropocene defaunation*" (Dirzo et al. 2014). Examples of defaunation are numerous across the world, yet the relative contribution of hunting versus other drivers such as climate change, habitat alteration (i.e. land-use changes, destruction, fragmentation), and impact of invasive species (Hoffmann et al. 2010; Wilkie et al. 2011; Roberts et al. 2013; Simberloff et al. 2013; Dirzo et al. 2014), makes it difficult attribute causation to hunting alone. Data from African sites indicate significantly higher mammal densities in un-hunted versus hunted sites; 13–42 % in Democratic Republic of Congo (Hart 2000), 44 % in Central African Republic (Noss 1995) and 43–100 % in Gabon (Lahm 1994; van Vliet 2008). As hunting pressure becomes heavier, primate numbers may drop to less than a tenth of their original densities (Oates 1996) and carnivores are significantly affected (Henschel 2009). Hunting may also be the cause of a reported 50 % decline in apes in Gabon within two decades (Walsh et al. 2003). The black colobus (*Colobus satanas*) was found to be more vulnerable to over-hunting in Equatorial Guinea (Kümpel et al. 2008) perhaps because it is an easy target owing to their relative inactivity and large body size (Brugiere 1998). In South America, hunted populations of spider (*Ateles sp.*) and woolly monkeys (*Lagothrix sp.*) in the Amazon basin have declined precipitously probably because of the over-hunting (Bodmer et al. 1994; Robinson and Redford 1994). Similar patterns have been recorded in the Amazon with declining white-lipped peccary (*Tayassu pecari*) populations being accompanied by increasing density and larger group sizes for collared peccaries (*Pecari tajacu*) (Fragoso 1994). There are also many examples of defaunation of large mammals in African savannahs, including in protected areas (Craigie et al. 2010). In Zambia, for example, wildlife populations in protected areas occur at just 6–26 % of their predicted carrying capacities due largely to the impacts of excessive bushmeat poaching (Lindsey et al. 2014).

Picture 7 Tourist hunter in Niger (Sophie Molia)

Picture 8 Tourist hunters in Nazinga, Burkina Faso (Daniel Cornelis)

Picture 9 Regulation hunting of red deer population though driven hunts in Ardennes, Belgium (Daniel Cornelis)

Yet, hunting does not always necessarily lead to defaunation. Species are impacted by hunting pressure to different extents. How populations respond to harvest can vary greatly depending on their social structure, reproductive strategies, dispersal patterns and intactness of habitats. Small species are typically more resilient to hunting than larger species, due to their higher reproductive rates (Cowlishaw et al. 2005). Dispersal, in particular, can have significant ramifications (both stabilizing and destabilizing) on population dynamics. Density-dependent dispersal may stabilize populations as immigration and emigration counterbalance between hunted (sink) and non-hunted (source). Cougar removal in small game management areas (about 1000 km^2) in Washington state, increases immigration and recruitment of younger animals from adjacent areas, resulting in little or no reduction in local cougar densities and a shift in population structure toward younger animals (Robinson et al. 2008). In areas where populations of larger species have been significantly depressed, abundance of small and medium-sized species can remain unaffected or even increase. For example, the small blue duiker is significantly less abundant in remote forests inside the Ivindo National Park (Gabon) than in hunted areas close to Makokou with similar vegetation cover (van Vliet et al. 2007). The explanation may be that abundance of resilient species may rise if their competitors are harvested, an ecosystem characteristic known as density compensation (or under-compensation) (Peres and Dolman 2000). Suggestions of density compensation have been made in Korup forest monkey communities (Cameroon) where putty-nosed guenons (*Cercopithecus nictitans*) densities increase in heavily hunted sites (Linder 2008). Source-sink effects (Novaro et al. 2000; Salas and Kim 2002), spatial heterogeneity (Kümpel et al. 2010a; van Vliet et al. 2010a, b) or high dispersal (Hart 2000) can also help maintain populations in hunted areas, masking or compensating for hunting driven population decline.

Long Term Impacts on Ecosystems

Defaunation may generate trophic cascades that alter ecological processes, that lead to changes in community composition and diversity loss (Dirzo et al. 2014; Muller-Landau 2007). In many ecosystems, the larger vertebrate fauna, especially frugivorous birds, primates, ungulates, and mammalian carnivores, have been extirpated or reduced in number. As these large animals vanish, so do their myriad (often non-redundant), ecological interactions and processes they generate, foremost trampling, ecosystem engineering, herbivory, seed predation, and dispersal (Beck et al. 2013; Dirzo and Mendoza 2007; Dirzo et al. 2014; Keesing and Young 2014; Stoner et al. 2007). Therefore, activities such as hunting have the potential to impact not only targeted species but the ecosystem more broadly. 'Keystone species', 'ecosystem engineers', or organisms with high community importance value are groups whose loss is expected to have a disproportionate impact on ecosystems compared to the loss of other species. Top predators (e.g. large cats, raptors, crocodiles) may impact biodiversity by providing resources that would otherwise be unavailable or rare for other species (e.g. carrion, safe breeding sites) (Terborgh and Feeley 2010). Local extinction of these predators can trigger large changes in prey populations, which in turn dramatically alters browsing or grazing to the point where large regime shifts or ecosystem collapse happen. For example, elephants can play a major role in modifying vegetation structure and composition through their feeding habits (including differential herbivory and seed dispersal) and movements in the forest (killing a large number of small trees). Ungulates such as wild pigs and duikers are among the most active seed dispersers or predators; thus, a significant change in their population densities will have a major effect on seedling survival and forest regeneration. In defaunated areas, studies found wide-ranging changes in plant physiognomy, recruitment, species composition, community changes, and declining in tree species diversity (Emmons 1989; Harrison et al. 2013; Keesing and Young 2014; Wilkie et al. 2011). In addition, plant species with autochorous and abiotic seed-dispersal syndrome increase in numbers (Corlett 2007; Emmons 1989; Terborgh et al. 2008).

On the other hand, numerous smaller species, primarily rodents, may increase in numbers due to a lack of predators or competitors (Terborgh and Feeley 2010.). Rodents typically affect different plant species, resulting in higher seed predation of small-seeded species (Emmons 1989; Terborgh et al. 2008; Wright 2003). In many temperate and boreal regions, populations crashes of apex predators (e.g. wolves, lynx, tigers, cougars and bears) along with land use change and behaviour change in humans has contributed to hyper-abundances of ungulates in North America, Eurasia, and eastern Asia (Côté et al. 2004; Martin et al. 2010; Ripple et al. 2010), which can trigger large-scale declines in forest ecosystems (Estes et al. 2011; Gill and Fuller 2007). Other studies have used the re-introduction of apex predator to re-establish ecological interactions. For example, 15 years after the re-introduction of grey wolf (*Canis lupus*) into the Yellowstone National Park, Ripple and Beschta (2012) found strong tri-trophic cascading effects involving wolf, elk (*Cervus elaphus*), and several plant species. Predators control the herbivore population in a strong top-down fashion, which reduces over-browsing and allows the recovery and succession of the plant community (Ripple and Beschta 2012; Ripple et al. 2010). Direct and indirect positive effects of the wolf re-introduction have also been recorded for other species, such as ravens (*Corvus corax*), bald eagles (*Haliaeetus leucocephalus*) (Wilmers et al. 2003), bison (*Bison bison*), and beavers (*Castor canadensis*) (Ripple and Beschta 2012).

External Drivers of Unsustainable Use

Wildlife populations worldwide are affected by a variety of sources, which may influence the sustainability of extractive use. Knowledge on how these different source influence wildlife populations is key to identifying management and policy measures that could help reduce negative impacts. Scholte (2011) described a series of proximate and underlying factors driving change in wildlife populations. Underlying drivers may not themselves cause change, but may act indirectly to contribute to change. Identifying drivers and, where possible, quantifying their impact, facilitates the formulation of appropriate management guidelines for extractive use.

The main drivers of change may be summarised as follows:

Habitat loss and degradation Hunter (2002) defines three forms of habitat destruction (viz. degradation, fragmentation and outright loss). Habitat loss has emerged in the twenty-first century as the most severe threat to biodiversity worldwide (Brooks et al. 2002; Baillie et al. 2004; Naeem et al. 1999; Smith and Smith 2003), threatening some 85 % of all species classified as "threatened" on the IUCN Red List (Baillie et al. 2004).

Large-scale extractive and production projects Many countries worldwide have allocated a large part of their territories to formal sector oil, mining, agriculture and extensive timber use (Walsh et al. 2003). For example, in central Africa selective logging is the most extensive extractive industry, with logging concessions occupying 30–45 % of all tropical forests and over 70 % of forests in some countries (Table 3) (Global Forest Watch 2002; Laporte et al. 2007). In many countries, the mineral boom is contributing to the emergence of "growth corridors" where infrastructure upgrades will improve the competitiveness of agriculture and other economic activities (Delgado et al. 1998) which impact wildlife habitats and disturb wildlife populations (noise, pollution etc…).

Conflict and war Wars have multiple impacts on biodiversity and protected areas, and livelihoods of local people dependent on natural resources. Impacts can be highly variable, and may be positive in some areas and negative in others (McNeely 1998). Very often, though, war has serious negative effects directly or indirectly on conservation (IUCN 2004). Modern wars and civil strife are typically associated with detrimental effects on wildlife and wildlife habitats (Fig. 5) (Dudley et al. 2002; Hatton et al. 2001; Said et al. 1995; Hart and Hall 1996; Hall et al. 1997; Plumptre et al. 1997, 2000; Vogel 2000; de Merode et al. 2004).

Table 3 Impacts of anthropogenic disturbance on wildlife in logging concessions

Major cause	Guild	Species or guild	Impact on species abundance	Country	Study
Disturbed habitat (logging)	Duikers		(+)	Congo	Clark et al. (2009)
	Elephant		(–)	Cameroon	Matthews and Matthews (2002)
	Great Apes	Chimpanzees	(–)	Cameroon	Matthews and Matthews (2004)
			(–)	Gabon	White and Edouards (2001)
	Rodent	Brush tailed porcupine	(+)	Gabon	Laurance et al. (2008)
		Murid rodents	(+)	Gabon	Laurance et al. (2008)
	Small monkeys	Collared mongabey	(–)	Cameroon	Matthews and Matthews (2002)
		Guenons	Not affected	Cameroon	Matthews and Matthews (2002)
Hunting	Duikers	Red duikers	(–)	Gabon	van Vliet and Nasi (2008a, b)
		Yellow back duiker	(–)	Gabon	van Vliet and Nasi (2008a, b)
		Elephants	(–)	Congo	Clark et al. (2009)
			Not affected	Gabon	van Vliet and Nasi (2008a, b)
	Great Apes	Chimpanzees	(–)	Cameroon	Matthews and Matthews (2004)
		Gorilla	(–)	Cameroon	Matthews and Matthews (2004)
	Rodent	Brush tailed porcupine	(+)	Gabon	Laurance et al. (2008)
		Murid rodents	(+)	Gabon	Laurance et al. (2008)
Proximity to big villages and towns	Great Apes	Chimpanzees	(–)	Congo	Clark et al. (2009)
	Small monkeys	Guenons	(–)	Congo	Clark et al. (2009)
	Duikers		(–)	Congo	Clark et al. (2009)
Proximity to small village	Duikers		(+)	Congo	Clark et al. (2009)
	Elephant		(+)	Congo	Clark et al. (2009)
	Forest buffalo		Not affected	Gabon	van Vliet and Nasi (2008a, b)
	Great Apes	Chimpanzees	Not affected	Gabon	van Vliet and Nasi (2008a, b)
		Gorilla	Not affected	Gabon	van Vliet and Nasi (2008a, b)
	Small monkeys		(–)	Gabon	van Vliet and Nasi (2008a, b)

Table 3 (continued)

Major cause	Guild	Species or guild	Impact on species abundance	Country	Study
Roads	Carnivores		Not affected	Cameroon	Van der Hoeven et al. (2010)
	Duikers	Bay duiker	Not affected	Gabon	Laurance et al. (2008)
		Blue duiker	(–)	Gabon	van Vliet and Nasi (2008a, b)
		Blue duiker	(–)	Gabon	Laurance et al. (2008)
		Duikers	(+)	Congo	Clark et al. (2009)
		Ogylbi duiker	(–)	Gabon	Laurance et al. (2008)
		Peter's duiker	Not affected	Gabon	Laurance et al. (2008)
		Red duikers	(–)	Gabon	van Vliet and Nasi (2008a, b)

Source: Nasi et al. (2011)

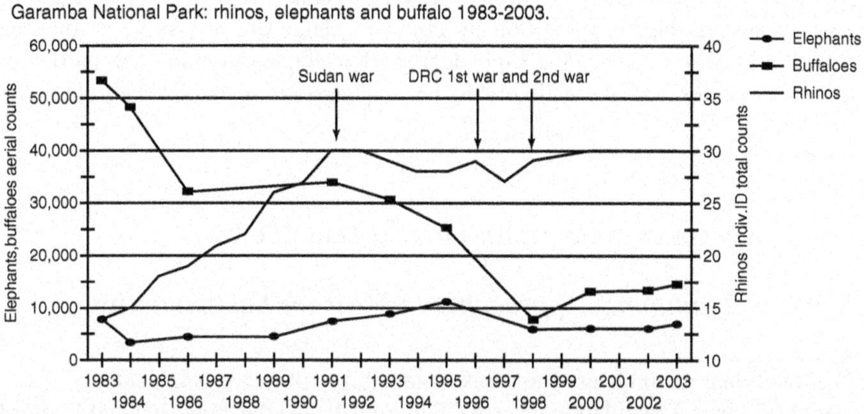

Fig. 5 Impact of war on mammal species in Garamba National Park, Democratic Republic of Congo (Source: Hanson et al. 2009)

Population growth The impacts that human population growth has on natural resources is the subject of much debate. While neo-Malthusian theories place population growth in a vicious circle of destruction, others suggest that such theories oversimplify the issue of environmental degradation (Sunderlin and Resosudarno 1999; Leach and Fairhead 2000). According to neo-malthusian theory, population growth may cause intensified pressures on natural habitats and resources to satisfy the growing demand for space, housing, food and water for drinking and sanitation. However, in Boserup's theory, when population density increases, people adapt to the constraint through innovative technologies that reduce pressure on natural resources.

Wildlife diseases Ecological disturbances can also influence on the emergence and proliferation of wildlife diseases. Each environmental change, whether occurring as a natural phenomenon or through human intervention (deforestation, changes in land use, human settlement, commercial development, road construction, water control systems), changes the ecological balance and context within which disease hosts or vectors and parasites breed, develop, and transmit disease. The global trade in wildlife provides disease transmission mechanisms (Smith et al. 2012; Walsh et al. 1993; Leroy et al. 2011; Bell et al. 2004; Guarner et al. 2004; Weldon et al. 2004; Pence and Ueckermann 2002; Kilonzo et al. 2013) that not only cause human disease outbreaks but also threaten livestock, international trade, rural livelihoods, native wildlife populations, and the health of ecosystems.

Climate change Climate change might have diverse indirect effect on wildlife depending on the characteristics of the species (Foden et al. 2013; Kaeslin et al. 2012). Species with generalised and unspecialised habitat requirements are likely to be able to tolerate a greater level of climatic and ecosystem change than specialised species. However, many species rely on environmental triggers or cues for migration, breeding, egg laying, seed germination, hibernation, spring emergence and a range of other essential processes. Species dependent on interactions that are susceptible to disruption by climate change are at risk of extinction, particularly where they have high degree of specialization for the particular resource species and are unlikely to be able to switch to or substitute other species.

Challenges and Opportunities for Sustainable Use

Limits of Traditional Approaches to Measure Sustainability

The traditional methods used to asses sustainability of harvests include (1) demographic models of population growth ('Full model') (2) the Robinson and Redford (1991) model for assessing Maximum Sustainable Yields, (3) population trend methods; (4) harvest-based indicators and (5) comparisons of demographic parameters between sites ('Compare sites'). Until the early 2000, the most commonly used model was the Robinson and Redford's model (1991), which has its origin in fisheries and has been the most popular in Africa and the Neotropics. In Central Africa for example, out of 17 publications dealing with the estimation of hunting sustainability, 13 have used the popular Robinson and Redford model (1991) (van Vliet and Nasi 2008a, b). This approach is based on the simple assumption that hunting remains sustainable as long as the amount harvested per year does not exceed annual recruitment. Key to the use of these models is our capacity to estimate offtakes, prey densities and our knowledge on biological parameters such as age at first/last reproduction and fecundity rate.

While the Robinson and Redford model is a simple algorithm that provides a crude estimate of sustainability, there is wide spread agreement that this model is plagued with different levels of errors. Although all indicators will have trade-offs in terms of effort required for data collection, scale of coverage, timeliness, accuracy and precision, some of the commonly used indicators have weaker theoretical support and thus may provide only very coarse-scale information of questionable reliability. Static, one-off indicators cannot ultimately predict sustainability; for example, it has been shown that in a sustainable system, half of a random sample of sustainability indicator evaluations would indicate unsustainability due to stochastic processes alone (Ling and Milner-Gulland 2006). Milner-Gulland and Akcakaya (2001) and van Vliet and Nasi (2008a, b) show that major problems related to the use of simple biological models are the paucity of available biological data even for the most common species and the difficulty of collecting the data required for a full sustainability assessment.

Besides the uncertainty caused by the inherent variability of natural systems and observational uncertainty arising from methodological shortfalls for assessing the variables of a system, there is an additional level of uncertainty that reflects our ignorance about the complexity of natural systems (Milner-Gulland and Akcakaya 2001). Recognition of the importance of uncertainty and of complexities of ecological systems is growing in all fields of theoretical ecology, including conservation. One issue that is difficult to address with simple biological models but which is increasingly recognized as being crucial for the sustainability of bushmeat hunting, is spatial heterogeneity. The emergence of geographic information systems now permits the taking into account of spatial effects on wildlife populations. Studies on sustainable hunting using spatially explicit individual based models (Salas and Kim 2002; Novaro et al. 2000; Siren et al. 2004), have tested the role of landscape structure and dispersal characteristics that might influence the sustainability of hunting. Salas and Kim (2002) suggest that spatial factors, such as shape of the hunted area and the size of the surrounding population, may be important in determining the sustainability of extraction. Novaro et al. (2000) found that dispersal could have a key role in rebuilding animal populations depleted by hunting. Thus, factors that strongly affect dispersal such as spatial distribution and size of areas with and without hunting population size in source areas, and social behaviour, should be considered when sustainability of hunting is evaluated in areas with heterogeneous hunting pressure (Novaro et al. 2000). Ling and Milner-Gulland (2006) consider the animal-hunter couple, as a dynamic system governed by the responses of hunters as well as the population dynamics of prey species. Seasonality in hunting activity, related to socioeconomic drivers (van Vliet et al. 2010a, b), to prey dynamics, to climate or to food availability, may require further consideration since the degree of seasonality in one or both of these factors could have considerable impact on sustainability predictions. Another important area for future development is the treatment of hunters' prey choice. In previous models, exploited populations are considered in isolation while, in most instances in which the indices are applied, the prey base consists of many different species (Rowcliffe et al. 2003).

Because of the difficulties in assessing sustainability with one-off indicators, Weinbaum et al. (2013) propose the monitoring of harvested populations through time as one of the gold standards in sustainability monitoring. Ideally, population monitoring is an ongoing process and is accompanied by adaptive harvesting strategies (Johnson et al. 2002).

From One-Off Indicators of Sustainability to Resilience Analysis

Simplistic models to asses ecological sustainability ignore important determinants of human behaviour (Peterson 2000), which may cause scientists to provide advice or formulate policy that is either inadequate, or open to misuse (Ludwig et al. 1993; Gunderson 1999). Indeed, assessing sustainability of hunting, entails the recognition that we are dealing with complex systems and that the sustainability of hunting may depend on exogenous factors other than hunting, such as habitat or climatic changes, or unmonitored harvests elsewhere in the population (Hill et al. 2003). Besides, sustainability needs to be understood within its three main pillars: economic, ecological and social sustainability. The links between hunting and livelihoods, health, culture and local economy (CBD 2008) are still poorly understood or not properly taken into account, but recent efforts have been made to understand the multifunctionality of hunting, and therefore seek sustainability taking into account the multiple roles that hunting plays (Fisher et al. 2013).

Sustainability, hinges on the feedbacks and balances between social and ecological systems, and should be investigated with a holistic framework (Ostrom 2007; Iwaruma et al. 2013). For example, habitat fragmentation can cause the sudden decline of animal abundance around villages, and lead to agricultural expansion to compensate for food loss due to unsuccessful hunting (Bennett 2002; Damania et al. 2005). Hunting systems may be understood as socio-ecological systems as defined by Gallopin et al. (1989), in which the focus is not on the impacts of hunting on prey populations, but rather on the complex and dynamic relationships between the territory, it's resources, the stakeholders at play (e.g. hunters, consumers, traders), and the different exogenous drivers of change that either affect the social or the ecological components of the system. The implications of this interpretation for sustainability science include changing the focus from seeking optimal states and the determinants of maximum sustainable yield (the MSY paradigm), to resilience analysis, adaptive resource management, and adaptive governance (Walker et al. 2004). The concept of a social-ecological system reflects "the idea that human action and social structures are integral to nature and hence any distinction between social and natural systems is arbitrary" (Berkes and Folke 1998). Clearly natural systems refer to biological and biophysical processes while social systems are made up of rules and institutions that mediate human use of resources (Berkes and Folke 1998). In the context of the concept of social-ecological systems, measuring vulnerability refers to identifying the degree to which a system is susceptible to cope with adverse effects. In all formulations, the key parameters of vulnerability and resilience are exposure (the stress to which a system is exposed), sensitivity, and adaptive capacity. It is crucial to recognize that the social ecological system is not stable, but dynamic: what is vulnerable in one period is not necessarily vulnerable (or vulnerable in the same way) in the next, and some new exposures and sensitivities arise over time (Smit and Wandel 2006). Those processes are constantly changing and, hence, must be constantly probed. Therefore, analysing the resilience of a system requires a monitoring system that analyses changes over time. It is also clear that we must seek more integrative approaches, because focusing on one scale and narrow goal-seeking (such as optimizing ecological sustainability) are likely to be maladaptive (Gunderson 1999) or lead to un-desired outcomes.

Alternatives to Extractive Use: Wildlife Production

The ever-increasing human population and high demand for game meat justifies exploring opportunities for the production of game meat from wildlife species. This is particularly justified in areas of the planet that are not suitable for crop or domestic livestock production due to their extreme climatic conditions such as tropical forests, arid regions or arctic areas. Animals can be produced in extensive ranging systems (game ranching), which usually includes several wildlife species, exploited for different purposes (sport hunting, tourism, live game sales and/or game meat production) or in more intensive conditions (game farming). The production is aimed to fulfil local or national markets but also, if well organized, international markets for which the demand of game meat is increasing. Only in the EU where game meat is far from being the main source of animal protein, the demand for game meat is currently achieving 200,000 tons per year. In countries typified by large and unsustainable bushmeat trades, legal wildlife-based land uses offer a potentially viable and sustainable alternative that contrasts with the lose-lose scenario that poaching offers (wildlife population declines (except weed species like cane rats) with no long-term livelihood benefits). In Africa for example, given the right legislative environment, legal wildlife-based land uses have potential to create vastly more jobs, meat and income than informal (and usually illegal) bushmeat harvesting.

Game ranching Game ranching generally occurs on a relatively extensive scale with relatively low intensity management. Wildlife is often provided with supplementary water in dry areas, but other than during extreme drought periods is usually not provided with additional food. Forms of wildlife use on game ranches and game farms are varied and include sport hunting, live animal sales, ecotourism and game meat production, among others. Wildlife ranching is especially common in southern Africa, with notably large industries in South Africa, Namibia (and previously Zimbabwe) and smaller industries in Botswana, Zambia and Mozambique

Picture 10 Russa deer (*Cervus rusa timorensis*) ranched for venison production in Mauritius (Ferran Jori)

Picture 11 Capybaras (*Hydrochoerus hydrochaeris*) in farmed in extyensive condition in Brasil (Ferran Jori)

(Cousins et al. 2008; Lindsey et al. 2013b). It is known that in semi-arid lands, wildlife based land uses are commonly more profitable than livestock, generates foreign currency incomes, is less susceptible to drought and climate change and contributes to food security and income generation (Bond et al. 2004). In the last

Picture 12 Intensive breeding of colared peccaries (*Tayassu tajacu*) in French Guyana (Ferran Jori)

15 years, game ranching has been one of the fastest growing agricultural industry in South Africa with currently more than 12 000 game farms covering at least 205.000 km^2, encompassing a total of 16–20 million heads of wild species in private lands (Dry 2014). Game meat produced in ranches, originates from individual hunting campaigns or from organized commercial culling operations culled and processed annually. Approximately 100.000 animals (including springbok (*Antidorcas marsupialis*), blesbok (*Damaliscus pygargus*), impala (*Aepyceros melampus*) and kudu (*Tragelaphus strepsiceros*)) are exported to the EU and only a minor proportion is consumed in South Africa. Game ranching is also expanding in Namibia, where there is ~287.000 km^2 (more than 15 % of private farmland) dedicated to this activity and where its economic outputs are exceeding those generated by domestic livestock production, showing important benefits for wildlife populations and food security of local populations (Lindsey et al. 2013a; Magwedere et al. 2012). Indeed, between 16 000 and 26 000 tons of game meat from African ungulates are produced annually in Namibian farmlands for local, regional and international export markets (Lindsey et al. 2013a), and demand seems is increasing (Hoffmann et al. 2010)

The spread of wildlife ranching in Africa is limited by three key factors (Lindsey et al. 2013a). Firstly, most governments continue to fail to devolve sufficient user rights and/or ownership over wildlife to land owners and communities. Secondly, on community lands, establishing game ranches on communal lands is often difficult due to vague land tenure, lack of capital and lack of expertise. Thirdly, legal wildlife production is often threatened by a failure of governments to treat wildlife poaching with anything near the severity with which livestock theft is granted.

However, the exponential spread of this model has also some shortcomings from the conservation and social perspective. On one side, the ecosystems on some private lands are often unbalanced and biased towards high densities of the most valuable species, elimination of predators and introduction of exotic species which are detrimental to the conservation natural ecosystems in Southern Africa (Cousins et al. 2008; Lindsey et al. 2009). Many of these problems fall away, however, when adjacent wildlife ranches are combined into a single larger management unit or conservancy (Lindsey et al. 2009). In addition, there is a need to seek ways in which game ranching can be used to integrate poor rural communities. One possibility that has not been adequately explored is the development of community owned wildlife ranches (Le Bel et al. 2013), or joint ventures between communities and the private sector. Some such joint ventures have been explored in South Africa. At Phinda Resource Reserve in South Africa, for example, the private owners of the land did not contest a land claim over the property from neighbouring communities, but rather chose to accept a government pay out for the property and to enter into a long term lease-agreement with the new community owners of the land. Similarly, negotiations are underway in Savé Valley Conservancy in Zimbabwe to achieve a community shareholding of the privately owned and run protected area.

Game farming Game farming is the term used to define animal production in more intensive conditions, and in some contexts involves the production of a single or a limited suite of species. In southern Africa, a substantial industry has developed around the breeding and trade of rare or high trophy value species, such as sable antelope. Elsewhere, game farming is conducted primarily to produce venison. For example, various deer species are commonly farmed in many parts of the world using extensive and intensive production systems (Bertolini et al. 2011). Since 1970, the New Zealand deer industry has grown exponentially and in 2013 it included 2800 farmers and produced approximately 1.1 million farmed deer, and the country became the major supplier of venison, deer velvet and other deer products in the world (Bertolini et al. 2011). More than 90 % of the venison production is exported. In 2013, total revenues for export of deer meat equalled US$ 132 million to European countries (75 % of the total production). The species most commonly farmed in New Zealand and throughout the world is the red deer (*Cervus elaphus*). However, other deer species are also being farmed successfully such as the reindeer in the Northern hemisphere and the rusa deer (*Cervus timorensis rusa*) in Eastern tropical countries (Dahlan 2009; Jori et al. 2013), New Caledonia hosts a huge feral population of deer after the introduction of rusa in the late 1800s. Reindeer and caribou comprise an integral part of the diet of local inhabitants of the Northern Hemisphere in Europe and Canada (Rincker et al. 2006). The domestication of reindeer by nomadic tribes from northern Europe is thought to date back 3 000 years and nowadays this species accounts for more than 63 % of total numbers of deer reared in captive or semi-captive conditions (Chardonnet et al. 2002). However, despite a large number of benefits, the success of ungulate production also comes with certain constraints in terms of intensification, disease emergence and the availability of land and capital investment that are not accessible to small-scale farmers and not feasible in tropical forested environments, where bushmeat trade is more common and the demand for game meat is higher.

'Mini' livestock Several authors have promoted the production of small sized species of wildlife that can be reared on a small-scale for animal or human food production (Hardouin et al. 2003; Assan 2014). The term applies to different invertebrate species such as the breeding of manure worms or tropical snails for animal and food consumption and small or medium sized species of rodents, birds, reptiles, rodents or small antelopes. Among all these options, some species of rodents exhibit greater potential for captive rearing, due to their generally high rate of reproduction and widespread popularity in tropical areas of Africa (Jori et al. 2005) Latin America (Jori et al. 2001; Nogueira-Filho and Nogueira 2011) and Asia (Drury 2009). More generally, this kind of wildlife farming is only recommended for species that are not endangered and that are in high demand (Bulte and Damania 2005). One good example is the case of cane rat (*Thryonomys swinderianus*) production which has been extensively studied since the mid 1980s in West Africa (Jori et al. 1995) and represents a successful example of sustainable production of bushmeat. Its technical feasibility and economic potential having been extensively proven (Jori and Chardonnet 2001), cane rat farming is now a fully accepted small scale farming activity in Benin, Ghana and Nigeria, proposed as a sustainable and profitable alternative to wildlife exploitation by local and international development agencies (Aiyeloja and Ogunjinmi 2013; Anang et al. 2011). The main constraints identified for a wider adoption are access to dissemination and extension support, credit facilities for initial infrastructure, availability of grass for food during the dry season (Anang et al. 2011; Ogunjimi et al. 2012), and access to breeding stock adapted to captivity. However, when breeding stock is taken from the wild as occurs with other captive breeding programs of Asian porcupines (*Hystrix brachyura*), promoted in Vietnam, these systems might deplete natural populations and be of serious conservation concerns (Brooks et al. 2010).

The capybara (*Hydrochaerus hydrochaeris*), together with the collared peccary (*Tayassu tajacu*) and white lipped peccary (*Tayassu pecari*) are among the most commonly exploited mid-sized mammal species in Latin America for their meat and hides (Bodmer and Robinson 2004; Moreira et al. 2012). The first two have been extensively studied and exhaustive technical information has been produced to breed those species in captive conditions. However, in practice, economic viability is challenging since initial investment is high and commercialization and marketing are restricted to niche gourmet market of exotic meats in urban centres and production costs are high. Moreover, production costs are not negligible and whereas hunters can access the same meat without the production costs. Legal bottle-necks for the trade of wild animals (even when coming from farms) are probably the main the reason why farming of capybaras or collared peccaries and has never really taken off in South America, despite profitability and technical feasibility (except in Venezuela) (Le Pendu et al. 2011; Moreira et al. 2012; Nogueira-Filho and Nogueira 2004; Nogueira-Filho and Nogueira 2011).

Sustainable wildlife management Sustainable Wildlife Management (SWM) is the careful management of socially or economically important wildlife species, to sustain their populations and habitat over time. In view of its economic, ecological and social value, wildlife is an important renewable natural resource. If sustainably managed, these species can provide continuous nutrition and income and therefore contribute considerably to the poverty alleviation, food security, and ecosystem maintenance and services. Sustainable wildlife use is an optimal solution for maintaining natural habitats while benefitting local communities at minimal cost. Several examples exist in Africa, Latin America, Australia and Asia for the management of the different species including ungulates, rodents (Maldonado-Chaparro and Blumstein 2008; Moreira et al. 2012), macropods (Cooney 2009) and reptiles (Webb et al. 2004). Reptiles have the capacity to lay large numbers of eggs, many of which will not survive in the wild due to predation and other natural causes. From that perspective sustainable management programs of different species of crocodiles, marine turtles, tortoises and lizards have been implemented worldwide with different levels of success (Alves et al. 2012; Schlaepfer et al. 2005; Webb et al. 2004). In the case of capybara and white-lipped peccaries, natural populations are regularly harvested at sustainable levels in Venezuela (Maldonado-Chaparro and Blumstein 2008) and Peru (Bodmer and Robinson 2004). There have been significant efforts to integrate communities into sustainable wildlife management. For example, in Zimbabwe during early 1990s the implementation of the Communal Areas Management Programme for Indigenous Resources (CAMPFIRE) was established as a means of extending the benefits of wildlife use on community lands to the people occupying those areas. It suggests that community-based natural resource management (CBNRM) is a potential solution to solve the interlinked problems of poverty and conservation of wildlife (Child 1996). However, the key factor limiting community conservation efforts in Africa, as with game ranching, is failure to devolve user rights or ownership of wildlife sufficiently to communities, and the retention of too-high proportions of revenue by governments (Child 2008). The most successful example of community conservation in Africa is in Namibia where those constraints have been largely overcome: there, communities that form conservancies are entitled to retain 100 % of income from wildlife (Jones and Weaver 2008).

These initiatives work successfully as an alternative to non-regulated hunting as they are based on an adaptive management approach where monitoring take a key role to define new quotas (Maldonado-Chaparro and Blumstein 2008). The main risk often encountered with the sustainable use of wildlife is overharvesting. This has been observed with the Saiga antelope (*Saiga tartarica*) in Central Asia (Berger et al. 2008) or some species of riverine turtles (De Souza Alcantara 2014). Therefore, a detailed baseline of population sizes and a good knowledge of the biological parameters of the species is needed before implementing extractive activities. Monitoring tools need to be developed in order to adapt harvesting strategies to unpredicted events (Letnic and Crowther 2013) or environmental changes (Mawdsley et al. 2009).

Conclusion and Prospects

Wildlife constitutes a renewable resource that generates a wide range of benefits worldwide. Extractive use of wildlife concerns numerous species and ecosystems, and involves a wide typology of actors, purposes, and extraction modes.

In our changing world, one global challenge facing humanity is to balance space and food needs of human populations and the maintenance of our biological heritage. As regards more particularly the consumption of renewable resources, the question arises of how to develop the sustainable use of wildlife, for the mutual benefit of biota, man and ecosystems.

This is a particularly hot issue in parts of the world where man has not yet completed its demographic transition (e.g. tropical biodiversity hotspots) and where unprotected natural ecosystems are being gradually replaced by agro-ecosystems. So far, humans have modified more than 50 % of Earth's land surface and since human population is projected to double in the next 40 years (Hooke et al. 2012), hunting will occur in ecosystems that are increasingly anthropomorphised. Many species are thus likely to decline over the next century as a result of land conversion and overexploitation, particularly specialist and non-resilient ones (Milner-Gulland and Bennett 2003). In contexts where hunting is practised for livelihood and wild meat consumption still firmly rooted in rural cultures, the challenge of the next decades is twofold: (i) maintain full assemblages of wildlife species within a network of protected areas and (ii) meeting the rural demand for wild meat through the sustainable harvest or production of resilient and productive wild species in non-protected areas. At the same time, we need to raise awareness and improve education to curve the demand for protected species and develop solutions to mitigate human-wildlife conflicts. In agro-industrial landscapes (e.g. North America, Europe) where pristine ecosystems and natural processes (e.g. predation) have been wiped out, sustainability issues relate to the maintenance of large ungulate populations at levels compatibles with a multifunctional use of space (agriculture, domestic stock raising, production forests, nature tourism, etc).

Within this global context, further research is needed focusing on the production systems of non-endangered species (in open, semi-open or fenced spaces) for which demand is popular. As regards to subsistence hunting, models to assess the sustainability of harvests still need further development; for example the model developed by Iwaruma et al. (2014) holds gret promise for the sustainable harvesting of wildlife in peopled forests. This type pf model may eventually be easily used to facilitate management decision making. For most common game species in tropical areas, zootechnical parameters remain poorly investigated, mainly because research has focused so far on emblematic and endangered species. Although poorly investigated, the transformation of natural habitats to degraded forests (e.g. through logging, shifting cultivation, timber/oil plantations) in tropical landscapes may increase the ecological balance to the benefit of resilient game species, thus providing future opportunities for sustainable harvesting models. For example, in South-East Asia where plantation crops generate high deforestation rates (Sayer et al. 2012), the emergence of commercial hunting practices of wild boar in oil palm plantation has recently been described (Luskin et al. 2013; Pangau-Adam et al. 2012). If sustainably managed, hunting in multifunctional spaces may thus be a source of wild meat and income and alleviate the pressure on threatened species in protected areas.

Recreational or game hunting in Africa was shown to play an important role in conserving natural ecosystems and buffering human pressure on protected areas. However, game hunting remains an exclusive use mode that brings few benefits to local people compared to mass tourism, thus hardly compatible with high human densities. In contrast, recreational hunters in Europe and North America are benefiting from the growth of large ungulate populations. In a context where the number of hunters is declining and the return of large predators is very controversial, the question arises as to how to manage the growing ungulate populations in a few decades. Equally, good practices in terms of governance, processes (hunting rules) or products (meat) should be promoted through the implementation of certification systems in the recreational hunting business.

In the case of wildlife ranching, research has been developed for many years and a technical guidelines are available, although marginally applied (Lindsey et al. 2013a). For wildlife ranching to flourish in the savannahs of southern and East Africa, for example, governments need to take the necessary steps and devolve user rights over wildlife to land owners and communities, encourage joint ventures between communities and the private sector, and treat wildlife poaching as a serious crime comparable to livestock theft. In that way, community benefits from sustainable legal wildlife production would replace the unsustainable and marginal benefits from illegal wildlife harvesting. In this context, research should investigate options to better integrate local rural communities in the process of managing wildlife on farms. For species that breed well in captivity (game farming and mini-livestock), the focus should be on fulfilling some basic knowledge gaps and reducing production costs. One major shortcoming with most of the wildlife species under production is the lack of research and knowledge on the pathogens their hosts which can affect their productivity and the one from the producers and consumers.

Overall, success stories of sustainable management modes of wildlife populations should be further promoted and tested elsewhere together with enough law enforcement to prevent illegal exploitation. In that sense, exchange of experiences at international level can be highly beneficial.

Several health issues also need consideration when managing and rearing wildlife species, are transversal to most production modes and require investigations, in the light of recent sanitary crisis linked with wildlife reservoirs such as SARS or Ebola (Jones et al. 2013; Kock 2014). More emphasis should be focused on the investigation and knowledge of pathogens circulating in exploitable wildlife populations for the benefit of the health of animals being produced and their consumers.

Finally yet importantly, managing wildlife effectively requires appropriate policies, social acceptability, good governance, and a degree of decentralization congruent with scales of wildlife management. The legal bottlenecks need to be addressed to allow innovations in terms of sustainable extractive use. For the moment, our knowledge has been generated either by research and theoretical models or by small scale/short term projects, without support by legal frameworks that allow scaling them up to national or regional levels. Holistic support is needed from local and national governments and international organisations and research and academic institutions to drive changes at all levels (legal, administrative, rural extension, training, credit availability).

Acknowledgements This work was possible thanks to funding from USAID and UKAID, through the Bushmeat Research Initiative from CIFOR

References

Abernethy K, Coad L, Taylor G et al (2013) Extent and ecological consequences of hunting in Central African rainforest in the 21st century. Phil Trans R Soc B 368:20130494

Agnew DJ, Pearce J, Pramod G et al (2009) Estimating the worldwide extent of illegal fishing. PLoS One 4:e4570. doi:10.1371/journal.pone.0004570

Aiyeloja A, Ogunjinmi A (2013) Economic aspects of Grasscutter Farming in Southwest Nigeria: implications for sustainable adoption and conservation. IJSER 4(10):17–23

Altrichter M (2006) Wildlife in the life of local people of the semi-arid Argentine Chaco. Biodivers Conserv 15:2719–2736. doi:10.1007/s10531-005-0307-5

Alves R, Alves HN (2011) The faunal drugstore: animal-based remedies used in traditiontal medicines in Latin America. J Ethnobiol Ethnomed 7(9). doi:10.1007/s10661-011-2465-0

Alves R, Vieira K, Santana G et al (2012) A review on human attitudes towards reptiles in Brazil. Environ Monit Assess 184(11):6877–6901

Alves R, de Lima R, Araujo HFP (2013) The live bird trade in Brazil and its conservation implications: an overview. Bird Conserv Int 23(1):53–65. doi:10.1017/S095927091200010X

Anang B, Awuni J, Mustapha A (2011) Factors affecting the adoption of grasscutter (Thryonomys swinderianus) farming in Sunyani Municipality, Brong Ahafo Region, Ghana. J Agric Bio Sci 2(3):43–48

Assan H (2014) Micro-Livestock Farming and Food Security in Sub Saharan Africa. J Anim Prod Adv 4:374–387

Auzel P, Wilkie DS (2000) Wildlife use in Northern Congo: hunting in a commercial logging concession. In: Robinson JG, Bennet EL (eds) Hunting for sustainability in tropical forests. Columbia University Press, New York, pp 413–454

Baillie JEM, Hilton-Taylor C, Stuart SN (eds) (2004) IUCN red list of threatened species. A global species assessment. IUCN, Gland, Switzerland and Cambridge, UK

Bair-Brake H, Bell T, Higgins A et al (2014) Is that a rodent in your luggage? A mixed method approach to describe bushmeat importation into the United States. Zoonoses Public Health 61(2):97–104

Bartel RA, Brunson MW (2003) Effects of Utah's coyote bounty program on harvester behavior. Wildl Soc Bull 31:738–743

Beck H, Snodgrass JW, Thebpanya P (2013) Long-term exclosure of large terrestrial vertebrates: Implications of defaunation for seedling demographics in the Amazon rainforest. Biol Conserv 163:115–121

Bell D, Roberton S, Hunter PR (2004) Animal origins of SARScoronavirus: possible links with the international trade in small carnivores. Philos Trans R Soc Lond B Biol Sci 359:1107–1114

Bennett EL (2002) Is there a link between wild meat and food security? Conserv Biol 16(3):590–592

Berger J, Young JK, Berger KM (2008) Protecting migration corridors: challenges and optimism for Mongolian Saiga. PLoS Biol 6(e165). doi:Protecting Migration Corridors: Challenges and Optimism for Mongolian Saiga

Berkes FJ, Folke C (eds) (1998) Linking social and ecological systems: management practices and social mechanisms for building resilience. Cambridge University Press, New York

Bertolini R, Zgrablic G, Cuffolo E (2005) Wild game meat: products, market, legislation, and processing controls. Vet Res Commun 29:97–100

Bodmer R, Robinson J (2004) Evaluating the sustainability of hunting in the Neotropics. In: Silvius K, Bodmer R, Fragoso J (eds) People in nature: wildlife conservation in South and Central America. Columbia University Press, New York, pp 299–323

Bodmer RE, Fang TG, Moya I (1994) Managing wildlife to conserve Amazonian forests: population biology and economic considerations of game hunting. Biol Conserv 67(1):29–35. doi:10.1016/0006-3207(94)90005-1

Bond I, Child B, de la Harpe D et al (2004) Private-land contribution to conservation in South Africa. In: Child D (ed) Parks in transition. Earthscan, London, pp 29–62

Bourdieu P (1977) Outline of a theory of practice. Cambridge University Press, Cambridge

Brooks TM, Mittermeier RA, Mittermeier CG (2002) Habitat loss and extinction in the hotspots of biodiversity. Conserv Biol 16(4):909–923

Brooks EGE, Roberton SI, Bell DJ (2010) The conservation impact of commercial wildlife farming of porcupines in Vietnam. Biol Conserv 143(11):2808–2814

Brown D (2003) Bushmeat and poverty alleviation: implications for development policy. ODI Wildlife Policy Briefing 2. ODI, London

Brugiere D (1998) Population size of the black colobus monkey *Colobus satanas* and the impact of logging in the Lopé Reserve, central Gabon. Biol Conserv 86(1):15–20

Bulte EH, Damania R (2005) An economic assessment of wildlife farming and conservation. Conserv Biol 19(4):1222–1233

Campbell M, Mackay KJ (2009) Communicating the role of hunting for wildlife management. Hum Dimens Wildl 14:21–36

CBD (2008) Conservation and use of wildlife-based resources: the bushmeat crisis. Technical Series No. 33. Secretary of the Convention on Biological Diversity and CIFOR, Montreal, Quebec

Chaber AL, Allebone-Webb S, Lignereux Y et al (2010) The scale of illegal meat importation from Africa to Europe via Paris. Conserv Lett 3(5):317–321

Chardonnet Ph, des Clers B, Fischer J et al (2002) The value of wildlife. Rev Sci Tech Off Int Epiz 21(1):15–51

Child G (1996) The practice and principles of community-based wildlife management in Zimbabwe: the CAMFIRE programme. Biodivers Conserv 5(3):369–398

Child B (2008) Community conservation in southern Africa: rights based natural resource management. In: Suich H, Child B, Spenceley A (eds) Evolution and innovation in wildlife conservation. Earthscan, London, pp 187–200

Clark C, Poulsen J, Malonga R et al (2009) Logging concessions can extend the conservation estate for Central African tropical forests. Conserv Biol 23(5):1281–1293

Coad L (2007) Bushmeat hunting in Gabon: socioeconomics and hunter behavior. University of Cambridge, Cambridge

Cooney R (2009) Landholder collaboration in wildlife management – models for landholders to share benefits from kangaroo harvesting. Barton, A.C.T.: Rural Industries Research and Development Corporation, Canberra, Australia

Corlett R (2007) The impact of hunting on the mammalian fauna of tropical Asian forests. Biotropica 39(3):292–303

Côté SD, Rooney TP, Tremblay JP et al (2004) Ecological impacts of deer overabundance. Annu Rev Ecol Evol Syst 35:113–147

Cousins JA, Sadler JP, Evans J (2008) Exploring the role of private wildlife ranching as a conservation tool in South Africa: stakeholder perspectives. Ecol Soc 13(2):43

Cowlishaw G, Mendelson S, Rowcliffe JM (2005) Evidence for post-depletion sustainability in a mature bushmeat market. J Appl Ecol 42:460–468
Craigie I, Baillie JEM, Balmford A, Carbone C, Collen B, Green RE, Hutton JM (2010) Large mammal population declines in Africa's protected areas. Biol Conserv 143(2010):2221–2228
Dahlan I (2009) Characteristics and Cutability of Farmed Rusa Deer (*Cervus timorensis*) Carcasses for Marketing of Venison. Asian-Aust J Anim Sci 22(5):740–746
Dale VH, Brown S, Haeuber R et al (2000) Ecological principles and guidelines for managing the use of land. Ecol Appl 10:639–670
Damania R, Milner-Gulland EJ, Crookes DJ (2005) A bioeconomic analysis of bushmeat hunting. Proc R Soc B 272(1560):259–266
de Merode E, Homewood K, Cowlishaw G (2004) The value of bushmeat and other wild foods to rural households living in extreme poverty in Democratic Republic of Congo. Biol Conserv 118:573–581
De Souza Alcantara A (2014) Conservation of freshwater turtles in Amazonia: retrospective and future prospects. JCLM 2:666–672
Delgado CL, Hopkins J, Kelly VA (1998) Agricultural Growt Linkages in Sub-Saharan Africa, IFPRI Reserac Report. International Food Policy Research Institute, Washington, DC
Delvingt W, Dethier M, Auzel P et al (1997) La chasse villageaoise Badjoué, gestion coutumière durable ou pillage de la resource gibier? Presses Agronomiques de Gembloux, Gembloux *Colobus satanas*
Dethier M (1995) Etude chasse. Rapport ECOFAC, Composante Cameroun. AGRECO-CTFT, Bruxelles
Dirzo R (2001) Tropical Forests. In: Chapin FS, Sala OE, Huber-Sannwald E (eds) Global biodiversity in a changing environment: scenarios for the 21st century. Springer, New York, pp 251–276
Dirzo R, Mendoza EPO (2007) Size-related differential seed predation in a heavily defaunated neotropical rain forest. Biotropica 39(3):355–362. doi:10.1111/j.1744-7429.2007.00274.x
Dirzo R, Young HS, Galetti M et al (2014) Defaunation in the anthropocene. Science 345(6195):401–406. doi:10.1126/science.1251817
Douglas LR, Alie K (2014) High-value natural resources: linking wildlife conservation to international conflict, insecurity, and development concerns. Biol Conserv 171:270–277. doi:10.1016/j.biocon.2014.01.031
Drury R (2009) Reducing urban demand for wild animals in Vietnam: examining the potential of wildlife farming as a conservation tool. Conserv Lett 2(6):263–270
Drury R (2011) Hungry for success: urban consumer demand for wild animal products in Vietnam. Conserv Soc 9(3):247–257
Dry G (2014) Wildlife Ranching Industry: the South African flagship of sustainable green economy. In 8th International Wildlife Ranching Symposium. Estes Park, Colorado
Dudley JP, Ginsberg JR, Plumptre AJ et al (2002) Effects of war and civil strife on wildlife and wildlife habitats. Conserv Biol 16(2):319–329
Emmons LH (1989) Tropical rain forests: why they have so many species and how we may lose this biodiversity without cutting a single tree. Orion Nat Quart 8:8–14
Estes JA, Terborgh J, Brashares JS et al (2011) Trophic downgrading of Planet Earth. Science 333(6040):301–306. doi:10.1126/science.1205106
Fa JE, Yuste J, Perez del Val J et al (1995) Impact of market hunting on mammalian species of Equatorial Guinea. Conserv Biol 9(5):1107–1115
Fa J, Yuste JEG (2001) Commercial bushmeat hunting in the Monte Mitra forests, Equatorial Guinea: extent and impact. Anim Biodivers Conserv 24(1):31–52
Fargeot C (2009) La viande de chasse en Afrique central: un PFNL essential. Le Flamboyant 65:13–18
Fisher A, Sandström C, Delibes-Mateos M (2013) On the multifunctionality of hunting – an institutional analysis of eight cases from Europe and Africa. J Environ Plann Manag 56(4):531–552

Foden WB, Butchart SHM, Stuart SN et al (2013) Identifying the world's most climate change vulnerable species: a systematic trait-based assessment of all birds, amphibians and corals. PLoS One 8(e65427). doi:10.1371/journal.pone.0065427
Fragoso JMV (1994) Large mammals and the community dynamics of an Amazonian rain forest. Zoology. University of Florida, Gainesville
Fulton EA, Smith ADM, Smith DC et al (2011) Human behaviour: the key source of uncertainty in fisheries management. Fish Fish 12(1):2–17. doi:10.1111/j.1467-2979.2010.00371.x
Gallopin GC, Gutman P, Maletta H (1989) Global impoverishment, sustainable development and the environment: a conceptual approach. Int Soc Sci J 121:375–397
Gally M, Jeanmart P (1996) Etude de la chasse villageoise en forêt dense humide d'Afrique centrale (Cameroun, Congo, République Centrafricaine). Mémoire de fin d'études, Faculté Universitaire des Sciences agronomiques de Gembloux, p. 142
Gavin MC, Solomon JN, Blank SG (2010) Measuring and monitoring illegal use of natural resources. Conserv Biol 24(1):89–100. doi:10.1111/j.1523-1739.2009.01387.x
Gill RMA, Fuller RJ (2007) The effects of deer browsing on woodland structure and songbirds in lowland Britain. Ibis 149:119–127
Global Forest Watch (2002) An analysis of access into Central Africa's rainforest. World Resource Institute, Washington, DC
Gordon IJ, Hester AJ, Festa-Bianchet M (2004) The management of wild large herbivores to meet economic, conservation and environmental objectives. J Appl Ecol 41:1021–1031
Gratwicke B, Mills J, Dutton A et al (2008) Attitudes toward consumption and conservation of tigers in China. PLoS One 3(e2544). doi:10.1371/journal.pone.0002544
Grayson DK (2001) The archaeological record of human impacts on animal populations. J World Prehistory 15(1):1–68
Guarner J, Johnson B, Paddock C et al (2004) Monkeypox transmission and pathogenesis in prairie dogs. Emerg Infect Dis 10:426–431
Gunderson L (1999) Resilient management: comments on – ecological and social dynamics in simple models of ecosystem management|| by S.R. Carpenter, W.A. Brock, P. Hanson. Conserv Ecol 3(2):7
Hall JS, Inogwabini BI, Williamson EA et al (1997) A survey of elephants (Loxodonta Africana) in te Kahuzi-Biega national Park lowland sector and adjacent forest in eastern Zaire. Afr J Ecol 35:213–223
Hanson T, Brooks TM, Da Fonseca GAB et al (2009) Warfare in biodiversity hotspots. Conserv Biol 23:578–587
Hardouin J, Thys É, Joiris V, Fielding D (2003) Mini-livestock breeding with indigenous species in the tropics. Livestock Research for Rural Development 15: Article #30. http://www.lrrd.org/lrrd15/4/hard154.htm
Harrison RD, Tan S, Plotkin JB et al (2013) Consequences of defaunation for a tropical tree community. Ecol Lett 16(5):687–694
Hart J (2000) Impact and sustainability of indigenous hunting in the Ituri Forest, Congo-Zaire: a comparison of unhunted and hunted duiker populations. In: Robinson JG, Bennet EL (eds) Hunting for sustainability in tropical forest. Columbia University Press, New York, pp 106–153
Hart J, Hall J (1996) Status of Eastern Zaires forest parks and reserves. Conserv Biol 10(2):316–324
Hatton J, Couto M, Oglethorpe J (2001) Biodiversity and war: a case study of Mozambique. WWF Biodiversity Support Programme, Washington, DC
Hayes R, Riffell S, Minnis R et al (2009) Survival and habitat use of feral hogs in Mississippi. Southeast Nat 8:411–426
Heberlein TA (2008) Open season? Should Wisconsin allow hunting to control the wolf population? Wis State J B4:1. http://dces.wisc.edu/wp-content/uploads/sites/30/2013/08/Heberlein-Yes-to-WI-Wolf-Hunt.pdf
Henschel P (2009) The status and conservation of leopards and other large carnivores in the Congo Basin, and the potential role of reintroduction. In: Hayward MW, Somers M (eds) Reintroduction of top-order predators. Blackwell Publishing, Oxford, pp 206–237
Hilborn R, Arcese P, Borner M et al (2006) Effective enforcement in a conservation area. Science 314(1266). doi:10.1126/science.1132780

Hill K, McMillan G, Fariña R (2003) Hunting-related changes in game encounter rates from 1994 to 2001 in the Mbaracayu Reserve, Paraguay. Conserv Biol 17(5):1312–1323

Hoffmann M, Hilton-Taylor C, Angulo A et al (2010) The impact of conservation on the status of the world's vertebrates. Science 330(6010):1503–1509. doi:10.1126/science.1194442

Hooke RL, Martín-Duque J, Pedraza J (2012) Land transformation by humans: a review. GSA Today 22(12):4–10

Hunter DM (2002) Landscape structure, hábitat fragmentation and the ecology of insects. Agric For Entomol 4:159–166. doi:10.1046/j.1461-9563.2002.00152.x

IUCN (2004) War and protected areas. Protected Areas Programme 14(1). NatureBureau, UK

IUCN/PACO (2009) Big game hunting in West Africa. What is its contribution to conservation? IUCN, Cambridge, UK

Iwaruma T, Guisan A, Wilson KA et al (2013) How robust are global conservation priorities to climate change? Glob Environ Change 23(5):1277–1284

Iwamura T, Lambin EF, Silviys KM et al (2014) Agent-based modeling of hunting and subsistence agriculture on indigenous lands: understanding interactions between social and ecological systems. Environ Model Software 58:109–127

Johnson FA, Kendall WL, Dubovsky JA (2002) Conditions and limitations on learning in the adaptative management of mallard harvest. Wildl Soc Bull 30(1):176–185

Jones B, Weaver C (2008) CBNRM in Namibia: growth, trends, lessons and constraints. In: Suich H, Child B, Spenceley A (eds) Evolution and innovation in wildlife conservation in southern Africa. Earthscan, London, pp 223–242

Jones BA, Grace D, Kock R et al (2013) Zoonosis emergence linked to agricultural intensification and environmental change. Proc Natl Acad Sci 110:8399–8404

Jorge AA, Vanak AT, Thaker M et al (2013) Costs and benefits of the presence of leopards to the sport-hunting industry and local communities in Niassa National Reserve, Mozambique. Conserv Biol 27:832–843

Jori F, Chardonnet P (2001) Cane rat farming in Gabon. Status and perspectives. In: Ebedes H, Reilly B, van Houven W et al (eds) Sustainable utilisation – conservation in practice. South African Game Ranchers' Organization, Pretoria, South Africa, pp 99–109

Jori F, Mensah GA, Adjanohoun E (1995) Grasscutter production: an example of rational exploitation of wildlife. Biodivers Conserv 4(3):257–265

Jori F, Cooper J, Casal J (2001) A survey of pathology on captive cane rats (Trynonomys swinderianus). Vet Rec 148:624–628

Jori F, Edderai D, Houben P (2005) Potential of rodents for minilivestok in Africa. In: Paoletti MG (ed) Ecological implications of minilivestok: potential of insects, rodents, frogs, and snails. Science Publishers, Enfield, pp 25–45

Jori F, Godfroid J, Michel AL et al (2013) An assessment of Zoonotic and Production Limiting Pathogens in Rusa Deer (Cervus timorensis rusa) from Mauritius. Transboun Emerg Dis 61:31.42. doi:10.1111/tbed.12206

Kaeslin E, Redmond I, Dudley N (eds) (2012) Wildlife in a changing climate. FAO Forestry Paper 167. Food and Agriculture Organization of the United Nations. Rome. http://www.fao.org/docrep/015/i2498e/i2498e00.htm

Keesing F, Young TP (2014) Cascading consequences of the loss of large mammals in an African savanna. Bioscience 64:487–495. doi:10.1093/biosci/biu059

Kilonzo C, Stopka TJ, Chomel B (2013) Illegal animal and (bush)meat trade associated risk of spread of viral infections. In: Singh SK (ed) Viral infections and global change. Wiley, Hoboken. doi:10.1002/9781118297469.ch10

Kock R (2014) Drivers of disease emergence and spread: is wildlife to blame? Onderstepoort J Vet Res 81(2):E1–E4. doi:10.4102/ojvr.v81i2.739

Kümpel NF (2006) Incentives for sustainable hunting of bushmeat in Río Muni, Equatorial Guinea. Institute of Zoology/University of London, London

Kümpel NF, Milner-Gulland EJ, Rowcliffe JM et al (2008) Impact of gun-hunting on diurnal primates in continental Equatorial Guinea. Int J Primatol 29(4):1065–1082

Kümpel N, Milner-Gulland E, Cowlisaw G et al (2010a) Assessing sustainability at multiple scales in a rotational bushmeat hunting system. Conserv Biol 24(3):861–871

Kümpel N, Milner-Gulland E, Cowlisaw G et al (2010a) Incentives for hunting: the role of bushmeat in the household economy in rural Equatorial Guinea. Hum Ecol 38(2):251–264
Lahm SA (1993) Ecology and economics of human/wildlife interaction in northeastern Gabon. New York University, New York
Lahm SA (1994) Impact of elephants on agricultura-Gabon. African Elephant Conservation Programme. Final report. Vol 4, European Commission, Ecology in Developing Countries Programme, Oxford, UK
Laporte NT, Stabach JA, Grosch RG et al (2007) Expansion of industrial logging in Central Africa. Science 316(5830):1451. doi:10.1126/science.1141057
Laurance WF, Croes BM, Guissouegou N et al (2008) Impacts of roads, hunting, and habitat alteration on nocturnal mammals in African rainforests. Conserv Biol 22(3):721–732
Le Bel S, Stansfield F, La Grange M et al (2013) Managing local overabundance of elephants through the supply of game meat: the case of Savé Valley Conservancy, Zimbabwe. South Afr J Wildl Res 43(2):103–119
Le Pendu Y, Guimaraes D, Linhares A (2011) Estado da arte sobre a criação comercial da fauna silvestre brasileira. Rev Bras Zootecn 40:52–59
Leach M, Fairhead J (2000) Challenging neo-Malthusian deforestation analyses in West Africa's dynamic forest landscapes. Popul Dev Rev 26:17–43
Leroy EM, Gonzalez JP, Baize S (2011) Ebola and Marburg haemorrhagic fever viruses: major scientific advances, but a relatively minor public health threat for Africa. Clin Microbiol Infect 17(7):964–976
Letnic M, Crowther MS (2013) Patterns in the abundance of kangaroo populations in arid Australia are consistent with the exploitation ecosystems hypothesis. Oikos 122(5):761–769
Linder JM (2008) The impact of hunting on primates in Korup National Park, Cameroon: implications for primate conservation. Department of Anthropology, City University of New York, New York
Lindsey P, Roulet P, Romanach S (2007) Economic and conservation significance of the trophy hunting industry in sub-Saharan Africa. Biol Conserv 134(4):455–469
Lindsey PA, Romañach SS, Davies-Mostert H (2009) The importance of conservancies for enhancing the value of game ranch land for large mammal conservation in southern Africa. J Zool 277(2):99–105
Lindsey P, Alexander R, Balme G et al (2012) Possible relationships between the South African captive-bred lion hunting industry and the hunting and conservation of lions elsewhere in Africa. South Afr J Wildl Res 42(1):11–22
Lindsey PA, Barnes J, Nyirenda V et al (2013a) The zambian wildlife ranching industry: scale, associated benefits, and limitations affecting its development. PLoS One 8(12). doi:10.1371/journal.pone.0081761
Lindsey PA, Havemann CP, Lines RM et al (2013b) Benefits of wildlife-based land uses on private lands in Namibia and limitations affecting their development. Oryx 47:41–53
Lindsey P, Nyirenda VR, Barnes JI et al (2014) Underperformance of African protected area networks and the case for new conservation models: insights from Zambia. PLoS One 9(5):e94109
Ling S, Milner-Gulland EJ (2006) Assessment of the sustainability of bushmeat hunting based on dynamic bioeconomic models. Conserv Biol 20(4):1294–1299
Lowassa A, Tadie D, Fisher A (2012) On the role of women in bushmeat hunting: insights from Tanzania and Ethiopia. J Rural Stud 28:622–630
Ludwig D, Hilborn R, Walters C (1993) Uncertainty, resource exploitation, and conservation: lessons from history. Science 260(5140):17–36
Luskin M, Christina E, Kelley L et al (2013) Modern hunting practices and wild meat trade in the oil palm plantation-dominated landscapes of Sumatra, Indonesia. Hum Ecol 42(1):35–45
MacMillan DC, Leitch K (2008) Conservation with a gun: understanding landowner attitudes to deer hunting in the Scottish highlands. Hum Ecol 36(4):473–484
Magwedere K, Hemberger M, Hoffman L et al (2012) Zoonoses: a potential obstacle to the growing wildlife industry of Namibia. Infect Ecol Epidemiol 2:1–16

Mahoney SP (2009) Recreational hunting and sustainable wildlife use in North America. In: Dickson B, Hutton J, Adams WM (eds) Recreational hunting, conservation and rural livelihoods. Wiley-Blackwell, Oxford, UK, pp 266–281

Maisels F, Strindberg S, Blake S et al (2013) Devastating decline of Forest Elephants in Central Africa. PLoS One 8(e 59469). doi:10.1371/journal.pone.0059469

Maldonado-Chaparro A, Blumstein D (2008) Management implications of capybara (Hydrochoerus hydrochaeris) social behavior. Biol Conserv 141(8):1945–1952

Marlowe FW (2005) Hunter-gatherers and human evolution. Evol Anthropol 14:54–67

Martin JL, Stockton SA, Allombert S (2010) Top-down and bottom- up consequences of unchecked ungulate browsing on plant and animal diversity in temperate forests: lessons from a deer introduction. Biol Invasions 12:353–371

Matthews A, Matthews A (2002) Distribution, population density, and status of sympatric cercopithecids in the Campo-Ma'an area, Southwestern Cameroon. Primates 43:155–168

Matthews A, Matthews A (2004) Survey of gorillas (*Gorilla gorilla gorilla*) and chimpanzees (*Pan* troglodytes troglodytes) in Southwestern Cameroon. Primates 45(1):15–24

Mawdsley JR, O'Malley R, Ojima DS (2009) A review of climate-change adaptation strategies for wildlife management and biodiversity conservation. Conserv Biol 23(5):1080–1089

McCorquodale SM (1997) Cultural contexts of recreational hunting and native subsistence and ceremonial hunting: their significance for wildlife management. Wildl Soc Bull 25: 568–573

McNeely J (1998) War and biodiversity: an assessment of impacts. First international conference on addressing environmental consequences of war: legal, economic, and scientific perspectives, Washington, DC 10–12 June 1998

Milner-Gulland EJ, Akcakaya HR (2001) Sustainability indices for exploited populations. Trends Ecol Evol 16:686–692

Milner-Gulland EJ, Bennett EL (2003) Wild meat: the bigger picture. Trends Ecol Evol 18: 351–357

Mincher BJ (2002) Harvest as a component of Greater Yellowstone Ecosystem grizzly bear management. Wild Soc Bull 30(4):1287–1292

Moreira J, Ferraz K, Herrera EA et al (eds) (2012) Capybara: biology, use and conservation of an exceptional Neotropical Species. Springer Science Business Media, New York

Muller-Landau HC (2007) Predicting the long-term effects of hunting on plant species composition and diversity in tropical forests. Biotropica 39(3):372–384

Naeem S, Byers D, Tjossem SF et al (1999) Plant neighborhood diversity and production. Ecoscience 6:355–365

Nasi R, Brown D, Wilkie D et al (2008) Conservation and use of wildlife-based resources: the bushmeat crisis. Secretariat of the Convention on Biological Diversity and Center for International Forestry Research (CIFOR), Bogor, Indonesia and Montreal, Canada

Nasi R, Taber A, van Vliet N (2011) Empty forests, empty stomachs? Bushmeat and livelihoods in the Congo and Amazon Basins. Int For Rev 13(3):355–368

Ngnegueu PR, et Fotso RC (1996) Chasse villageoise et conséquences pour la conservation de la biodiversité dans la réserve de biosphère du Dja. Rapport, ECOFAC/Cameroun

Nogueira-Filho S, Nogueira S (2004) Captive breeding programs as an alternative for wildlife conservation in Brazil. In: Silvius K, Bodmer R, Fragoso J (eds) People in nature: wildlife conservation in South and Central America. Columbia University Press, New York, pp 171–190

Nogueira-Filho S, Nogueira S (2011) Wildlife farming: an alternative to unsustainable hunting and deforestation in Neotropical forests? Biodivers Conserv 20(7):1385–1397

Noss AJ (1995) Duikers, cables and nets: a cultural ecology of hunting in a central African forest. Ph.D. thesis, University of Florida, Gainesville

Novaro AJ, Redford KH, Bodmer RE (2000) Effect of hunting in source-sink systems in the neotropics. Conserv Biol 14:713–721

Oates J (1996) African primates: status survey and conservation action plan. IUCN, Gland

Ogunjimi S, Yusuf OJ, Ajala A (2012) Factors influencing involvement of peri-urban farmers in mini-livestock farming in south-western Nigeria. Sci J Rev 1:58–69

Ostrom E (2007) Sustainable social-ecological systems: an impossibility?. Paper presented at the 2007 Annual Meetings of the American Association for the Advancement of Science, "Science and Technology for Sustainable Well-Being", San Francisco 15–19 February 1978
PACEC (2006) Economic and environmental impact of sporting shooting in the UK. Public and Corporate Economic Consultants, Cambridge
Packer C, Brink H, Kissui BM et al (2010) Effects of trophy hunting on lion and leopard populations in Tanzania. Conserv Biol 25(2):142–153
Pangau-Adam M, Noske R, Muehlenberg M (2012) Wildmeat or bushmeat? Subsistence hunting and commercial harvesting in Papua (West New Guinea), Indonesia. Hum Ecol 40: 611–621
Pence DB, Ueckermann E (2002) Sarcoptic mange in wildlige. Rev Sci Tech Off Int Epizoot 21(2):385–398
Peres CA, Dolman P (2000) Density compensation in neotropical primate communities: evidence from 56 hunted and nonhunted Amazonian forest of varying productivity. Oecologia 122:175–189
Peres CA, Nascimento HS (2006) Impact of game hunting by the Kayapó of south-eastern Amazonia: implications for wildlife conservation in tropical forest indigenous reserves. Biodivers Conserv 15:2626–2653
Peterson GD (2000) Political ecology and ecological resilience: an integration of human and ecological dynamics. Ecol Econ 35:323–336
Plumptre AJ, Bizumuremi JB, Uwimana F et al (1997) The effects of the Rwandan civil war on the poaching of ungulates in the Parc National des Volcans. Oryx 31(4):265–273. doi:10.1126/science.288.5466.617c
Plumptre AJ, Hart T, Vedder A et al (2000) Support for Congolese conservationists. Science 288(5466):617. doi:10.1126/science.288.5466.617c
Primack RB (2002) Essentials of conservation biology. Sinauer Associates Press, Sunderland
Putnam RD (2000) Bowling alone: the collapse and revival of American Community. Simon & Schuster, New York
Redford KH (1992) The empty forest. Bioscience 42(6):412–422
Redpath S, Thirgood S (2009) Hen harriers and red grouse: moving towards consensus?. J Appl Ecol 46:961–963. doi:10.1111/j.1365-2664.2009.01702.x
Richerson P, Borgeroff-Mulder M, Vila B (1996) Principles of human ecology. Simon and Schuster, New York
Rincker PJ, Bechtel PJ, Finstadt G et al (2006) Similarities and differences in composition and selected sensory attributes of reindeer, caribou and beef. J Muscle Foods 17:65–78
Ripple WJ, Beschta RL (2012) Trophic cascades in Yellowstone: the first 15 years after wolf reintroduction. Biol Conserv 145:205–213
Ripple WJ, Rooney TP, Beschta RL (2010) Large predators, deer, and trophic cascades in boreal and temperate ecosystems. In: Terborgh J, Estes JA (eds) Trophic cascades: predators, prey, and the changing dynamics of nature. Island Press, Washington, DC, pp 141–162
Roberts PD, Diaz-Soltero H, Hemming DJ et al (2013) What is the evidence that invasive species are a significant contributor to the decline or loss of threatened species? A systematic review map. Environ Evidence 2:2–7
Robinson JG, Redford KH (1991) Sustainable harvest of neotropical forest mammals. In: Robinson JG, Redford KH (eds) Neotropical wildlife use and conservation. University of Chicago Press, Chicago, pp 415–429
Robinson JG, Redford KH (1994) Measuring the sustainability of hunting in tropical forests. Oryx 28:249–256
Robinson HS, Wielgus RB, Cooley HS et al (2008) Sink populations in carnivore management: cougar demography and immigration in a hunted population. Ecol Appl 18(4):1028–1037
Rossing WAH, Zander P, Josien JCJ et al (2007) Integrative modelling approaches for analysis of impact of multifunctional agriculture: a review for France, Germany and The Netherlands. Agr Ecosyst Environ 120(1):41–57

Roulet PAR (2004) Chasseur blanc, cœur noir? La chasse sportive en Afrique Centrale. Une analyse de son rôle dans la conservation de la faune sauvage et le développement rural au travers des programmes de gestion communautaire. Les cas du nord RCA et du sud-est Cameroun. Dissertation PhD, Université d'Orléans, Orléans

Rowcliffe JM, Cowlishaw G, Long JA (2003) model of human hunting impacts in multi-prey communities. J Appl Ecol 40:872–889

Said MY, Chunge RN, Craig GC et al (1995) African elephants database. Occasional paper 11. Species Survival Commission, World Conservation Union, Gland, Switzerland

Salas LA, Kim JB (2002) Spatial factors and stochasticity in the evaluation of sustainable hunting of tapirs. Conserv Biol 16:86–96

Sayer J, Ghazoul J, Nelson P, Klintuni Boedhihartono A (2012) Oil palm expansion transforms tropical landscapes and livelihoods. Global Food Security 1:114–119

Scherrer S (2002) Les pertes d'usage récréatif du patrimoine forestier après les tempêtes de 1999: le cas de la forêt de Fontainebleau. Econ Stat 357–358:153–171

Schlaepfer MA, Hoover C, Dodd CK (2005) Challenges in evaluating the impact of the trade in amphibians and reptiles on wild populations. BioScience 55:256–264

Scholte P (2011) Towards understanding large mammal population declines in Africa's protected areas: a West-Central African perspective. Trop Conserv Sci 4(1):1–11

Schulp C, Thuiller W, Verburg P (2014) Wild food in Europe: a synthesis of knowledge and data of terrestrial wild food as an ecosystem service. Ecol Econ 105:292–305

Simberloff D, Martin J-L, Genovesi P et al (2013) Impacts of biological invasions: what's what and the way forward. Trends Ecol Evol 28(1):58–66

Siren AH (2012) Festival hunting by the kichwa people in the Ecuadorean Amazon. J Ethnobiol 32(1):30–50. doi:10.2993/0278-0771-32.1.30

Siren A, Hamback P, Machoa J (2004) Including spatial heterogeneity and animal dispersal when evaluating hunting: a model analysis and an empirical assessment in an amazonian community. Conserv Biol 18:1315–1329

Smit B, Wandel J (2006) Adaptation, adaptive capacity and vulnerability. Global Environ Change 16-282-292. doi:10.1016/j.gloenvcha.2006.03.008

Smith RL, Smith TM (2003) Elements of ecology. Benjamin Cumming, San Francisco

Smith KM, Anthony SJ, Switzer WM (2012) Zoonotic viruses associated with illegally imported wildlife products. PLoS One 7(1), e29505. doi:10.1371/journal.pone.0029505

Sonricker ALH, Li A, Joly D et al (2012) Digital surveillance: a novel approach to monitoring the illegal wildlife trade. PLoS One 7(12):e51156 10.1016/0006-3207(94)90005-1

Starkey M (2004) Commerce and subsistence: the hunting, sale and consumption of bushmeat in Gabon. Fitzwilliam College/Cambridge University, Cambridge, UK

Stoner KE, Vulinec K, Wright SJ et al (2007) Hunting and plant community dynamics in tropical forests: a synthesis and future directions. Biotropica 39(3):385–392

Sunderlin W, Resosudarno IAP (1999) The effect of population and migration on forest cover in Indonesia. J Environ Dev 8(2):152–169

Terborgh J, Feeley K (2010) Propagation of trophic cascades via multiple pathways in tropical forests. In: Terborgh J, Estes JA (eds) Trophic cascades: predators, prey and the changing dynamics of nature. Island Press, Washington, DC

Terborgh J, Nunes-Iturri G, Pitman NCA et al (2008) Tree recruitment in an empty forest. Ecology 89(6):1757–1768

TRAFFIC (2008) "What's driving the wildlife trade? A review of expert opinion on economic and social drivers of the wildlife trade and trade control efforts in Cambodia, Indonesia, Lao PDR and Vietnam". East Asia and Pacific Region Sustainable Development Discussion Papers. East Asia and Pacific Region Sustainable Development Department, World Bank, Washington, DC

Treves A, Karanth KU (2003) Human–Carnivore conflict and perspectives on carnivore management worldwide. Conserv Biol 17(6):1491–1499. doi:10.1111/j.1523-1739.2003.00059.x

U.S. Fish and Wildlife Service (2004) What do hunters do for conservation?. Available DIALOG. http://www.fws.gov/hunting/whatdo.html. Accessed 15 Sep 2014

U.S. Fish and Wildlife Service, U.S. Department of the Interior, U.S. Department of Commerce et al (2011). National survey of fishing, hunting, and wildlife-associated recreation. U.S. Fish & Wildlife Service. https://www.census.gov/prod/2012pubs/fhw11-nat.pdf

UICN (2009) Big game hunting in West Africa. What is its contribution to conservation? PAPACO Studies n°2, 110 pp

UICN/PACO (2009) Big game hunting in West Africa. What is its contribution to conservation?. IUCN, Gland, Switzerland and Cambridge, UK

Van Der Hoeven CA, De Boer WF, Prins HHT (2010) Roadside conditions as predictor for wildlife crossing probability in a Central African rainforest. Afr J Ecol 48(2):368–377

van Vliet N (2008) Variabilité spatiale et temporelle au sein système “chasseur – animal – territoire de chasse villageois” – pour une approche géographique de la durabilité de la chasse en Afrique Centrale. Université Toulouse le Mirail, Toulouse

van Vliet N, Nasi R (2008a) Why biological models do fail to assess the sustainability of duiker hunting in Central Africa? Oryx 42(3):392–399

van Vliet N, Nasi R (2008b) Mammal distribution in a Central African logging concession area. Biodivers Conserv 17(5):1241–1249

van Vliet N, Nasi R, Emmons L et al (2007) Evidence for the local depletion of bay duiker Cephalophus dorsalis, within the Ipassa Man and Biosphere Reserve, north-east Gabon. Afr J Ecol 45(3):440–443

van Vliet N, Nebesse C, Nasi R (2010a) The dynamics of bushmeat trade in the market of Kisangani. Paper presented at the XXIII IUFRO Congress, Seoul, Korea, 23–28 August 2010

van Vliet N, Milner-Gulland E, Bousquet F et al (2010b) Effect of small-scale heterogeneity of prey and hunter distributions on the sustainability of bushmeat hunting. Conserv Biol 24(5):1327–1337

van Vliet N, Quiceno-Mesa MP, Cruz-Antia D et al (2014) Carne de monte y seguridad alimentaria en la zona trifronteriza amazónica (Colombia, Perú y Brasil). CGIAR, USAID, CIFOR, Fundación SI, UFAM, Fundación Omacha, Bogotá DC

Vogel G (2000) Conflict in Congo threatens bonobos and rare gorillas. Science 287(5462): 2386–2387

Walker B, Holling CS, Carpenter SR et al (2004) Resilience, adaptability and transformability in social-ecological systems. Ecol Soc 9(2):5

Walsh JF, Molyneux DH, Birley MH (1993) Deforestation: effects on vector borne disease. Parasitology 106:S55–S75

Walsh P, Abernethy K, Bermejo M et al (2003) Catastrophic ape decline in western equatorial Africa. Nature 422(6932):611–614. doi:10.1038/nature01566

Walters GG, Touladjan S, Makouka L (2014) Integrating cultural and conservation contexts of hunting: the case of the Plateau Bateke Savannas of Gabon. Afr Study Monogr 35(2):99–128

Warchol GL (2004) The transnational illegal wildlife trade. Crim Justice Stud Crit J Crime J Law Soc 17(1):57–73

Webb GJW, Brook B, Whitehead P et al (2004) Wildlife management principles and practices in crocodile conservation and sustainable use. Paper presented at the 17th working meeting of the IUCN-SSC Crocodile Specialist Group. IUCN, Gland, Switzerland

Weinbaum KZ, Brashares JS, Golden CD et al (2013) Searching for sustainability: are assessments of wildlife harvests behind the times? Ecol Lett 16(1):99–111

Weldon C, du Preez LH, Hyatt AD et al (2004) Origin of the amphibian chytrid fungus. Emerg Infect Dis 10(12):2100–2105

White L, Edwards A (2000) Methods for assessing the status of animal populations. In: White L, Edwards A (eds) Conservation research in the African rain forests: a technical handbook. Wildlife Conservation Society, New York, pp 225–275

Wilkie D, Carpenter J (1999) Bushmeat hunting in the Congo Basin: an assessment of impacts and options for mitigation. Biodivers Conserv 8(7):927–955

Wilkie DS, Bennett EL, Peres CA et al (2011) The empty forest revisited. Ann N Y Acad Sci 1223:120–128. doi:10.1111/j.1749-6632.2010.05908.x

Wilmers CC, Crabtree RL, Smith DW et al (2003) Trophic facilitation by introduced top predators: grey wolf subsidies to scavengers in Yellowstone National Park. J Anim Ecol 72(6):909–916

Winkler R, Warnke K (2013) The future of hunting: an age-period-cohort analysis of deer hunter decline. Popul Env 34:460–480

Woodroffe R, Lindsey PA, Romañach S et al (2005) Livestock predation by endangered African wild dogs (Lycaon pictus) in northern Kenya. Biol Conserv 124:225–234

Wright SJ (2003) The myriad consequences of hunting for vertebrates and plants in tropical forests. Perspect Plant Ecol Evol Syst 6(1):73–86

Wright JH, Priston N (2010) Hunting and trapping in Lebialem Division, Cameroon: bushmeat harvesting practices and human reliance. Endang Species Res 11:1–12

WWF (2014) Living Planet Report 2014. Available via DIALOG. http://wwf.panda.org/about_our_earth/all_publications/living_planet_report/. Accessed 21 Sept 2014

Conservation Conflicts: Future Research Challenges

R.J. Gutiérrez, Kevin A. Wood, Stephen M. Redpath, and Juliette C. Young

Introduction

It has been long known that wildlife management is often more about working with people than with wildlife (Gilbert 1971). Much of this people management occurs as top down conservation measures codified in rules, regulations, and laws. When such measures are enforced, or when there is an absence thereof, both people and wildlife can be affected; this has the potential to give rise to conflict about these measures to conserve wildlife. Such conservation conflicts have the potential to be as widespread as they are varied, and have proven difficult to resolve. It is also clear that conservation conflicts that involve wildlife will escalate both in number and intensity in the future because of rising human population and resource use (Redpath et al. 2015).

The urgency inherent in conservation conflicts often leads directly to action upon identification of the issue (i.e., searching for solutions to ameliorate the situation). The idea of resolving conflicts is embedded in the literature; however, conflicts are often not resolved but rather managed. That is, there can be efforts to prevent disagreements escalating toward conflict, they can be mediated, and they can be managed to achieve various outcomes. We thus use the term "conflict management" to mean any process that attempts to lead parties in conflict towards shared solutions.

R.J. Gutiérrez (✉)
Department of Fisheries, Wildlife, and Conservation Biology, University of Minnesota, St. Paul, MN, USA
e-mail: gutie012@umn.edu

K.A. Wood
Conservation Science, Wildfowl and Wetlands Trust, Slimbridge, Gloucestershire GL2 7BT, UK

S.M. Redpath
Institute of Biological and Environmental Sciences, Aberdeen University, 23 St Machar Drive, Aberdeen, UK

J.C. Young
NERC Centre for Ecology and Hydrology, Edinburgh, UK

R. Mateo et al. (eds.), *Current Trends in Wildlife Research*,
Wildlife Research Monographs 1, DOI 10.1007/978-3-319-27912-1_11

Whilst the logical priority of many wildlife organisations or interest groups is to manage conflicts immediately, research also has a crucial role, specifically to evaluate the efficacy of many aspects that practitioners normally employ when managing conflicts. In this chapter we therefore have focused on key areas which have seen recent advances and we identify future research priorities. The areas we discuss are conflict identification, conflict management in adversarial systems, integrating non-traditional fields, evaluating conflicts, and adaptive management and monitoring of conflicts. We conclude this chapter with some final thoughts on the key challenges associated with conservation conflicts in the future.

Conflict Identification

Conservation conflicts have recently been described as between those who wish to conserve biodiversity (or specific wildlife or habitats), and those whose activities may impinge on such activities (Young et al. 2010). All conservation conflicts are therefore between people who have different views about conservation (Young et al. 2010).

Effective management approaches will vary depending on the nature of the problem and how it will affect the people involved. The relevance of deciding whether a situation is an impact or a conflict is that each usually requires a different course of action. In the former case, a range of technical solutions or strategies (e.g., fencing livestock or using dogs to reduce livestock predation) can be explored to ameliorate or stop the impact. However, in conflict situations an in-depth understanding of stakeholders' perceptions of the issues, their values and goals will be required before a joint understanding of the conflict can be achieved and then solutions proposed and implemented. An example is taken from the Bladnoch region of Scotland, where interviews with local stakeholders showed very different understandings about the impact of forestry practices on salmon fisheries (Young 2010). While government agencies and foresters perceived the situation as an impact requiring technical solutions such as releasing fish after capture, fishermen perceived it as a conflict between fishermen and foresters (Young 2010). The fishermen involved felt frustrated that their concerns were not being heard, and were increasingly adversarial towards government agencies (Young et al. 2013). This situation highlighted both the natural tendency to shy away from conflict when possible and the potential negative repercussions of ignoring stakeholders' concerns and worldviews, such as growing mistrust (e.g., Wynne 1992).

The above considerations highlight the need and challenge in conservation conflicts of a clear process of conflict mapping and management (Young et al. in press). Acknowledging the views and values of all stakeholders in a conflict is an essential step in such conflict mapping and management and requires resources and will to engage with all relevant stakeholders. Despite what may seem obvious in this discussion (i.e., facilitating a collective learning process) there are people and organisations whose goals are antagonistic to environmental concerns who purposely obfuscate both the state of scientific knowledge and the historical facts of situations (Oreskes and Conway 2010). Of course, the opposite is true (environmentalists or others can employ the same strategy). Research is needed to assess the undermining effects these contrarians have on conflict management processes. Moreover, we need to improve our capacity to evaluate effective ways of determining the cost and benefits of such people or organisations engaging with conflict management processes.

Conflict Management in an Adversarial System

It is axiomatic that conflicts are adversarial, but how they are dealt with is a key issue for both researcher and practitioners. The conflict management literature often advocates the need for stakeholders involved in a conflict to be open, transparent and clear about their objectives, so that both joint understanding and joint approaches can be adopted (Young et al. 2010). But "openness" can be exploited if other stakeholders engage in adversarial or strategic positioning. This dilemma not only identifies an area that requires further investigation, but suggests wildlife professionals draw upon fields with great experience in conflict management such as peace studies (Rogers 2015).

An important initial issue here is to understand why stakeholders might be adversarial and escalate a conflict, rather than seeking shared solutions through dialogue and discussion. Certain contexts may not lend themselves to dialogue, for example in situations where collaboration is limited or where imbalances of power occur. In these situations, stakeholders may turn to litigation and legislation as means to resolution rather than dialogue (McCool et al. 2000). In other situations, histories of distrust and confrontation between stakeholders may preclude dialogue (Hemmati 2002). In these situations, the questions are whether and how a system can be created where the risks of being adversarial outweigh the benefits.

For example, in the conflict between seal conservationists and salmon fishermen in the Moray Firth in Scotland, a top-down ban on seal shooting served as a catalyst to stop salmon fishermen from adopting adversarial positions (Young et al. 2012). Prior to the ban, the risks to fishermen who shot seals of being caught and either fined or prosecuted were low if non-existent. The benefits of shooting seals were perceived as high by the fisherman because they perceived that shooting seals increased the probability of salmon returning to the rivers, which, in turn, would assure their livelihoods. Once there was a ban, the risks of killing seals outweighed the benefits, thereby encouraging fishermen to engage in dialogue to find a shared solution, which emerged as a management plan with quotas on seal culling and the identification of seals more likely to predate salmon (Butler et al. 2008; Graham et al. 2011). Whereas a top-down approach, such as this shooting ban, provided an environment where dialogue became possible, it could also have had the opposite effect of emphasising the imbalance of power (with fishermen powerless against the government) and reinforcing strategic positioning. Future research is critical to understand the conditions required to create a system or environment in which the risks of being adversarial outweigh the benefits, and to ascertain those situations where dialogue is not a suitable option such as when there is intense distrust or unwillingness to share views with other stakeholders (Hemmati 2002).

When dialogue is established between relevant stakeholders, effective conflict management processes will depend on whether there is sufficient flexibility in positions held by conflict parties to find acceptable solutions or goals. Such mutually acceptable goals may often be very different from stakeholders' ideal outcome (McCool et al. 2000) and will require willingness to compromise. In adversarial situations, stakeholders may brand some of their values as "non-negotiable" – i.e. issues they feel they cannot compromise on. Research is lacking however on the nature of non-negotiable versus desired (valued) goals, and how such goals can be identified. We feel this is a critical area of research because parties declaring some element as non-negotiable have the potential to easily end or stall almost all conflict management efforts.

Integrating Non-traditional Fields

Over the past 100 years there has been a profound rise in human population and resource use, which have put increasing pressure on wildlife and its conservation. Traditionally, wildlife researchers have responded to conservation conflicts in two ways. They have conducted ecological research to increase our understanding of wildlife at the centre of conflicts (Woodroffe et al. 2005). Then they have used this knowledge to help environmental managers and policy makers design and evaluate management strategies (e.g., Wood et al. 2013).

Currently, the majority of wildlife research is carried out within a single discipline. However, there is a small but growing awareness within the wildlife research community that the management of conflicts, as well as the attainment of broader conservation objectives, are likely to benefit from a better integration of social, ecological, and economic realities because people are the reason conservation conflicts emerge (Dickman 2010; Wood et al. 2015). Thus, increasing only the available natural science information and technical solutions is usually insufficient to lead to successful management of conservation conflicts (Kirkpatrick and Turner 1997; Johns 2009). Indeed, an incomplete understanding of the causes of a conflict may actually exacerbate it, for example by marginalising stakeholders or proposing management solutions that are inappropriate for social, economic or logistical reasons (White and Ward 2010). Conflicts that are identified as conservation issues often serve as proxies for conflicts between people over less tangible social and psychological issues, including identity, status, values, power and beliefs – none of which is typically addressed by current approaches to resolving conflicts among people about wildlife (Madden and McQuinn 2014). To understand conflicts better so that they can be more completely addressed, the insights and approaches used by practitioners in other disciplines as well as the natural sciences are needed. Redpath et al. (2015b) have provided some of the potential contributions that other fields with experience of conflict management, from humanities to natural and social sciences, could make to conservation conflicts. For example, besides the necessary ecological information, the majority of conservation conflicts are likely to require information and expertise to understand the positions, goals, values and relations of the stakeholders involved, the politics overshadowing the situation, the history of the conflict, the law that provides a framework for actions and the ethical arguments to guide subsequent conflict management (Redpath et al. 2015b). In particular, much insight and research exists from fields that routinely consider the social underpinnings of human conflicts, including psychology, sociology, and peace studies. These fields can inform how we understand and address conservation conflicts, and so help us to improve the effectiveness of conservation efforts (Madden and McQuinn 2014). We currently lack an understanding about how well lessons from these other disciplines will transpose to conservation conflicts.

Whilst scientific information can inform decision-making processes, there are socio-political processes that include conflicts among people with different values, power, history and outlook; these differences among people will have a strong influence on the development and acceptance of potential solutions (Peterson et al. 2013). Scientific knowledge is therefore only part of the relevant knowledge needed to address conflicts effectively as part of evidence-informed conservation (*sensu* Adams and Sandbrook 2013). Other forms of knowledge, such as indigenous knowledge or that derived from stakeholders' experiences working in a conflict can also be valid and relevant. Integrating these disparate forms of knowledge can provide beneficial insight toward resolving conservation conflicts (Adams and Sandbrook 2013). Yet, two major barriers to such integration have been identified. Firstly, it has proven challenging for quantitatively-trained natural scientists to accept and interpret these non-traditional sources of knowledge (Endter-Wada et al. 1998). Secondly, researchers have found it difficult to combine these non-traditional sources of knowledge, which are typically qualitative, with traditional quantitative scientific data. To overcome such barriers, researchers are slowly developing a growing number of frameworks designed to integrate scientific and non-traditional knowledge. For example, an attempt to address a conflict over conservation and resource use in upland regions within the United Kingdom engaged different types of knowledge through an iterative combination of qualitative and quantitative methods (Raymond et al. 2010). Multiple semi-structured interviews, structured questionnaires, workshops and site visits allowed local knowledge to be gained from stakeholders such as farmers, game keepers, grouse moor owners, water companies, recreationalists and government officials. Such local knowledge was analysed using grounded theory analysis and incorporated into conceptual models of the upland socio-ecological system (Raymond et al. 2010). Knowledge from natural and social sciences was integrated through a literature review and quantitative socio-ecological modelling. Whilst Raymond et al. (2010) describe this and two other projects which aimed to integrate scientific and local knowledge, no indication is given as to whether such projects were successful in transforming conflicts. There is a clear need for more examples of such frameworks actually being used to address conflicts, together with evaluations of their success.

The effort to combine knowledge from different fields of study reflects the suggestion to shift from "evidence-based" to "evidence-informed" decision-making in conservation (Adams and Sandbrook 2013). Such a shift may be particularly urgent in conflict situations, where a shared understanding of the evidence amongst scientists, policy makers and other stakeholders is necessary to make progress toward resolving conservation conflicts. However, a major barrier to further progress in developing effective approaches is that, to date, there have been few attempts to develop such frameworks, let alone evaluate the effectiveness of such multidisciplinary approaches to conflicts.

Evaluating Conservation Conflict Interventions

Conflict occurs when those representing conservation pursue incompatible goals from other groups, be they local communities, farmers, hunters, businesses or others. Given the negative consequences of these conservation conflicts for biodiversity, as well as for human livelihoods and wellbeing, there is an urgent need to evaluate the effectiveness of different interventions in conflict situations, so that we can learn from the vast array of case studies being conducted around the world (Wood et al. 2014).

The need to improve the way we evaluate conservation interventions has been increasingly recognised (Sutherland et al. 2004; Waylen et al. 2010; Howe and Milner-Gulland 2012). Of particular concern in conflict situations is the argument that we need to move away from thinking about biodiversity outcomes in isolation, to recognising that conservation is also a social and political process so the approaches and evaluation needs to reflect this (Brechin et al. 2002; Waylen et al. 2010).

There have been several attempts to evaluate the success of conflict interventions (e.g., Parker and Osborn 2006; Gore et al. 2006, 2008). Inevitably, however, interventions focus on reducing *impact* to improve conservation success rather than focusing on managing *conflict* or evaluating the interventions. Increasingly, evaluations try to measure through attitudinal surveys the efficacy of interventions and responses by local people (Washington et al. 2014 and references therein), but evaluations of interventions need to be more comprehensive to judge their efficacy.

So far, the aim of wildlife researchers has naturally been to focus on short-term conservation success. However, in conflict situations, the social conflicts are often dominant, so there is a strong argument to switch the focus from outcomes relative to a single party (i.e., conservation) to reducing conflict among all conflict parties in the long-term. As discussed earlier, such a change requires different approaches to interventions because the focus is not on getting a conservation "win" irrespective of the cost to some, but on the need for parties to acknowledge alternative viewpoints and then to agree on interventions and outcomes that are, hopefully, beneficial to all (Redpath et al. 2013). This will require conservationists to reflect on their role in these issues (Madden and McQuinn 2014; Redpath et al. 2015a). Naturally, however, conservationists are likely to engage in such processes only if there are benefits to biodiversity or particular species. So, as conservationists there is a need to understand how biodiversity outcomes relate to processes of genuine engagement in conflict management. We hypothesise that engaging with stakeholders in conflict management approaches will lead to better biodiversity and social outcomes. However, we simply do not yet know how engagement affects conservation outcomes, because of the dearth of evaluations accounting for both social and ecological outcomes (Young et al. 2013).

To illustrate these issues, consider the on-going and often discussed conflict over birds of prey in the United Kingdom (UK; Thirgood and Redpath 2008; Sotherton et al. 2009; Thompson et al. 2009). Essentially this conflict revolves around the fact that game managers seek to maximise numbers of gamebirds so they kill predators to help them achieve this goal. Some of these predators are protected, leading to conflict with conservation organisations. One of the main aims of conservationists here is to conserve protected predators by preventing illegal killing, but what is the most effective way of doing that? One approach is to enforce existing legislation, increase penalties and force game managers to adhere to the law, regardless of the cost to hunters. Another approach is to focus on engagement and conflict management, through bringing the parties together to develop shared solutions that ensure that illegal killing is stopped, but at the same time minimise the costs that hunters will incur with additional predators. Many on the conservation side are nervous about engagement and compromise. On the one hand we might predict that attempts to enforce legislation may lead to more predators (a conservation win) but heightened conflict, as hunters feel their way of life is being threatened. On the other hand, an agreed plan for the legal management of predator populations may result in fewer predators, but reduced conflict if parties agree to it. From a conservation perspective, the question is: will going for the win with heightened conflict deliver more robust long-term conservation outcomes than accepting a compromise with reduced conflict? Arguments are ongoing across both social media and the national press, but which approach will deliver better conservation outcomes and how do these relate to social outcomes?

Wildlife biologists lack a systematic approach to evaluating conflict management processes, professional training, and an appropriate database of interventions and outcomes. To evaluate properly the effectiveness of alternative approaches for reducing conflict and delivering long-term sustainable outcomes we therefore need to consider not just the environmental indicators, but also the social ones and the process of engagement. More specifically, we need to monitor conservation measures (such as the number of predators and the extent of illegal activity), social measures (such as the types and levels of conflict among the groups), and the type of processes that has been used to resolve conflict (including enforcement, dialogue, and the involvement of policy makers and mediators). Such understanding is critical to the long-term, sustainable management of conflict.

We consider that there are four questions that need to be addressed to help us deal with emerging conflicts more effectively: (1) When is it better to approach conflicts by actively enforcing conservation legislation, focusing on technical solutions to reduce impact, rather than engaging in conflict management approaches (Redpath et al. 2013)? (2) At what stage in the process should engagement occur and with whom? (3) Which conflict management approaches are most effective? (4) How sensitive are these approaches to local cultures and contexts (Waylen et al. 2010)?

Ultimately, we need to monitor the huge range of conflict situations around the world, and build a robust evidence base to support a detailed evaluation of the efficacy of interventions as well as the environmental and social outcomes those interventions deliver. We need to develop a consistent framework to allow us to evaluate their success. One of the many challenges of developing such a framework is that the interventions and evaluations are often done by conservation organisations that may lack expertise and resources to evaluate social outcomes. This will require more inter-disciplinary partnerships between natural and other sciences and the need for research funders to support robust, systematic evaluation, such as that supported by the UK's Darwin Initiative (Howe and Milner-Gulland 2012).

Adaptive Management and Monitoring: Mechanisms for Building Trust in Conflict Situations

People engaged in conflict typically do not trust their adversaries. Hence, trust is a key element underpinning the perceptions of people who are engaged in conflict with others over conservation issues (e.g., Redpath et al. 2013). Mapping conflicts and exploring the knowledge, views and beliefs held by stakeholders are foundational elements to conflict management and provide opportunities to identify key issues of the conflict about which people are mistrustful. For example, if stakeholders believe that a management strategy designed to exploit the habitat of a species will result in irreparable harm to the species, it provides the opportunity to develop a mechanism to demonstrate either the efficacy of the strategy or the validity of the belief. One way of evaluating efficacy of management strategies that has received substantial attention in the past few decades is an approach called adaptive management (Holling 1978; Walters 1986). A related paradigm is "adaptive monitoring" (Lindenmayer et al. 2011), but in our context monitoring the effects on wildlife (or habitat) of some management action is an issue central to both adaptive management and trust. Thus, adaptive management programmes can be used to build trust because many issues leading to distrust, such as assertions that certain management interventions will have little or no impact on wildlife, can be tested explicitly. Hence, we predict that adaptive management also has the potential to assist in conflict management involving wildlife conservation.

A Cursory Overview of Adaptive Management

Adaptive management is a process predicated on determining the efficacy of management interventions in an environment of uncertainty (Walters 1986; Walters and Holling 1990). It presumes an organisation creates a structure and operational framework that allows specific information to be gathered about the explicit management actions being conducted (e.g., effects of logging on wildlife habitat), which can then be analysed to determine the effects of the actions on the element(s) of interest. Once this analysis has been done and a conclusion about these objectives relative to the original predictions is reached, a feedback structure within the organisational structure facilitates modification or cessation of future actions based on this analysis of effects. Thus, it is a way to determine if management interventions are deleterious and, if they are, learn why they are and then either stop or modify actions before there is irreparable harm. Walters (1986) discusses three ways this learning can occur as part of adaptive management: trial and error; passive adaptive management informed by existing data; and active adaptive management where data are gathered (ideally in the form of management experiments) and then used to evaluate trade-offs for a range of alternatives. In turn each of these learning methods is more complex than the preceding, but the latter two imply a sophisticated operational structure because there has to be an explicit design, predictions based on that design, monitoring to gather data for analyses (or the analysis of historical data), feedback loops to inform managers, revisions based on analyses of data and interaction among a variety of people, including stakeholders, scientists, managers, decision makers and policy makers (Colyvan et al. 2011). The last element is often at odds with the typical command and control management structures so common to natural resource and wildlife management agencies (Holling and Meffee 1996).

Substantial research and management effort has been devoted to adaptive management in the past four decades, and it has been applied to many ecosystem and conservation problems (e.g., Schreiber et al. 2004; Gregory et al. 2006; Conroy et al. 2011). The reason for this wide interest is because adaptive management encourages management to continue while presumably containing mechanisms to stop or modify management actions should they be found faulty. Adaptive management has been applied less often to wildlife management, but there are notable examples of its use in wildlife conservation, such as for game harvest strategies (e.g., Williams and Johnson 1995; Nielsen et al. 1997; Johnson and Williams 1999), large mammal monitoring (Boddicker et al. 2002) and endangered species management (e.g., Briceño-Linares et al. 2011; Runge 2011; Runge et al. 2011; Swaisgood et al. 2011). Even less attention has been paid to employing adaptive management to deal with the social context related to conservation conflicts.

Because adaptive management is conceptually straightforward and encourages management interventions, its acceptability to both decision makers and scientists is enhanced (Gregory et al. 2006). However, adaptive management can also be used as a way to defer decisions or otherwise undermine the need to understand the effects of interventions (Gregory et al. 2006). While elegant in its conceptual simplicity, it is, by the nature of its promise (eliminating uncertainty and guarding against harmful practices), a complicated process. Adaptive management, particularly active adaptive management, often involves sophisticated analyses (Williams 2012), and hierarchical planning that can often exceed either the expertise or the worldview of stakeholders who are engaged in a conservation conflict centered on wildlife. It is because of this complexity and analytical approach that some have questioned its value, particularly as it relates to social outcomes (McLane and Lee 1996; Lee 1999).

Social Context of Adaptive Management

Adaptive management provides an opportunity for participation and learning among different people or stakeholders, which is often a central issue of conflicts (Walters 1986; Armitage et al. 2008). For example, in a conflict situation a classic top down or command and control structure would be antithetical to successful dialogue among parties in conflict (Holling and Meffee 1996; Reed and Sidoli del Ceno 2015).

The reasons why people engage in conflict about wildlife are many, and vary for a wide range of reasons. Yet they fall under at least three general categories of philosophy, tradition, and expectation. Often these are linked or interrelated. The basic philosophy of participants in conflict has much bearing on whether progress can be made, but determining what is core philosophy or current belief(s) of a person or group rather than simply a psychological positioning at the onset is essential to understanding whether progress can be made toward dialogue or management of a conflict (see also above). Tradition covers, at minimum, that which owes to history, social structure of stakeholder groups, taboos, ways of making a living, and belief systems. Expectation covers issues such as beliefs (or scientific predictions) about effects of management (i.e. whether you think [or have data to suggest] that an activity is either harmful or not harmful [in terms of negative impacts] vs. whether you think an activity is simply untenable as a matter of core philosophy/social system [see above]), desire to have access to use resources, and whether one sees others negatively (e.g., as exploiters, paternalistic, or prejudicial). An adaptive management process has the potential to evaluate all these elements as part of its structure.

Adaptive Management and Conflict Management

Because adaptive management can accommodate input of stakeholders, it holds great promise to ameliorate one of the fundamental blocks to conflict management – lack of trust (Reed 2008). This is because parties in conflict who also participate in an adaptive management process can identify those elements of concern to them, advocate for variance in methods that might affect those elements, provide input to monitoring strategies, observe the data collection under a monitoring scheme, scrutinize the analysis and conclusions about effects on management interventions on those elements of interest, and then review responses by the management agencies after such conclusions are drawn. Most importantly, they can demand, if the process is truly participatory with sufficient power allocated to participating parties, that management be modified if there are negative effects. Thus, trust can be established if transparency is established and power is allocated among all involved in the conflict.

However, adaptive management is often antithetical to conflict management because one group of stakeholders (e.g., preservationists) may be asked to allow exploitation of a species or habitat that they want to protect (Lee 1999). Yet the promise is that adaptive management can determine whether or not the species or habitat will not be harmed as a result of management. This is a difficult premise because one group may have a philosophical position that opposes exploitation of individual animals (Holland 2015), while wildlife managers only consider population as the object of management. The challenge for researchers is to define the limits of philosophical position versus psychological disposition among stakeholders. For example, does an individual or group really abhor the idea of any animal dying as part of their core philosophy or, when confronted with the alternatives of either having people incur harm from species, such as lions killing cattle or people, or allowing offtake to reduce impact from a species, would they accept the lesser of the two "evils"? The latter option represents changing a psychological preference or position. These kinds of dichotomies are often couched under the speculative hypothesis of potential short-term harm versus long-term gain (Runge et al. 2011).

Of Trojan Horses and the Promise of Adaptive Management

Despite the increasing literature mentioning adaptive management, there have been few reported examples of successful adaptive management programmes (Keith et al. 2011; Rist et al. 2013a,b). The lack of successful examples coupled with the high visibility of adaptive management suggests either that adaptive management is prohibitively complex and costly for most wildlife conservation organisations to undertake or they are simply not reported (see Lee 1999; McLane and Lee 1996; Moir and Block 2001). As Rist et al. (2013b) point out the technical, social and political appropriateness of adaptive management often become confused, which reduces clarity about the effectiveness or success of adaptive management programmes. It is in the context of the latter that we think adaptive management has potential to be used in conflict management. If stakeholders are allowed to participate in a meaningful way, the ambiguity about the social/technical application needs to be ameliorated. Yet participation could threaten those who hold power (McLane and Lee 1996), but the rewards of inclusive participation and empowerment may outweigh the risks (Reed 2008). Alternatively, adaptive management in many instances may have become analogous to a "Trojan Horse" strategy. That is, invoking adaptive management to facilitate or continue management (e.g., exploitation of a resource) is simply a ruse (Lee 1999; Gregory et al. 2006). Whilst such a label seems harsh, the increasing indicated of adaptive management as a component of management programmes, but for which there is no real commitment, seems to support this "dark side" of managers or administrators invoking adaptive management (Moir and Block 2001; Stankey et al. 2003; Gregory et al. 2006). However, there are ways of identifying whether an adaptive management proposal is too complicated or a Trojan Horse. In the first case, the complexity of an issue can be evaluated relative to the capabilities of the parties in conflict (or those who are charged with ameliorating or managing the conflict). For example, an adaptive management programme can be evaluated to determine if it has all the elements of adaptive management and, if so, whether there are sufficient resources (technical and monetary) for the programme to be successfully implemented. If the answer is "no" for any stage (e.g., incomplete design or insufficient resources), adaptive management should not be invoked as a component of a conflict management process. In the case of a Trojan Horse situation, there are some obvious indicators of the veracity of a proposed adaptive management programme. For example, all adaptive management programmes must have explicit goals (or objectives), explicit designs for achieving those goals, explicit predictions for management outcomes, an adequate monitoring programme to gather information on effects of the management interventions, thresholds or criteria that trigger a cessation or modification of management activities if harmful effects are discovered, and a feedback loop(s) that allow management to be changed if analysis indicates that change is needed. Moreover, there should be opportunities for stakeholders to participate in decision-making in light of the findings of the process. These are some of the key elements that must be present for a process to be considered adaptive management (Walters 1986; Moir and Block 2001; Gregory et al. 2006; Colyvan et al. 2011), and they can serve as benchmarks to participants in conflict management strategies. It is not our intention here to disparage the use of adaptive management by creating the label of Trojan Horse, quite the contrary. Rather we think that adaptive management can aid people in managing or resolving conflicts if done honestly and correctly, but used as ruse it will almost certainly lead to greater conflict and distrust.

Conclusions

Conflicts over the conservation of wildlife will continue to grow as finite wildlife habitats shrink in the face of growing human populations and rising resource use. Our challenge as wildlife biologists and managers is to help manage those conflicts. In our view conflicts, in general, are counterproductive to wildlife conservation so we think greater awareness and effort needs to be focused not only on techniques or processes to manage conflicts, but also to learn which techniques are most effective and we can do the latter through research. Yet, we are often ill equipped by training to engage in conflict management processes (Young and Redpath 2014). Therefore, we need to expand the research and education dimensions to include the factors relevant to the broader realities of conflict management in wildlife management. As wildlife managers and researchers, we are at a nascent stage our understanding of conflict management, but we have been long aware of the relationship humans play in wildlife conservation as evidenced by the emergence of the field of human dimensions, first as a component of The Wildlife Society and now with its own journal, *Human Dimensions of Wildlife*. Ecologists have expanded the study of human dimensions to general ecology with the online journal *Ecology and Society* and most other relevant scientific, conservation, and professional societies also recognize human dimensions in some form so our understanding should grow rapidly.

In this chapter we have provided an overview of key areas relevant to conservation conflicts that, in our experience with such conflicts, are particularly relevant to framing conflict situations, but these represent just a few of the relevant themes. Indeed, it would be impossible to provide a comprehensive treatment of all of the relevant areas of a field of study which is relatively new, complex, and rapidly evolving. We have simply focused on what we consider to be important questions about our uncertainty in these areas. In addition, there are many questions that need to be considered by wildlife biologists and managers, which we have started to pose in our chapter, and which will need to be addressed as we become increasingly involved in conflict management.

By its nature, conflict management is a hands-on practice and as such the approaches used and their efficacy in the short and long-term are often not subjected to analysis and published. Thus, we believe that the understanding and management of conflicts will benefit greatly from rigorous analysis and reporting of them by a wide range of stakeholders. There is much that remains to be learned from existing conservation conflicts that will help us to better understand and manage conservation conflicts in the future. We think wildlife researchers across the globe strive to evaluate data and draw inference from such data objectively. As such, working within conservation conflicts can sometimes be disconcerting when stakeholders seemingly refuse to accept results derived from robust empirical research. Some reasons why people do not accept scientifically derived knowledge is because they feel their customs, ideologies, and beliefs are threatened by such knowledge and its implications (Nyhan and Reifler 2015). Yet it is incumbent upon us to engage ourselves because it is in the best interests of wildlife conservation to do so. Thus, we are challenged with finding ways to create conflict management environments and approaches that do not threaten stakeholders but empower them to seek shared solutions with their adversaries.

Acknowledgements The University of Minnesota Agricultural Experiment Station (AES PROJECT MIN-41-036) provided financial support to RJG.

References

Adams WM, Sandbrook C (2013) Conservation, evidence and policy. Oryx 47:329–335

Armitage D, Marschke M, Plummer R (2008) Adaptive co-management and the paradox of learning. Glob Environ Chang 18:86–98

Boddicker M, Rodriguez JJ, Amanzo J (2002) Indices for assessment and monitoring of large mammals within an adaptive management framework. Environ Monit Assess 76:05–123

Brechin SR, Wilshusen PR, Fortwangler CL, West PC (2002) Beyond the square wheel: toward a more comprehensive understanding of biodiversity conservation as social and political process. Soc Nat Res 15:41–64

Briceño-Linares JM, Rodríguez JP, Rodríguez-Clark KM, Rojas-Suárez F, Millán PA, Vittori EG, Carrasco-Muñoz M (2011) Adapting to changing poaching intensity of yellow-shouldered parrot (*Amazona barbadensis*) nestlings in Margarita Island, Venezuela. Biol Conserv 144:1188–1193

Butler JRA, Middlemas SJ, McKelvey SA, McMyn I, Leyshon B, Walker I, Thompson PM, Boyd IL, Duck C, Armstrong JD, Graham IM, Baxter JM (2008) The Moray Firth Seal Management Plan: an adaptive framework for balancing the conservation of seals, salmon, fisheries and wildlife tourism in the UK. Aquat Conserv Mar Freshw Ecosyst 18:1025–1038

Colyvan M, Justus J, Regan HM (2011) The conservation game. Biol Conserv 144:1246–1253

Conroy MJ, Runge MC, Nichols JD, Stodola KW, Cooper RJ (2011) Conservation in the face of climate change: the roles of alternative models, monitoring, and adaptation in confronting and reducing uncertainty. Biol Conserv 144:1204–1213

Dickman AJ (2010) Complexities of conflict: the importance of considering social factors for effectively resolving human–wildlife conflict. Animal Conserv 13:458–466

Endter-Wada J, Blahna D, Krannich R, Brunson M (1998) A framework for understanding social science contributions to ecosystem management. Ecol Appl 8:891–904

Gilbert DL (1971) Natural resources and public relations. The Wildlife Society, Washington, DC

Gore ML, Knuth BA, Curtis PD, Shanahan JE (2006) Education programs for reducing American black bear-human conflict: indicators of success? Ursus 17(1):75–80

Gore ML, Knuth BA, Scherer CW, Curtis PD (2008) Evaluating a conservation investment designed to reduce human–wildlife conflict. Conserv Lett 1(3):136–145

Graham IM, Harris RN, Matejusová I, Middlemas SJ (2011) Do 'rogue' seals exist? Implications for seal conservation in the UK. Animal Conserv 14:587–598

Gregory R, Ohlson D, Arvai J (2006) Deconstructing adaptive management: criteria for applications to environmental management. Ecol Appl 16:2411–2425

Hemmati M (2002) Multi-stakeholder processes for governance and sustainability: beyond deadlock and conflict. Earthscan, London

Holland A (2015) Philosophy, conflict and conservation. In: Redpath SM, Gutierrez RJ, Wood KA, Young JC (eds), Conflicts in Conservation: Navigating towards solutions, Cambridge University Press, Cambridge, pp. 19–29

Holling CS (ed) (1978) Adaptive environmental assessment and management. Wiley, New York

Holling CS, Meffee GK (1996) Command and control and the pathology of natural resource management. Conserv Biol 10:328–337

Howe C, Milner-Gulland EJ (2012) Evaluating indices of conservation success: a comparative analysis of outcome-and output-based indices. Animal Conserv 15(3):217–226

Johns D (2009) A new conservation politics: power, organisation building, and effectiveness. Wiley-Blackwell, Chichester

Johnson F, Williams K (1999) Protocol and practice in the adaptive management of waterfowl harvests. Conserv Ecol 3(1):8. [online] URL: http://www.consecol.org/vol3/iss1/art8/

Keith DA, Martin TG, McDonald-Madden E, Walters C (2011) Uncertainty and adaptive management for biodiversity conservation. Biol Conserv 144:1175–1178

Kirkpatrick JF, Turner JW (1997) Urban deer contraception: the seven stages of grief. Wildl Soc Bull 25:515–519

Lee KN (1999) Appraising adaptive management. Conserv Ecol 3(2): 3. [online] URL: http://www.consecol.org/vol3/iss2/art3

Lindenmayer DB, Likens GE, Haywood A, Miezis L (2011) Adaptive monitoring in the real world: proof of concept. Trends Ecol Evol 26:641–646

McCool SF, Guthrie K, Smith JK (2000) Building consensus: legitimate hope or seductive paradox? Res. Pap. RMRS-RP-25. US Department of Agriculture, Forest Service, Rocky Mountain Research Station, Fort Collins

Madden F, McQuinn B (2014) Conservation's blind spot: the case for conflict transformation in wildlife conservation. Biol Conserv 178:97–106

McLane RJ, Lee RG (1996) Adaptive management: promises and pitfalls. Environ Manage 20:437–448

Moir WH, Block WM (2001) Adaptive management on public lands in the United States: commitment or rhetoric? Environ Manage 28:141–148

Nielsen CK, Porter WF, Underwood HB (1997) An adaptive management approach to controlling suburban deer. Wildl Soc Bull 25:470–477

Nyhan B, Reifler J (2015) Does correcting myths about the flu vaccine work? An experimental evaluation of the effects of corrective information. Vaccine 33:459–464

Oreskes N, Conway EM (2010) Merchants of doubt: how a handful of scientists obscured the truth on issues from tobacco smoke to global warming. Bloomsbury Publishing, New York

Parker GE, Osborn FV (2006) Investigating the potential for chilli Capsicum spp. to reduce human-wildlife conflict in Zimbabwe. Oryx 40:343–346

Peterson MN, Peterson MJ, Peterson TR, Leong K (2013) Why transforming biodiversity conservation conflict is essential and how to begin. Pacif Cons Biol 19:94–103

Raymond CM, Fazey I, Reed MS, Stringer LC, Robinson GM, Evely AC (2010) Integrating local and scientific knowledge for environmental management. J Environ Manage 91:1766–1777

Redpath SM, Young J, Evely A, Adams WM, Sutherland WJ, Whitehouse A, Amar A, Lambert R, Linnell J, Watt A, Gutiérrez RJ (2013) Understanding and managing conservation conflicts. Trends Ecol Evol 28:100–109

Redpath SM, Bhatia S, Young J (2015a) Tilting at wildlife: reconsidering human–wildlife conflict. Oryx 49:222–225

Redpath SM, Gutiérrez RJ, Wood KA, Young JC (eds) (2015b) Conflicts in conservation: navigating towards solutions. Cambridge University Press, Cambridge

Reed MS (2008) Stakeholder participation for environmental management: a literature review. Biol Conserv 141:2417–2431

Reed MS, Sidoli del Ceno J (2015) Mediation and conservation conflicts: from top-down to bottom-up. In: Redpath SM, Gutiérrez RJ, Wood KA, Young JC (eds) Conflicts in conservation: navigating towards solutions. Cambridge University Press, Cambridge, pp 222–232

Rist L, Cambell BM, Frost P (2013a) Adaptive management: where are we now? Environ Conserv 40:5–18

Rist L, Felton A, Samuelsson AL, Sandström C, Rosvall O (2013b) A new paradigm for adaptive management. Ecol Soc 18:63. http://dx.doi.org/10,5751/ES-06183-1800463

Rogers P (2015) Peace research and conservation conflicts. In: Redpath SM, Gutiérrez RJ, Wood KA, Young JC (eds) Conflicts in conservation: navigating towards solutions. Cambridge University Press, Cambridge, pp 164–172

Runge MC (2011) Adaptive management for threatened and endangered species. J Fish Wildlife Manage 2:220–233

Runge MC, Converse SJ, Lyons JE (2011) Which uncertainty? Using expert elicitation and expected value of information to design an adaptive program. Biol Conserv 144:1214–1223

Schreiber ESG, Bearlin AR, Nicol SJ, Todd CR (2004) Adaptive management: a synthesis of current understanding and effective application. Ecolog Manag Restorat 5:177–182

Sotherton N, Tapper S, Smith A (2009) Hen harriers and red grouse: economic aspects of red grouse shooting and the implications for moorland conservation. J Appl Ecol 46(5):955–960

Stankey GH, Borman BT, Ryan C, Shindler B, Sturtevant V, Clark RN, Philpot C (2003) Adaptive management and the Northwest Forest Plan: rhetoric or reality? J Forest 101:40–46

Sutherland WJ, Pullin AS, Dolman PM, Knight TM (2004) The need for evidence-based conservation. Trends Ecol Evol 19(6):305–308

Swaisgood RR, Fuwen W, McShea WJ, Wildt DE, Kouba AJ, Zhang Z (2011) Can science save the giant panda (Ailuropoda melanoleuca)? Unifying science and policy in an adaptive management paradigm. Integr Zool 6:290–296

Thirgood S, Redpath S (2008) Hen harriers and red grouse: science, politics and human–wildlife conflict. J Appl Ecol 45(5):1550–1554

Thompson PS, Amar A, Hoccom DG, Knott J, Wilson JD (2009) Resolving the conflict between driven-grouse shooting and conservation of hen harriers. J Appl Ecol 46(5):950–954

Walters CJ (1986) Adaptive management of renewable re-sources. McGraw-Hill, New York
Walters CJ, Holling CS (1990) Large-scale management experiments and learning by doing. Ecology 71:2060–2068
Washington H et al (2014) A framework for evaluating the effectiveness of conservation attention at the species level. Oryx. doi:10.1017/S0030605314000763
Waylen KA, Fischer A, McGowan PJ, Thirgood SJ, Milner-Gulland EJ (2010) Effect of local cultural context on the success of community-based conservation interventions. Conserv Biol 24:1119–1129
Williams BK (2012) Reducing uncertainty about objective functions in adaptive management. Ecol Model 225:61–65
Williams BK, Johnson FA (1995) Adaptive management and the regulation of waterfowl harvests. Wildl Soc Bull 23:430–436
White PCL, Ward AI (2010) Interdisciplinary approaches for the management of existing and emerging human–wildlife conflicts. Wildl Res 37:623–629
Wood KA, Stillman RA, Daunt F, O'Hare MT (2013) Evaluating the effects of population management on a herbivore grazing conflict. PLoS One 8:e56287
Wood KA, Stillman RA, Daunt F, O'Hare MT (2014) Can sacrificial feeding areas protect aquatic plants from herbivore grazing? Using behavioural ecology to inform wildlife management. PLoS One 9:e104034
Wood KA, Stillman RA, Goss-Custard JD (2015) Co-creation of individual-based models by practitioners and modellers to inform environmental decision-making. J Appl Ecol. doi:10.1111/1365-2664.12419
Woodroffe R, Thirgood S, Rabinowitz A (eds) (2005) People and wildlife: conflict or coexistence? Cambridge University Press, Cambridge
Wynne B (1992) Misunderstood misunderstanding: social identities and public uptake of science. Public Understand Sci 1:281–304
Young JC (2010) Analyzing the implementation of Natura 2000 in Scotland: an evaluation of processes and outcomes of stakeholder involvement in management plans. PhD thesis submitted to the University of East Anglia
Young JC, Marzano M, White RM, McCracken DI, Redpath SM, Carss DN, Quine CP, Watt AD (2010) The emergence of biodiversity conflicts from biodiversity impacts: characteristics and management strategies. Biodivers Conserv 19:3973–3990
Young JC, Butler JRA, Jordan A, Watt AD (2012) Less government intervention in biodiversity management: risks and opportunities. Biodivers Conserv 21(4):1095–1100
Young JC, Jordan AR, Searle K, Butler A, Chapman DS, Simmons P, Watt AD (2013) Does stakeholder involvement really benefit biodiversity conservation? Biol Conserv 158:359–370
Young JC, Redpath SM (2014) Conflict management principles and guidance. Scottish Natural Heritage Commissioned Report (iBids and Project no13654)
Young JC, Thompson D, Moore P, MacGugan A, Watt AD, Redpath SM (In Press). A conflict management tool for conservation agencies. J Appl Ecol.

Wildlife Research: Towards a Better Coexistence Between People and Wildlife

Jesús T. García, José Jiménez, Rafael Mateo, and Beatriz Arroyo

The rate at which humans transform the world is accelerating and, perhaps as a result, the interest in wildlife appears to be also growing. Reconciling human development and wildlife conservation in current human-dominated ecosystems is not an easy task, but is imperative for both mankind and biodiversity.

Approximately 7.3 billion humans inhabit the earth in 2015 (United Nations and Social Affairs 2015). Notably, human population has grown about fourfold in the last century, multiplying exponentially our needs (as Thomas R. Malthus argued more than 200 years ago) and dramatically impacting the environment, which is currently considered, to all intents and purposes, as our farm: truly a place to build lodgings and a food factory. However, this is not the end of the story: according to projections, the global population will continue to grow reaching 9–11 billion by 2050. Meeting the food, water and energy needs of current and future generations globally affects all the earth's resources and involves numerous environmental problems (e.g. Chapter 6: Mateo et al. 2016); Chapter 5: Sánchez-Zapata et al. 2016). This emphasizes the urgent need to evaluate how to match available resources to the requirements of a rapidly growing human population.

In addition to the basic resources (land, water, crops and livestock species), humans depend on the presence and functioning of approximately ten million other species existing in nature, many of them currently unknown, and whose maintenance frequently collide with our growth. Besides the approximately 20,000 plant species that are used by humans for food and medicine purposes, thousands of species are involved in direct or indirect ecosystem services such as pollination, recycling our waste, degradation of chemical pollutants, purification of water and soil, or regulation of pests and diseases (e.g. Swift et al. 2004). In this context, one of the most important goals should be the conservation of natural resources for their

J.T. García (✉) • J. Jiménez • R. Mateo • B. Arroyo
Department of Ecología, Instituto de Investigación en Recursos Cinegéticos (IREC) CSIC-UCLM-JCCM, Ronda de Toledo s/n, 13071, Ciudad Real, Spain
e-mail: jesusgarcia.irec@gmail.com

R. Mateo et al. (eds.), *Current Trends in Wildlife Research*,
Wildlife Research Monographs 1, DOI 10.1007/978-3-319-27912-1_12

sustainable future use, and the necessary establishment of an appropriate framework, policies and rules governing relationships between humans and nature to meet this objective. Scientific knowledge in general, and particular disciplines such as wildlife research, can contribute successfully to decision-making in this regard.

In this chapter we briefly review how wildlife research has changed in the last few decades to address these new challenges, how policies for wildlife conservation have also changed concomitantly to science, and discuss some points that might be important to consider in the future.

Changes in Wildlife Research During the Last Decades

In every chapter of this book, there are mentions of noticeable research discoveries and advances happened during the last decades, and how these have not only contributed to the development of the discipline of wildlife research at large, but how they frequently led to practical applications. Overall, recent decades have been characterized by marked scientific progress in all wildlife research topics, some of them having particularly rapid increases, as highlighted by the metrics reviewed in Chapter 1 (Arroyo et al. 2016).

Moments of rapid acceleration of scientific progress have different origins. Some occur due to scientific and/or technological advances, such as the invention of Internet or the development of the Polymerase Chain Reaction (PCR) mentioned in Chapter 9 (Piertney 2016). Some occur due to conceptual advances, such as the integration between pattern-oriented and process-oriented approaches (e.g. Shröeder and Seppelt 2006; Turner 2005), or the shift from management decisions based on anecdotal information to evidence-based approaches in decision-making for wildlife conservation (Sutherland et al. 2004). Other advances are 'forced' by the appearance of new issues for which there was no previous knowledge, such as global change or the emergence of new wildlife diseases because of human actions. Overall, recent changes in wildlife research may be grouped in one of these categories.

Scientific and/or Technological Advances

Recent scientific and technological advances in a number of fields outside wildlife research have allowed unprecedented quantitative analyses of natural systems. For example, technological advances have strikingly changed the way that researchers monitor, track and locate animals (e.g. Bridge et al. 2011), since the first bird banding experiences in North America (tying strings around bird legs), through the use of VHF telemetry in the 1960s, to the current use of Geolocators, GPS (Global Positioning Systems), GSM (Global System for Mobile Communication), PSAT (Pop-up Satellite Archival Tags), or the recent use of nanothechnology (e.g. quantum dots) for tracking small organisms in marine environments (Ekvall et al. 2015).

Currently, wildlife researchers can work from a distance by the use of digital surveillance camera traps to monitor organisms, record electronically large amount of data thanks to dataloggers and have even access to these data via wireless communications (Yasuda and Kawakami 2002). Being able to handle and process these large datasets has been possible due to the huge and quick development of computers, computation and mathematics (e.g. multiple processors, data-parallel processing, complex mathematical models), together with the fact that access and storage of information has greatly improved by the invention of Internet. Can anyone imagine now doing research without Internet? Wildlife researchers, as scientists of all other disciplines, nowadays use Internet for communicating and sharing information with colleagues and the public, for publishing the result of their investigations (a process that has been extremely shortened) and for creating and accessing large databases that allow spatial and temporal analyses at scales that would have been impossible in earlier times (e.g. Chapter 7: Acevedo et al. 2016).

Another set of important advances that have marked the history of wildlife research were those related to DNA, such as the direct Sanger DNA sequencing in the 1970s, or the development of the Polymerase Chain Reaction (PCR) in the early 1990s, which has allowed to add a genetic dimension to the study of wildlife (Sarre and Georges 2009). The improvement in this field, together with the development of computers, the Internet and other online tools have all created new promising methods to assist in wildlife research and conservation, such as high-throughput DNA, DNA barcoding, or DNA and eDNA metabarcoding, for example (Chapter 9: Piertney 2016).

Similar progresses in other scientific disciplines have also allowed wildlife research to grow, like the contribution of analytical chemistry to the field of ecotoxicology and specifically the study of pollutants (Chapter 6: Mateo et al. 2016); or the development and application of complex mathematical models to the study of habitat (Chapter 4: Tellería 2016), animal and plant distribution (Chapter 7: Acevedo et al. 2016) or demography (Chapter 2: Lebreton and Gaillard 2016).

Last, but not least, it is fair to mention explicitly the recent and growing importance of the development of the open-source programming language R (www.r-project.org), one of the most popular tools for data analyses today that allows scientists to address problems that would have been unapproachable in the past. In our field, a large amount of specific packages on CRAN ("Comprehensive R Archive Network") and other repositories are available today, allowing a better understanding of the patterns and processes underlying human-wildlife interactions and facilitating predictions of how species and ecosystems will respond to future changes (see also Tufto and Cavallini 2005).

Emerging Issues in Human-Wildlife Interactions

As humans increase in numbers and needs, human-wildlife interactions multiply. There is little evidence to suggest that these types of problems will subside; on the contrary, it appears that they may intensify in the next decades. As these problems grow, the needs for applied solutions will also increase, and there is a raising interest within wildlife research in more specialized research disciplines dealing with these new challenges, as has been highlighted throughout the previous chapters.

Global trade, for example, has resulted in the movement of species across natural barriers to dispersal, and invasive species currently present a major challenge to conservationists (Chapter 8: Sol 2016). Research to provide a better understanding of the processes and consequences of invasions, as well as applied research to mitigate the adverse effects arising from this kind of human-related activities, are currently and will continue to be important developments within wildlife research. Similarly, global trade and increased human and animal movements have also led to emergence or re-emergence of important wildlife and zoonotic infections, and wildlife research has adapted to study, prevent and manage, these new challenges in an increasingly global world (Chapter 3: Gortazar et al. 2016).

As human development poses new challenges for wildlife at many levels, several branches within wildlife research have evolved in recent decades, and others have opened up to focus on these issues. Specific research on the potential conflicts between wildlife and the development of alternative or renewable energy systems (Chapter 5: Sánchez-Zapata et al. 2016); the impact of global change on wildlife (Chapter 7: Acevedo et al. 2016); or the impact on wildlife of new pollutants arising from human activities (Chapter 6: Mateo et al. 2016) are some examples of it. Additionally, there is increasing interest in research on wildlife tourism (Reynolds and Braithwaite 2001), or urban wildlife (Magle et al. 2012), which also respond to the growing weight that cities and humans as tourists have (and will continue to have) on wildlife.

Conceptual Advances in Wildlife Research

The important scientific and technological developments occurred during the last decades and the advance in our knowledge of natural patterns and processes has caused a real revolution in the field of wildlife research, largely due to all this progress being properly absorbed (e.g. Robinson et al. 2015). Notable changes in the conceptual frameworks about how to define and face the problems have played a crucial role.

One prime change has been that wildlife researchers have taken advantage of developments from other fields to assist in research and conservation management. Throughout the last half of the twentieth century, wildlife researchers have increasingly included approaches from various disciplines in their work. These include, for example, the incorporation of multidimensionality (human, social, politic and economic dimensions) to the study of human-wildlife conflicts (Chapter 11: Gutierrez et al. 2016); the synergy between veterinary sciences, molecular biology, genetics, human medicine and ecology to investigate the impact of infections on wild fauna, livestock and humans (Chapter 3: Gortazar et al. 2016); or the integration of chemistry, ecotoxicology and ecophysiology to study the impact of chemical substances

(Chapter 6: Mateo et al. 2016). In summary, wildlife research has adopted a clear interdisciplinary character, becoming a bridge-building discipline, which is evidenced by the large impact of pluridisciplinary studies and the transversality of topics such as biodiversity conservation, demography, pollution or human-wildlife conflicts highlighted in Chapter 1 (Arroyo et al. 2016).

In addition, important changes in conceptual frameworks have emerged from the need to address new challenges, and by recently developed analytical methods. There have been indisputably great progresses in this regard, adding new concepts and revisiting old ones for the study of habitat-wildlife interactions and management (e.g. Chapter 4: Tellería 2016); incorporating the scale-dependence on both research studies and management solutions (see Beever et al. 2006); the change from pattern-oriented towards process-oriented approaches in demography (Chapter 2: Lebreton and Gaillard 2016); the inclusion and increasing use of projection of future environmental conditions and environmental predictability to deal with global change impact (e.g. Bierwagen et al. 2010); or an increasing awareness of how data quality and precision, as well as appropriateness of analyses under uncertainty, influence robustness of inferences and thus management guidelines (e.g. Barker and Link 2015; Nuno et al. 2013; see also Chapter 7: Acevedo et al. 2016; Chapter 2: Lebreton and Gaillard 2016). It is noteworthy also the increasing importance of citizen science as a tool for both monitoring and managing wildlife in our changing world (e.g. Lee et al. 2006; Weckel et al. 2010).

Remarkable conceptual changes have also occurred in relation to wildlife conservation, including a shift to evidence-based decision-making (Dicks et al. 2013; Sutherland et al. 2004). This has led to an increasing focus on evaluating the cost-effectiveness of management actions (Shwiff et al. 2013). Similarly relevant has been the development of the concept of *resilience* in conservation (Brand and Jax 2007), linked to the debate on whether biodiversity conservation should be better served through policies that encourage multifunctional landscapes ("land sharing") rather than protected areas ("land sparing") (Fischer et al. 2008; Green et al. 2005). The growing influence of the "land sharing" concept gives increasing importance to research focusing on the relationship between wildlife and human activities, as a means for sustainable coexistence. The high impact and rising trends of studies dealing with these issues, identified in Chapter 1 (Arroyo et al. 2016), indicates that wildlife research is contributing timely to this scientific and societal debate. Similarly, the importance of conservation on private lands has also been recently emphasized, and this has triggered a paradigm shift, from top-down to bottom-up approaches (Knight 1999; Miller et al. 2011), and a growing importance of social sciences in wildlife research and conservation (e.g. Bennet and Roth 2015).

Finally, there has also been a conceptual shift in the idea of wildlife as a renewable resource, as new alternatives to extractive use are proposed as means for sustainability (Chapter 10: Van Vliet et al. 2016). Ideas like wildlife production or game farming are highly debated (see also below), but represent an example of how these new concepts may shape wildlife research in the future.

Changes in Environmental Policies and Public Awareness

Wildlife research has, as emphasized in Chapter 1 (Arroyo et al. 2016), a very strong applied focus, and thus an ultimate aim is to influence management actions or conservation policies. In that sense, it is important to highlight that conservation policies have changed in the last years, sometimes reflecting and sometimes guiding changes in conceptual approaches in wildlife research.

Much has been done in wildlife conservation since the World Conservation Strategy, developed by the IUCN (in collaboration with UNEP, WWF, FAO and UNESCO), was adopted in 1980. The strategy itself has undergone its own historic change, from the first versions focused on the protection of nature to the later ones, expanded to include both social and economic issues. The strategy is intended to "*stimulate a more focussed approach to the management of living resources and to provide policy guidance on how this can be carried out by three main groups: (i) government policy makers and their advisers; (ii) conservationists and others directly concerned with living resources; (iii) development practitioners, including development agencies, industry and commerce, and trade unions*". In 1988, the IPCC (Intergovernmental Panel on Climate Change; www.ipcc.ch) was created, focusing its activity on global assessment of scientific, technical and socioeconomic knowledge on climate change. More recently (2012), the IPBES (Intergovernmental Platform on Biodiversity and Ecosystem Services; www.ipbes.net) incorporated additional functions of assessment, such as knowledge generation, capacity building, and policy support, and was more focused on biodiversity and ecosystem services, and represented yet an inflexion point in the relationship between science and policies. These two international bodies thus exemplify not only current governmental concerns about global destruction of biodiversity, but also the importance they attach to scientific knowledge in terms of organising a global response to it, as well as the necessary communication with society.

Indeed, effective interactions of the triangle "science-policy-people" are the basis for successful wildlife conservation, and in that sense it is worrying that there is still in many countries and societies a poor understanding of, or apathy concerning, environmental practices (e.g. Peterson et al. 2005). Thus, many regional, national and international conservation strategies include the implementation of guidelines to increase the information transferred to the public and people's education on environmental issues and wildlife, with the aim of increasing the public willingness to engage with nature conservation.

Recent changes in education policies of some countries have not been too helpful in that regard. For example, many universities have changed in recent years their science programmes by dropping traditional disciplines with a focus on a sound knowledge of organismal biology, which has been substituted with more general "soft" environmental studies. This has entailed, in many countries, a gradual lack of knowledge on e.g. ecology, zoology, botanic, taxonomy, that may compromise the design of effective wildlife management and conservation strategies.

At another level of interactions, there is still a polarization of actions and knowledge use between (wildlife) researchers and other stakeholders (wildlife managers, farmers, hunters, politicians, business people). However, efforts are being made to shorten this distance between the scientific and the non-expert communities (Treves et al. 2006). In that sense, it is particularly important the growing development of interactions between social and ecological sciences for the study and management of wildlife conservation issues (Chapter 11: Gutierrez et al. 2016), including the explicit involvement of all players in identified conflicts to develop effective management approaches.

Perspectives on Wildlife Research

In spite of the noticeable progress and modernization of wildlife research in the last decades, the rate at which humans transform the planet is even faster, so we need to rethink if the advances in our field of research are enough and, if not, how to face the near future.

Modifications induced by humans will be the most harmful to wildlife and overall biodiversity. For example, by using a sample covering 20 % of terrestrial surface, Thomas et al. (Thomas et al. 2004) estimated, on the basis of mid-range climate-warming scenarios for 2050, that 15–37 % of species within the sample will be "committed to extinction". Indeed, predictable changes in abundance, distribution, interaction and phenology of those species not necessarily threatened with extinction will impact both ecosystem functionality and ecosystem services. The net loss of total forest area, including logging, change in land uses, degradation, fragmentation and transformation of primary forests to agriculture, will worsen the consequences of climate change (Hansen et al. 2013) and consequently biodiversity loss. With much biodiversity conservation policy decoupled from forestry and woodland policy at many levels (governmental and supranational), the capacity to address those threats is technically constrained. It is probably one of the areas where it is more needed to collect, analyse and synthetize information and scientific knowledge to assist in decision-making.

Furthermore, the impact and direction of the changes ahead will not be the same all over the world, but different on different geopolitical regions (Jenkins 2003). At one extreme, developed countries are suffering in certain areas human depopulation, abandonment of agricultural land, and re-naturalization of environment. In Europe, for example, the predicted cessation of farming in many high nature value areas together with the loss of low-intensity farmland (the areas perhaps most at risk of land abandonment) is likely to lead to a loss in farmland biodiversity, particularly in rural areas (Renwick et al. 2013). The opposing pressures between land abandonment and land intensification in developed countries, and how this will affect wildlife and human-wildlife interactions, will be a key topic to address in the future.

In developed countries, change in conservation policies has also led to population increases of large predators, for which conservation cannot be successfully focused on protected areas, which are often not large enough to contain viable populations of the target species. Coexistence between humans and large predators (mainly carnivores) in densely populated areas will need the continued development of research approaches including the involvement of psychology, sociology and education, in addition to ecology.

In the opposite extreme, in developing countries, where the great majority of the world's terrestrial biological diversity is found, the economic and social pressures (including poverty) erode their capacity to develop efficiently the appropriate policies to preserve biodiversity. Defaunation in those regions (e.g. tropics) is closely linked to forest ecosystems, particularly to deforestation and hunting activities therein ("empty forest syndrome"; e.g. Wilkie et al. 2011). This is even more worrying due to the lack of scientific background about many species inhabiting those high-diversity environments, which makes it harder for us evaluating the potential magnitude of biodiversity loss processes and the search for appropriate solutions.

Matters in the oceans are not much better, having decreased biological diversity continuously since industrialization in the nineteenth century because of practices such as trawler fishing, but also through pollution, eutrophication and overexploitation of the seas, as well as the prevalent effect of climate change. We have a clear outstanding debt with the research on seas and oceans, where only a small fraction of species have so far been identified, making the loss of biodiversity much more difficult to record and evaluate than on land (McCauley et al. 2015). Even if the concept of "wildlife" does exclude, in many definitions, fish and ocean fauna, we believe that it should be included within wildlife research, or at least that further interactions should exist between marine biologists and wildlife researchers, as problems, processes and approaches are similar (e.g. extractive use, pollution, epizootics, etc).

In the near future, as human population will continue to expand, there will be growing competition for space between wildlife and humans, increasing the incidence of conservation conflicts (Chapter 11: Gutierrez et al. 2016). In this context, protected areas deserve specific attention. Traditionally viewed as the solution to all conservation problems, it is clear that they will be necessary but insufficient to protect biodiversity. In addition to the reasons already mentioned (e.g. problems associated to size of these areas), there is no opportunity for populations to survive under geographic isolation, so one of our priorities regarding protected areas will be defining, creating and managing extensive corridor networks to connect protected areas at global scale. Additionally, it will be increasingly necessary to develop conservation strategies that are integrated with human activities and the economic use of natural resources, and even where the presence or diversity of wildlife provides additional value to the economic use of land. Some current conservation strategies go in this direction, such as the Natura 2000 network in EU; this approach is meant to create a complex network of areas with different levels of preservation, from strict protection to intensive economic use. Beyond this, it is clear that research focusing on the efficacy for wildlife conservation of multifunctional landscapes outside protected areas, and on the socio-economic and ecological consequences of different management systems in human-dominated landscapes should be also necessary in the near future.

Additionally, at some point over the next few years, it will be necessary to tackle an in-depth debate about hunting as a recreational activity. This has been a foundation stone of wildlife research as a scientific discipline, but it is also undeniable that there is increasing societal debate about it. This debate has many roots that confront, sometimes actively, several groups with disparate interests. Beyond the moral debates associated to this activity, it is unquestionable that due to human action there is currently overabundance of certain species, which compromises sometimes other species and ecosystems, may have agronomic (crop losses) or health (infections) risks, and which may be controlled by hunting. It is also true that an important part of natural and semi-natural areas in many regions of the planet are maintained due to commercial hunting interests. But it is also true that hunting and hunting management may lead to problems like poaching; illegal persecution of predators; poisoning; genetic, sanitary and ecological problems associated to overabundance, translocations, and releases of farm-reared animals for hunting, all of which also have negative consequences for wildlife and ecosystem conservation. More research will be needed to quantify when and how extractive use of wildlife is compatible with population and ecosystem conservation. But, additionally, other aspects influence biodiversity conservation and are severely under-researched: the potential impacts of management to boost game populations to sustain economic activities on other fauna; the potential impacts of the economic benefits provided by recreational hunting on conservation; the relationship between hunting activities (or anti-hunting attitudes) and engagement and commitment to conservation policies, etc.

Probably, we also need in the next years to develop research on questions, issues, organisms or regions that may report low scientific –and economic- benefits before it is too late. It is necessary to alleviate some existing bias in wildlife research through an increasing investment in studies on poorly known organisms and taxa. This should be carried out by a good planning of public founded research not guided by economic convenience. Additionally, although there is -and will be- a clear need for quick answers in today's crisis-to-crisis management environment, investment in long-term research is also necessary. Neither researchers nor the supporting agencies are eager to initiate research efforts that are slow to produce rewards. However, long-term studies are required to several areas of wildlife research and the evaluation of management decisions, so we need to combine both short and long-term studies, as the latter can yield high-quality results important to develop accurate models and strategies of management. Probably something needs to change in the funding agencies to support long-term studies or studies on less "trendy" topics.

Finally, as wildlife researchers, a greater effort is needed to speak the same language than other stakeholders, particularly resource managers, but also founding bodies, policy makers and the wider public. We need to progress in communication skills to be more effective explaining the needs and results of our research and their practical applications to the non-scientific community, so that science, conservation and management can be more effectively integrated.

References

Acevedo P, Jiménez-Valverde A, Aragón P, Niamir A (2016) New developments in the study of species distribution. In: Mateo R, Arroyo B, García JT (eds) Current trends in wildlife research. Springer, Switzerland

Arroyo B, Mateo R, García JT (2016) Trends in wildlife research: a bibliometric approach. In: Mateo R, Arroyo B, García JT (eds) Current trends in wildlife research. Springer, Switzerland

Barker RJ, Link WA (2015) Truth, models, model sets, AIC and multimodel inference: a Bayesian perspective. J Wildlife Manage 79:730–738

Beever EA, Swihart RK, Bestelmeyer BT (2006) Linking the concept of scale to studies of biological diversity: evolving approaches and tolos. Diversity Distrib 12:229–235

Bennet N, Roth R (2015) The conservation social sciences: what? how? and why? Canadian Wildlife Federation and Institute for Resources, Environmentand Sustainability/University of British Columbia, Vancouver

Bierwagen BG, Theobald DM, Pyke CR, Choate A, Groth P, Thomas JV, Morefield P (2010) National housing and impervious surface scenarios for integrated climate impact assessments. Proc Natl Acad Sci U S A 107:20887–20892

Brand FS, Jax K (2007) Focusing the meaning(s) of resilience: resilience as a descriptive concept and a boundary object. Ecol Soc 12(1):23

Bridge ES, Thorup K, Bowlin MS, Chilson PB, Diehl RH, Fléron RW, Hartl P, Kays R, Kelly JF, Robinson WD, Wikelski M (2011) Technology on the move: recent and forthcoming innovations for tracking migratory birds. BioSci 61:689–697

Dicks L, Hodge I, Randall N, Scharlemann J, Siriwardena G, Smith H, Smith R, Sutherland W (2013) A transparent process for 'evidence-informed' policy making. Conserv Lett 7(2):119–125

Ekvall MT, Bianco G, Linse S, Linke H, Bäckman J, Hansson L-A (2015) Three-dimensional tracking of small aquatic organisms using fluorescent nanopoarticles. Plos One 8:e78498

Fischer J, Brosi B, Daily GC, Ehrlich PR, Goldman R, Goldstein J, Lindenmayer DB, Manning AD, Mooney HA, Pejchar L, Ranganathan J, Tallis H (2008) Should agricultural policies encourage land sparing or wildlife-friendly farming? Front Ecol Environ 6:380–385

Gortazar C, Vicente J, Villar M, Ruiz-Fons F, Höfle U de la Fuente JJ (2016) Research priorities and trends in infections shared with wildlife. In: Mateo R, Arroyo B, García JT (eds) Current trends in wildlife research. Springer, Switzerland

Green RE, Cornell SJ, Scharlemann JPW, Balmford A (2005) Farming and the fate of wild nature. Science 307:550–555

Gutierrez R, Wood K, Redpath S, Young Y (2016) Conservation conflicts: future research challenges. In: Mateo R, Arroyo B, García JT (eds) Current trends in wildlife research. Springer, Switzerland

Hansen MC, Potapov PV, Moore R, Hancher M, Turubanova SA, Tyukavina A, Thau D, Stehman SV, Goetz SJ, Loveland TR, Kommareddy A, Egorov A, Chini L, Justice CO, Townshend JRG (2013) High-resolution global maps of 21st-century forest cover change. Science 342(6160): 850–853

Jenkins M (2003) Prospects for biodiversity. Science 302:1175–1177

Knight RL (1999) Private lands: the neglected geography. Conserv Biol 13:223–224

Lebreton J-D, Gaillard J-M (2016) Wildlife demography: from population concepts to population management. In: Mateo R, Arroyo B, García JT (eds) Current trends in wildlife research. Springer, Switzerland

Lee T, Quinn MS, Duke D (2006) Citizen, science, highways, and wildlife: using a web-based GIS to engage citizens in collecting wildlife information. Ecol Soc 11:11

Magle SB, Hunt VM, Crooks KR (2012) Urban wildlife research: past, present, and future. Biol Conserv 155:23–32

Mateo R, Lacorte S, Taggart M (2016) An overview of recent trends in wildlife toxicology. In: Mateo R, Arroyo B, García JT (eds) Current trends in wildlife research. Springer, Switzerland

McCauley DJ, Pinsky ML, Palumbi SR, Estes JA, Joyce FH, Warner RR (2015) Marine defaunation: animal loss in the global ocean. Science 347:1255641

Miller TR, Minteer B, Malan LC (2011) The new conservation debate: the view from practical ethics. Biol Conserv 144:948–957
Nuno A, Bunnefeld N, Milner-Gulland EJ (2013) Matching observations and reality: using simulation models to improve monitoring under uncertainty in the Serengeti. J Appl Ecol 50:488–498
Peterson MN, Peterson MJ, Peterson TR (2005) Conservation and the myth of consensus. Conserv Biol 19:762–767
Piertney SB (2016) High-throughput DNA sequencing and the next generation of molecular markers in wildlife research. In: Mateo R, Arroyo B, García JT (eds) Current trends in wildlife research. Springer, Switzerland
Renwick A, Janssonb T, Verburgc PH, Revoredo-Gihaa C, Britz W, Gocht A, McCrackena D (2013) Policy reform and agricultural land abandonment in the EU. Land Use Policy 30:446–457
Reynolds PC, Braithwaite D (2001) Towards aconceptual framework for wildlife tourism. Tour Manage 22:31–42
Robinson WD, Bowlin MS, Bisson I, Shamoun-Baranes J, Thorup K, Diehl RH, Kunz TH, Mabey S, Winkler DW (2015) Integrating concepts and technologies to advance the study of bird migration. Front Ecol Environ 8(7):354–361
Sarre SD, Georges A (2009) Genetics in conservation and wildlife management: a revolution since Caughley. Wildl Res 36(1):70–80
Shröeder B, Seppelt R (2006) Analysis of pattern–process interactions based on landscape models -overview, general concepts, and methodological issues. Ecol Model 199:505–516
Shwiff SA, Anderson A, Cullen R, White PCL, Shwiff SS (2013) Assignment of measurable costs and benefits to wildlife conservation projects. Wildl Res 40:134–141
Sol D (2016) Progresses and controversies in invasion biology. In: Mateo R, Arroyo B, García JT (eds) Current trends in wildlife research. Springer, Switzerland
Sutherland WJ, Pullin AS, Dolman PM, Knight TM (2004) The need for evidence-based conservation. Trends Ecol Evol 19:305–308
Swift MJ, Izac A-MN, van Noordwijk M (2004) Biodiversity and ecosystem services in agricultural landscapes – we are asking the right questions? Agric Ecosyst Environ 104:113–134
Tellería JL (2016) Wildlife habitat requirements: concepts and research approaches. In: Mateo R, Arroyo B, García JT (eds) Current trends in wildlife research. Springer, Switzerland
Thomas CD, Cameron A, Green RE, Bakkenes M, Beaumont LJ, Collingham YC, Erasmus BF, De Siqueira MF, Grainger A, Hannah L, Hughes L, Huntley B, Van Jaarsveld AS, Midgley GF, Miles L, Ortega-Huerta MA, Peterson AT, Phillips OL, Williams SE (2004) Extinction risk from climate change. Nature 427(6970):145–148
Treves A, Wallace RB, Naughton-Treves L, Morales A (2006) Co-managing human-wildlife conflicts: a review. Hum Dimens Wildl 11:383–396
Tufto J, Cavallini P (2005) Should wildlife biologists use free software? Wildl Biol 11(1):67–76
Turner MG (2005) Landscape ecology in North America: past, present, and future. Ecology 86:1967–1974
United Nations, Department of Economic and Social Affairs (2015). The world population prospects: 2015. New York, United Nation. Revision 29 Jul 2015
Van Vliet N, Cornelis D, Beck H, Lindsey P, Nasi R, Lebeck S, Moreno J, Fragoso J, Jori F (2016) Meat from the wild: extractive uses of wildlife and alternatives for sustainability. In: Mateo R, Arroyo B, García JT (eds) Current trends in wildlife research. Springer, Switzerland
Weckel ME, MacK D, Nagy C, Christie R, Wincorn A (2010) Using citizen science to map human-coyote interaction in suburban New York, USA. J Wildl Man 74:1163–1171
Wilkie DS, Bennett EL, Peres CA, Cunningham AA (2011) The empty forest revisited. Ann N Y Acad Sci 1223:120–128
Yasuda M, Kawakami K (2002) New method of monitoring remote wildlife via the Internet. Ecol Res 17:119–124
Zapata JA, Clavero M, Carrete M, DeVault T, Hermoso V, Losada M, Polo MJ Sánchez-Navarro S, Pérez M, Botella F, Ibañez C, Donazar JA (2016) Effects of renewable energy production and infrastructure on wildlife. In: Mateo R, Arroyo B, García JT (eds) Current trends in wildlife research. Springer, Switzerland

The manufacturer's authorised representative in the EU is Springer Nature Customer Service Centre GmbH, Europaplatz 3, 69115 Heidelberg, Germany. If you have any concerns regarding our products, please contact ProductSafety@springernature.com

Printed and bound by CPI Group (UK) Ltd, Croydon, CR0 4YY
30/09/2026
02225454-0007